Applied Engineering Mathematics with MATLAB® and Python

This comprehensive new textbook bridges the gap between mathematical theory and practical engineering applications. Organized into eight chapters, it introduces programming skills in MATLAB and Python alongside essential mathematical concepts, progressing from basic computation and linear algebra to more advanced topics such as differential equations, Fourier series, and Laplace transforms.

The book develops both mathematical reasoning and computational proficiency, equipping students with essential tools for modern engineering practice. It includes numerous MATLAB and Python scripts, presented within the text and available through a dedicated GitHub repository for hands-on learning and reproducibility. Each chapter concludes with problems—many with solutions—that reinforce understanding and encourage practical application. By integrating theory with coding exercises, the book ensures that students not only grasp key concepts but also gain the skills to apply them in real-world contexts.

The book is intended for undergraduate engineering students in Civil, Mechanical, Electrical, and Chemical Engineering. It is particularly suited for modules in Engineering Mathematics, Programming, Computations, and Analytical Studies, and can also serve as a supplementary reference for postgraduate students seeking applied mathematical and computational skills. Instructors will find it a valuable teaching resource that integrates theory with practice, adaptable for both semester- and year-long courses.

The GitHub repository can be accesses at https://github.com/prog871/CRC-book2025.

Applied Engineering Mathematics with MATLAB® and Python

From Theory to Practice

Mohammad Heidarzadeh and
Payam Khazaeinejad

CRC Press
Taylor & Francis Group
Boca Raton London New York

CRC Press is an imprint of the
Taylor & Francis Group, an **informa** business

Cover image: Shutterstock

First edition published 2026
by CRC Press
4 Park Square, Milton Park, Abingdon, Oxon, OX14 4RN

and by CRC Press
2385 NW Executive Center Drive, Suite 320, Boca Raton FL 33431

CRC Press is an imprint of Informa UK Limited

British Library Cataloguing-in-Publication Data
A catalogue record for this book is available from the British Library

ISBN: 978-1-032-83635-5 (hbk)
ISBN: 978-1-032-84242-4 (pbk)
ISBN: 978-1-003-51185-4 (ebk)

DOI: 10.1201/9781003511854

Typeset in TeXGyreTermes-Regular font
by KnowledgeWorks Global Ltd.

Publisher's note: This book has been prepared from camera-ready copy provided by the authors

Contents

Preface

Innovations of this book

The book *Applied Engineering Mathematics with MATLAB and Python: From Theory to Practice* aims to address a well-recognized gap in the teaching of mathematics to undergraduate and postgraduate engineering students. While many existing textbooks in this field are rigorous, they often lack a strong emphasis on practical engineering applications, making it difficult for students to connect mathematical concepts to real-world engineering problems.

Through our teaching experience, we have observed that engineering students often become disengaged when faced with highly abstract material lacking tangible applications. Another common limitation of traditional textbooks is their focus on manual calculations, with minimal or no integration of computational tools. In an era where engineering practice relies heavily on software for analysis and design, this approach can leave students underprepared for professional work.

This book was developed to help bridge the gap between theoretical mathematics and practical engineering applications. It features numerous real-world engineering examples, accompanied by implementations in MATLAB and Python. This integrated method supports students in developing both mathematical reasoning and programming proficiency, preparing them for the demands of modern engineering practice.

The audience

This book is primarily intended for undergraduate engineering students across a wide range of disciplines, including Civil Engineering, Mechanical Engineering, Electrical Engineering, and Chemical Engineering. It is particularly well-suited for use in teaching modules such as *Engineering Mathematics*, *Programming*, *Engineering Computations*, and *Analytical Studies*. Instructors may find this book to be a valuable core resource for these subjects, as it integrates mathematical theory with practical applications using MATLAB and Python. In addition, the book can serve as a supplementary reference for some postgraduate engineering students, especially those looking to strengthen their computational and analytical skills in applied contexts.

Structure of the book

This book is structured into eight chapters, each building on essential programming skills in MATLAB and Python, followed by key mathematical concepts used in engineering. It begins with basic programming in Chapter 1, providing readers with the tools for computation. The book then covers core topics like matrices, vectors, and systems of equations, forming the foundation for more advanced concepts. As the chapters progress, readers delve into critical mathematical principles, including limits, differentiation, integration, and differential equations. Applied methods such as root finding, interpolation, and curve fitting are then introduced, followed by Fourier series and Laplace transforms—essential for analyzing engineering systems. The final chapter covers statistics and probabilities, completing the mathematical toolkit for engineering analysis.

Each chapter features a range of engineering applications, supported by MATLAB and Python scripts specifically designed to solve problems within each domain. In total, the book includes 124 MATLAB and Python scripts. These scripts are presented within the main body of the text, enclosed in boxes, and are also available for download from the book's dedicated GitHub page: `https://github.com/prog871/CRC-book2025`

Each chapter includes ten problems, many of which are accompanied by answers. These problems, with solutions provided at the end of each chapter, offer students the opportunity to reinforce their understanding and practical skills. This comprehensive approach ensures that students not only grasp the theoretical concepts but also gain hands-on experience in using programming tools to solve real-world engineering problems.

How to use the book?

The book is suitable for both semester-long and year-long modules, as its content can be effectively delivered within either timeframe. We recommend structuring the course with a combination of a two-hour lecture each week to consolidate theoretical knowledge, followed by a one-hour computer laboratory session for hands-on practice with MATLAB and Python codes. This dual approach ensures that students not only understand the theoretical foundations but also gain practical experience in applying their knowledge to real-world engineering problems.

Mohammad Heidarzadeh, PhD, CEng, MICE, C.WEM, MCIWEM, FHEA

Payam Khazaeinejad, PhD, CEng, CSci, SFHEA, FIMechE, SMAIAA, MInstP, MASME

London, UK

October 2025

Authors

Mohammad Heidarzadeh is an Associate Professor of Civil Engineering at the University of Bath (UK). He is a Chartered Civil Engineer (CEng) with the Institution of Civil Engineers (ICE), UK, and a Chartered Water and Environmental Manager (C.WEM) with the Chartered Institution of Water and Environmental Management (CIWEM), UK. His expertise lies in coastal and hydraulic engineering, with research focused on improving coastal resilience to natural hazards, including storms, tsunamis, landslides, and earthquakes.

Payam Khazaeinejad is an Associate Professor of Solid Mechanics at Kingston University London (UK). He is a Chartered Engineer, a Chartered Scientist, a Senior Fellow of Advance HE (formerly the Higher Education Academy), and a Fellow of the Institution of Mechanical Engineers (IMechE). He received his PhD in Engineering from the University of Edinburgh (UK). His research focuses on computational design as a transformative approach to materials science and structural engineering, establishing principles for creating materials and systems with exceptional performance and adaptability, from microstructural architectures to full-scale applications.

1 Basics of MATLAB and Python for Engineers

This chapter provides an introduction to MATLAB and Python, two of the most popular programming languages used in engineering applications. They are also highly useful for performing routine mathematical analyses and visualizing data and information (e.g., [2, 3]). MATLAB and Python are commonly employed as part of larger design toolsets used for calculating and developing various infrastructure projects, including bridges, sea defenses, railways, and skyscrapers, as shown in Figure 1.1.

In this chapter, we start by covering foundational topics such as installation details for both software packages and proceed to subjects including basic programming concepts, functions, plotting and visualization techniques, and debugging code. It is hoped that by reading this chapter, students will be able to install both software packages and perform basic mathematical analyses and visualizations using MATLAB and Python.

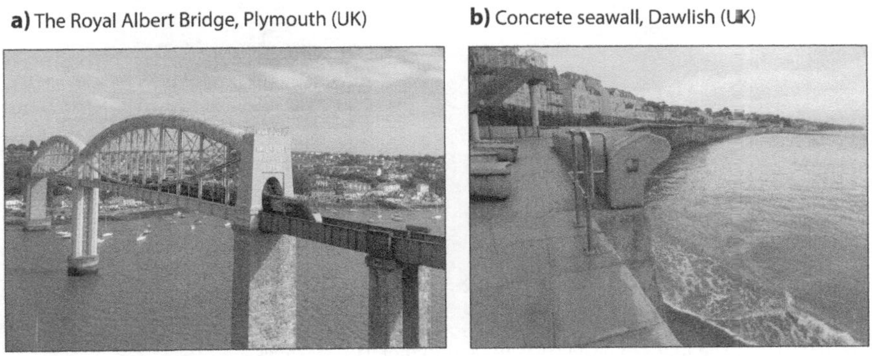

FIGURE 1.1 Examples of infrastructure systems where MATLAB and Python can be used as part of larger design tool sets. **a)** The Royal Albert bridge in Plymouth, UK. **b)** The seawall in Dawlish, UK.

1.1 INSTALLATIONS, BASIC SYNTAX AND DATA STRUCTURES

1.1.1 MATLAB

MATLAB, short for "Matrix Laboratory," is a versatile software package designed for programming and a wide range of mathematical analyses, spanning from basic computations to advanced operations such as signal processing and deep learning. MATLAB is developed by The MathWorks, Inc. and is a commercial software [6].

DOI: 10.1201/9781003511854-1

To install MATLAB on your PC, you need to create an account on the MathWorks website and obtain a valid license [6]. Numerous universities globally have obtained MATLAB subscriptions, offering institution-wide licenses to all students and staff. For information on accessing the MATLAB license, contact your university's IT department. Assuming that a MathWorks account has been created and a valid MATLAB license has been obtained, the following steps outline the installation process for MATLAB [6]:

i) Visit the MathWorks download center (https://www.mathworks.com/downloads/) and choose an appropriate MATLAB release and download the installer.

ii) After downloading the installer, double-click the file and follow the on-screen instructions to complete the installation process.

iii) Once the installation is complete, MATLAB can be launched by locating its icon (Figure 1.2a) on the desktop or in the Start Menu of your PC or laptop.

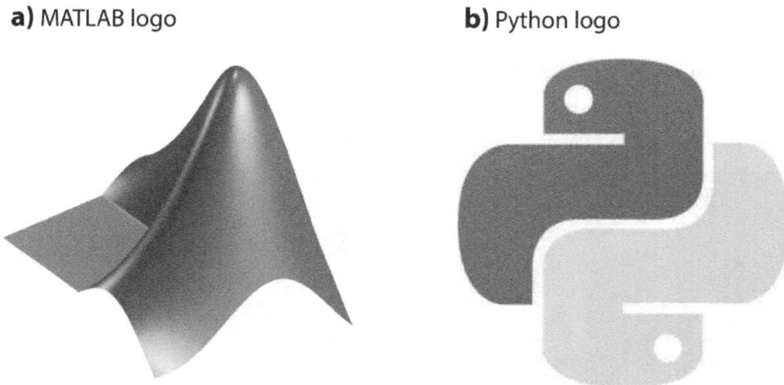

FIGURE 1.2 The logos of MATLAB (**a**) and Python (**b**) packages.

Upon launching, the MATLAB interface appears as depicted in Figure 1.3, consisting of several windows as named in the figure. At its most basic level, MATLAB functions like a pocket calculator, enabling various mathematical computations through its Command Window. Figure 1.3 illustrates how fundamental calculations, such as 2×2 or $2\cos(2\pi)$, are performed in MATLAB. The script `CH01Mcode1.m`, shown in Figure 1.3, together with all other code and scripts in this book, can be downloaded from: `https://github.com/prog871/CRC-book2025`

Basic arithmetic operations in MATLAB are introduced in Table 1.1. In MATLAB, values can be stored in variables for later use in calculations, a process known as variable assignment. For instance, we can assign a value of 5 to the variable x and a value of 10 to the variable y using the syntax shown in Example 1.1, as one would type in the MATLAB command window.

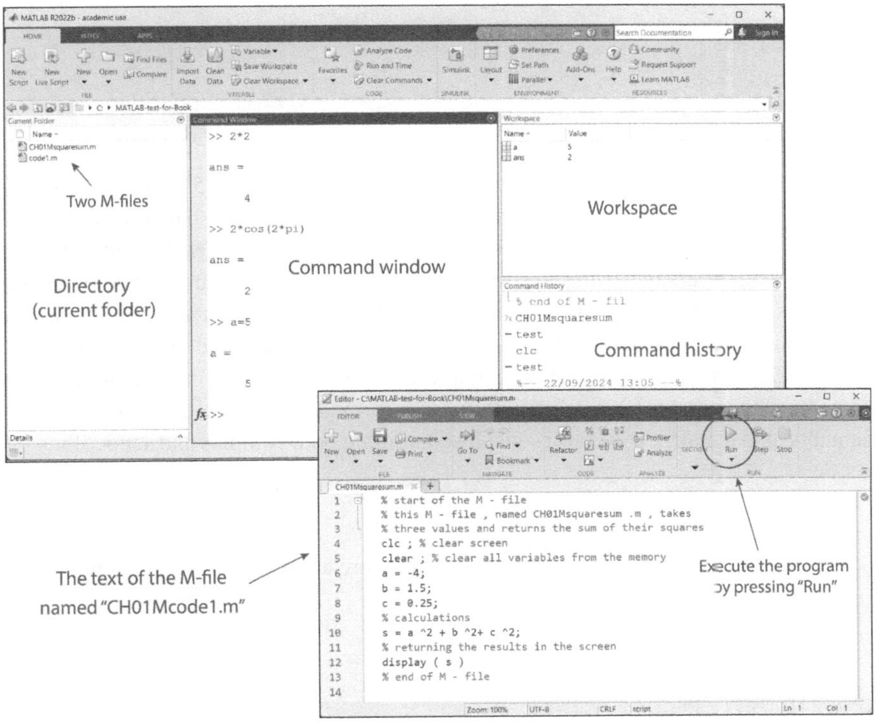

FIGURE 1.3 MATLAB interface showing various windows of the interface along with a sample M-file. The script `CH01Mcode1.m`, along with all other code and scripts presented in this book, can be downloaded from: `https://github.com/prog871/CRC-book2025`

TABLE 1.1
Basic arithmetic operations in MATLAB

Operation	Symbol
Addition	+
Subtraction	-
Multiplication	*
Division	/
Exponentiation	^

Example 1.1

Variable assignment in MATLAB. Here two values of 5 and 10 are assigned to x and y, respectively:

```
x =5;
y = 10;
z = x+y;
```

In the third line in Example 1.1, the term $z = x + y$ calculates 5 + 10 and stores the result, 15, in the variable z. MATLAB is particularly powerful in handling vectors and matrices, as its name—short for "Matrix Laboratory"—suggests. Vectors and matrices in MATLAB can be created using square brackets "[ ]." Row vectors are formed by separating values with spaces or commas, while column vectors are created using semicolons. Matrices are constructed by combining both spaces (or commas) and semicolons, as illustrated in Example 1.2. As shown in Example 1.2, a row vector v is created by separating elements with commas (,), a column vector w is formed by using semicolons (;) between elements, and a matrix p is constructed by combining both commas and semicolons.

Example 1.2

Making vectors and matrices in MATLAB. Here a row vector (v), a column vector (w) and a matrix (p) are made:

MATLAB Input:

```
v = [1, 2, 3];
w = [1; 2; 3];
p = [1, 2, 3; 4, 5, 6; 7, 8, 9];
```

MATLAB Output:

```
v = 1 2 3
w=
1
2
3
p =
1 2 3
4 5 6
7 8 9
```

A list of most useful MATLAB functions are provided in Table 1.2 which can be employed to execute various operations. Assuming a row vector $k = [1, -2, 3.5, 8, 11, -5.8]$, various operations using MATLAB functions are demonstrated in Example 1.3. The percent symbol "%" is used to add comments in MATLAB. Therefore, any text or command after "%" is not executed. Some other useful MATLAB syntax are:

clear—Removes all variables from the workspace.
clear x—Removes the variable.
clear all—Clears variables, functions, and files from the workspace.
clc—Clears the command window.
close—Closes the current figure window.
close all—Closes all figure windows.
display (x)—Displays text or variables in the command window.
run—Runs a specified script or function.

TABLE 1.2
Some basic MATLAB functions and their descriptions

Function	Description
sum()	Computes the sum of array elements
mean()	Calculates the average of array elements
max()	Finds the maximum value in an array
min()	Finds the minimum value in an array
std()	Computes the standard deviation of array elements
var()	Computes the variance of array elements
abs()	Computes the absolute value of each element
sqrt()	Computes the square root of each element
exp()	Computes the exponential of each element
log()	Computes the natural logarithm of each element
sin()	Computes the sine of each element
cosin()	Computes the cosine of each element
length()	Returns the number of elements in the largest dimension
size()	Returns the dimensions of an array
transpose()	Transposes a matrix
median()	Computes the median of array elements
plot()	Creates a 2D plot of data
hist()	Creates a histogram of data
scatter()	Creates a scatter plot

Example 1.3

Performing various operations using MATLAB functions. Here a row vector $k = [1, -2, 3.5, 8, 11, -5.8]$ is assumed:

MATLAB Input:

```
k = [1, -2, 3.5, 8, 11, -5.8];
s=sum(k);
m=max(k);
```

```
z=size(k);
a=abs(k);
```

MATLAB Output:

```
s= 15.7
m= 11
z= 1 6 (this means one row and six columns)
a= 1, 2, 3.5, 8, 11, 5.8
```

The MATLAB commands used previously can be accessed from the "Command History" panel located on the right side of the MATLAB interface (Figure 1.3). While this feature is useful, it allows for the reloading of only one command at a time. This limitation can be inconvenient if the user wishes to execute a series of commands to repeat a previous task, such as plotting a figure or conducting a specific analysis. In such situations, the best approach is to save the series of commands used to perform a specific task in a script or file known as an M-file in MATLAB. To create an M-file in MATLAB, open the editor by clicking on "New Script" and write your commands in the editor. Then save the file with a ".m" extension, using a name that starts with a letter. You can run the M-file script by typing its name in the command window or by simply pressing the "Run" button in the M-file editor.

An M-file named `CH01Mcode1.m` is presented in Example 1.4 and is shown in Figure 1.3, which takes three values, a, b, and c, as input from the user and returns the sum of their squares, displaying the result in the command window. This M-file, along with all other scripts mentioned in this book, is available for free download at the following website: https://github.com/prog871/CRC-book2025

In MATLAB, the colon operator, :, is used to generate sequences of numbers, as illustrated below:

- 1:5 generates the sequence 1, 2, 3, 4, 5.
- 4:2:10 generates the sequence 4, 6, 8, 10. In this notation, 4 is the starting number, 2 is the step size, and 10 is the end value.

You can use this syntax to quickly create arrays without having to type each number individually. This is especially useful for looping and plotting. In MATLAB, ".*" and "./" are element-wise operators. They allow you to perform multiplication or division on corresponding elements of two arrays (vectors or matrices), or between an array and a constant, rather than applying the operation to the entire arrays as a whole.

Example 1.4

MATLAB code (an M-file). This MATLAB script (M-file), named `CH01Mcode1.m`, along with all other scripts provided in this book, can be downloaded from the following website: https://github.com/prog871/CRC-book2025

MATLAB Input:

```
% start of the M - file
%this M-file, named CH01Mcode1.m,
%takes three values and returns
% sum of their squares
clc ; % clear screen
clear ;%clear all variables from memory
a = -4;
b = 1.5;
c = 0.25;
% calculations
s = a ^2 + b ^2+ c ^2;
% returning the results in the screen
display ( s )
% end of M - file
```

MATLAB Output:

```
s = 18.3125
```

1.1.2 PYTHON

Python is a widely used programming language that has gained popularity among both beginners and professionals in recent years, continuing to rise in prominence. A study by Amos et al. (2015) [1] revealed that Python has experienced significant growth compared to the flat or declining trends of other similar programming languages. Unlike MATLAB, Python is open-source software, meaning it is freely available to users worldwide. Python's logo is shown in Figure 1.2.

Python can be freely downloaded from www.python.org, where numerous resources and documentation, such as example programs and community news, are also available. First, download the appropriate Python version for your operating system—whether Windows, macOS, Linux, or others—from www.python.org/downloads/. If you are using a Windows system, Python can be installed by double-clicking the downloaded installation file and following the on-screen instructions. After you complete your installation of Python, the following components are typically installed on your Windows PC [5]:

- **Python interpreter**—The core component that allows you to run Python scripts and interact with Python code.
- **IDLE (Integrated Development and Learning Environment)**—This comes bundled with Python and provides a basic interface for writing, editing, and executing Python code.
- **Python documentation**—An offline version of the Python documentation is installed, which can be accessed using the pydoc module or through IDLE.

a) Python IDLE programming environment **b)** A sample Python script: "CH01Pcode1.py"

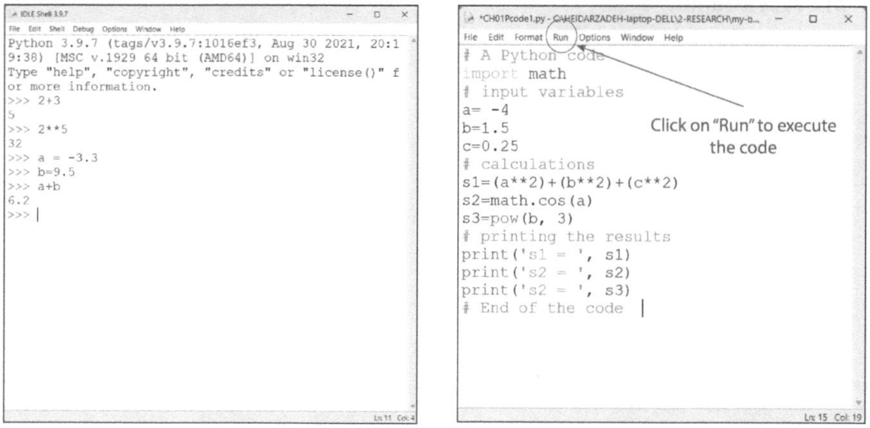

FIGURE 1.4 **a)** Python IDLE (ntegrated Development and Learning Environment). **b)** An example Python script.

TABLE 1.3

Basic arithmetic operations in Python

Operation	Symbol
Addition	+
Subtraction	-
Multiplication	*
Division	/
Floor Division	//
Modulus	%
Exponentiation	**

Python's IDLE is an ideal environment for Python programming and is used for the programming examples presented in this book. Python IDLE can function like a basic calculator for simple arithmetic operations, such as addition and multiplication as shown in Figure 1.4a. Variable assignment in Python functions similarly to MATLAB. In Figure 1.4a, the values -3.3 and 9.5 are assigned to the variables a and b, respectively, followed by a command to calculate $a + b$. Basic arithmetic operations in Python are presented in Table 1.3. Note the distinction between the exponentiation operators in MATLAB (^) and Python (**).

Similar to MATLAB, for more efficient Python programming, it is helpful to write all the commands for a specific task in a script and then run the script as a whole. A Python script consists of a ".py" file containing all the commands needed to perform a particular task, which can then be executed in one go. An example Python script is shown in Figure 1.4b, where values are assigned to three variables, a, b, and c, and various mathematical calculations are performed, with the results displayed on the screen. The electronic file of this Python script, named `CH01Pcode1.py`, can be downloaded from https://github.com/prog871/CRC-book2025

Example 1.5 gives this Python script (`CH01Pcode1.py`) and its outputs. It is important to note the second line of the code: `import math`. Unlike MATLAB, which has many built-in functions, Python requires you to import specific modules, such as `math`, to access additional functions. This means you need to include statements like `import math` at the beginning of your code, depending on the application (see Example 1.5 and Figure 1.4b). Note that in some cases, your Python code may not run if a required library is not installed on your system, and you may need to install it separately.

Example 1.5

Python code. This Python script, named `CH01Pcode1.py`, can be downloaded at the following website: https://github.com/prog871/CRC-book2025

Python Input:

```
# A Python code
import math
# input variables
a= -4
b=1.5
c=0.25
# calculations
s1=(a**2)+(b**2)+(c**2)
s2=math.cos(a)
s3=pow(b, 3)
# printing the results
print('s1 = ', s1)
print('s2 = ', s2)
print('s2 = ', s3)
# End of the code
```

Python Output:

```
s1 = 18.3125
s2 = -0.6536436208636119
s2 = 3.375
```

Python provides several basic mathematical functions, such as abs(), which returns the absolute value of a number, and round(), which rounds a floating-point number to a specified number of decimal places. For additional mathematical functions, users need to explicitly import the necessary functions from the appropriate modules. For example, the math module provides a range of functions for various mathematical operations. To view the functions available in the Python module `math`, enter the following commands consecutively in Python IDLE: `import math` and `dir(math)`. Example 1.6 shows the input and output of these Python commands, demonstrating the availability of various mathematical functions such as $log()$, $sin()$, and constants like pi, among others. A list of useful Python modules is presented in Table 1.4.

Example 1.6

Python code. This Python script displays the functions available in the `math` module.

Python Input:

```
import math
dir(math)
```

Python Output:

```
['__doc__', '__loader__', '__name__', '__package__', '__spec__
    ', 'acos', 'acosh', 'asin', 'asinh', 'atan', 'atan2', '
    atanh', 'ceil', 'comb', 'copysign', 'cos', 'cosh', 'degrees
    ', 'dist', 'e', 'erf', 'erfc', 'exp', 'expm1', 'fabs', '
    factorial', 'floor', 'fmod', 'frexp', 'fsum', 'gamma', 'gcd
    ', 'hypot', 'inf', 'isclose', 'isfinite', 'isinf', 'isnan',
     'isqrt', 'lcm', 'ldexp', 'lgamma', 'log', 'log10', 'log1p
    ', 'log2', 'modf', 'nan', 'nextafter', 'perm', 'pi', 'pow',
     'prod', 'radians', 'remainder', 'sin', 'sinh', 'sqrt', '
    tan', 'tanh', 'tau', 'trunc', 'ulp']
```

It is important to note that to use any of these functions, you must write the module name followed by a dot and then the function name. For example, to access the value of π, use the command `math.pi`. Similarly, to calculate $sin(1.6)$, type `math.sin(1.6)`. The Python code, named `CH01Pcode2.py`, shown in Example 1.7, provides several examples of using various functions from the `math` module. The electronic file of this Python script can be downloaded from: https://github.com/prog871/CRC-book2025

Example 1.7

Python code. This Python script, named `CH01Pcode2.py`, imports `math` module and conducts various calculations using functions embedded in this module.

Python Input:

TABLE 1.4

Basic Python modules for mathematics and visualization

Module name	Description
`math`	Provides mathematical functions like square root, and sine
`random`	Implements pseudo-random number generators
`sys`	Provides access to system-specific parameters and functions
`time`	Provides time-related functions, including sleep and time retrieval
`csv`	Facilitates reading from and writing to CSV files
`collections`	Implements specialized container data types like namedtuples and deque
`matplotlib`	A plotting library used for creating static, animated, and interactive visualizations
`numpy`	A powerful library for numerical computations, providing support for arrays and matrices, along with a collection of mathematical functions
`seaborn`	A statistical data visualization library built on top of Matplotlib, providing a high-level interface for drawing attractive statistical graphics
`plotly`	A library for creating interactive plots and dashboards, allowing for online and offline graphing
`scipy`	Builds on NumPy and provides additional functionality for scientific and technical computing, including optimization and integration
`sympy`	A library for symbolic mathematics for algebraic manipulations and solving equations

```
import math
# A Python code
import math
# input variable
a= 0.95
# calculations
c1= math.sin(a)
c2= math.cos(a)
c3= math.log(a)
c4= math.sqrt(a) #square root
c5= math.asin(a) #inverse sine (arcsine)
# printing the results
print('c1 = ', c1)
print('c2 = ', c2)
print('c3 = ', c3)
print('c4 = ', c4)
print('c5 = ', c5)
# End of the code
```

Python Output:

```
c1 = 0.8134155047893737
c2 = 0.5816830894638836
```

```
c3 = -0.05129329438755058
c4 = 0.9746794344808963
c5 = 1.253235897503375
```

1.2 CONTROL STRUCTURES AND LOOPS

Control structures are important for making decisions and repeating tasks in programming. They enable the programmer to dictate the order in which actions occur based on specific conditions. For instance, control structures allow the program to perform a task repeatedly until a specified stopping point is reached. The most important control structures are the "`if statement`," "`for loop`," and "`while loop`," which are discussed here with examples.

1.2.1 "IF STATEMENT"

The "`if statement`" allows the program to make a decision and perform a task based on a condition. For example, if it is a sunny day, you decide to wash your car. In this case, the condition is the weather being sunny, and the task is washing the car.

In MATLAB, the structure of a general "`if statement`" is as follows:

```
if condition1
   % Code to execute if condition1 is true
elseif condition2
   % Code to execute if condition2 is true
else
   % Code to execute if none of the conditions are true
end
```

In Python, the "`if statement`" is structured as follows:

```
if condition1:
   % Code to execute if condition1 is true
elif condition2:
   % Code to execute if condition2 is true
else:
   % Code to execute if none of the conditions are true
```

Engineering application: Application of the "`if statement`" is shown through an engineering example here. Stress (σ) in a structural column is calculated using the following equation (see Figure 1.5):

$$\sigma = \frac{F}{A} \tag{1.1}$$

where σ is stress in the column, F is axial force, and A is the cross-section area of the column (Figure 1.5). Assuming the bridge column has a cross-section with dimensions $1.5\,\text{m} \times 1.5\,\text{m}$, its cross-sectional area will be $A = 2.25\,\text{m}^2$, which is equivalent to 2,250,000 mm^2. The allowable stress for the column material is σ =30 megapascals (MPa, where 1 MPa = 1 N/mm^2). Given that the applied force is F = 42,000 kN, the question is whether the resulting stress exceeds the safe limit. The MATLAB and Python codes presented in Examples 1.8 (script `CH01Mcode2.m`) and 1.9 (script `CH01Pcode3.py`), respectively, use an "`if statement`" to determine this.

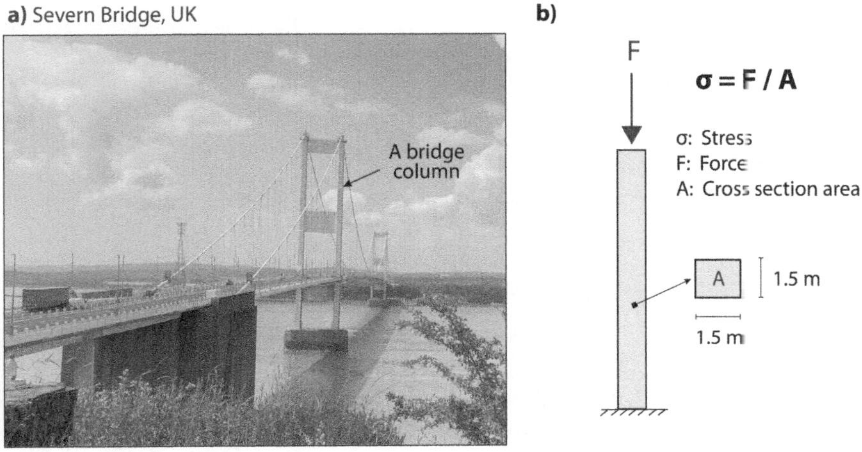

FIGURE 1.5 **a)** The Severn bridge in the UK. **b)** Diagram showing force (F) in a column of the bridge and the equation to calculate the stress σ.

Example 1.8

MATLAB code. This MATLAB script, named `CH01Mcode2.m`, calculates the stress inside a column and determines whether it exceeds the safe limit. This MATLAB script can be downloaded at the following website: https://github.com/prog871/CRC-book2025

MATLAB Input:

```
%This code, named 'CH01Mcode2.m'
% calculates stress inside a
%column and determines if it is safe
clc ; % clear screen
clear ;%clear ariables from the memory
A = 2250000; % cross section area (mm^2)
F = 42000*1000; % force in N
allow_str=30;%allowabale stress(MPa=N/mm^2)
% calculations
str= F/A; % calculate stress in MPa
display(str);
```

```
% Check if stress exceeds allowable limit
if str > allow_str
disp('str>allowabale stress;column fails');
else
disp('str<allowabale stress;column is safe');
end
% end of M - file
```

MATLAB Output:

```
str =
  18.6667
str < allowabale stress; column is safe
```

Example 1.9

Python code. This Python code, named `CH01Pcode3.py`, calculates the stress inside a column and determines whether it exceeds the safe limit.

Python Input:

```
# This Python code calculates stress inside a
# column and determines if it is safe
A = 2250000 # cross section area (mm^2)
F = 42000 * 1000 # force in N
allow_str = 30 # allowable stress (MPa = N/mm^2)
# Calculations
str = F / A # calculate stress in MPa
# Display the stress value
print(f'stress = {str}') #calculated stress
# Check if stress exceeds allowable limit
if str > allow_str:
   print('str > allowable stress; column fails')
else:
   print('str < allowable stress; column is safe')
# End of code.
```

Python Output:

```
stress = 18.666666666666668
str < allowable stress; column is safe
```

1.2.2 "FOR LOOP"

A "`for loop`" allows the repetition of a block of code multiple times. It is particularly useful for performing repetitive tasks without the need to rewrite the same code repeatedly.

In MATLAB, the basic structure of a "`for loop`" is:

```
for i= 1:n
  % Code to execute
end
```

In Python, a "`for loop`" is structured as follows:

```
for i in range(1, n+1):
  % Code to execute
```

Note that the range(1, n + 1) function in Python generates numbers starting from 1 up to n.

Engineering application: We utilize a "`for loop`" to calculate the deflection (δ) of a simply supported beam subjected to a point load F at its midpoint (see Figure 1.6b). We know that the deflection of the midpoint for the beam shown in Figure 1.6b can be calculated using the following equation:

$$\delta = \frac{FL^3}{48EI} \tag{1.2}$$

where L is the length of the beam (mm), δ is deflection of the midpoint (mm), F is point load (N), E is modulus of elasticity (N/mm^2), and I is moment of inertia (mm^4).

Assuming a simply-supported beam with $L = 5000$ mm, $I = 2.13 \times 10^9$ mm^4, and $E = 35$ kN/mm^2, the MATLAB and Python scripts presented in Examples 1.10 (script `CH01Mcode3.m`) and 1.11 (script `CH01Pcode4.py`), respectively, calculate the deflection at the beam's midpoint for various point loads of 50 kN, 60 kN, 70 kN, 80 kN, and 90 kN using a "`for loop`."

Example 1.10

MATLAB code. This MATLAB code, named `CH01Mcode3.m`, calculates the deflection of a beam under various point loads.

MATLAB Input:

```
%This M-file, named CH01Mcode3.m,
%calculates deflection of a
%simply-supported beam
% under point load
clc ; % clear screen
clear; % clear memory
L = 5000; % length in mm
```

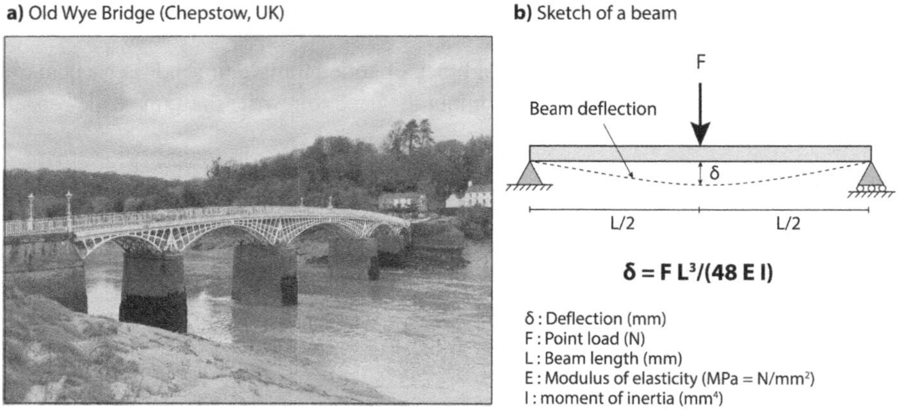

FIGURE 1.6 a) The Old Wye bridge in Chepstow, UK. b) Sketch showing the deflection of a simply-supported beam along with the equation to calculate the deflection of its midpoint subjected to a point load F.

```
I=2.13*(10^9);%moment of inertia (mm^4)
E=35000;%modulus of elasticity (N/mm^2)
F = [50,60,70,80,90].*1000;% force (N)
% calculating 'delta': deflection(mm)
for i=1:5
delta(i)= F(i).*(L^3)./(48*E*I);
end
% printing the reslts
display (delta, 'Deflection (mm)')
% end of M-file
```

MATLAB Output:

```
Deflection (mm) =
1.7466 2.0959 2.4452 2.7945 3.1439
```

Example 1.11

Python code. This Python code, named `CH01Pcode4.py`, calculates the deflection of a beam under various point loads.

Python Input:

```
# This Python script, named 'CH01Pcode4.py',
```

```
# calculates the deflection of a
# simply-supported beam under point load
import numpy as np # importing module 'numpy'
L = 5000 # length in mm
I = 2.13 * (10**9) # moment of inertia (mm^4)
E = 35000 # modulus of elasticity (N/mm^2)
F = np.array([50,60,70,80,90])*1000 #force(N)
# Calculations
delta = F * (L**3) / (48 * E * I) #deflection(mm)
# Printing the results
print("Deflection (mm):", delta)
# End of Python script
```

Python Output:

```
Deflection (mm): [1.74659066 2.09590879 2.44522692 2.79454505
    3.14386318]
```

1.2.3 "WHILE LOOP"

Previously, we discussed how to use the "`if statement`" to evaluate a condition and execute code when the condition is true. We also explored the "`for loop`," which is employed to repeat a task a specific number of times. In cases where the number of repetitions is not predetermined, the "`while loop`" is used. This section will focus on understanding the "`while loop`" and its applications.

The structure of the "`while loop`" in MATLAB is:

```
while condition
    % Code to execute while the condition is true
end
```

In Python, the "`while loop`" takes the following form:

```
while condition:
    # Code to execute while the condition is true
```

Engineering application: The "`while loop`" is applied here to solve an engineering problem related to dams and reservoirs. The Craig Goch Dam and Reservoir are shown in Figure 1.7. Considering a reservoir volume of $V = 9.2 \times 10^6$ m^3, and assuming a river inflow discharge of $q = 5$ m^3/s, our objective is to calculate the time required to fill the empty reservoir. The volume of water (ΔV) entering the reservoir over a time interval Δt can be calculated using the following equation:

$$\Delta V = q \times \Delta t \tag{1.3}$$

where q is river inflow discharge. The MATLAB and Python codes presented in Example 1.12 (script CH01Mcode4.m) and Example 1.13 (script CH01Pcode5.py), respectively, solve this problem, showing that it will take 21.3 days. Note that in these scripts, the time interval is set to 1 s ($\Delta t = 1$ s).

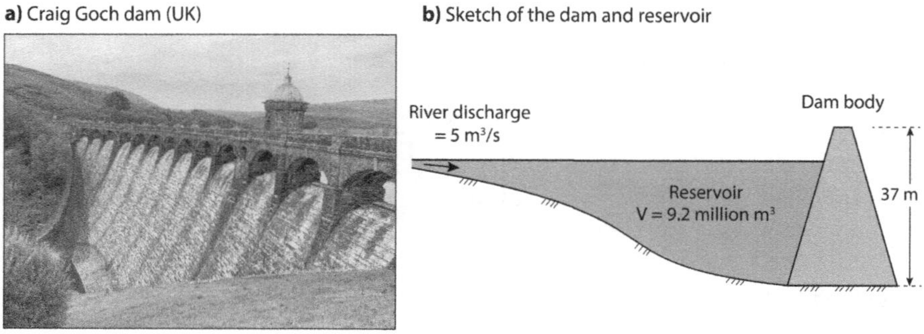

FIGURE 1.7 a) The Craig Goch dam, UK. b) Sketch showing the dam and reservoir, the river input discharge and the reservoir capacity.

Example 1.12

MATLAB code. This MATLAB script, named CH01Mcode4.m, calculates the time it takes to fill the Craig Goch reservoir in the UK.

MATLAB Input:

```
%this M-file, named CH01Mcode4.m,
% calculates the time taken to
% fill Craig Goch reservoir (UK)
clc ; % clear screen
clear ; % clear memory
V=9.2*(10^6);%reservoir volume (m3)
q=5; % river dischage (m3/s)
% Initialize variables
volume=0;%reservoir starting volume
time = 0; % Time (seconds)
% while loop, time taken to fill dam
while volume < V
 volume = volume + q;% volume added
 time = time + 1;%time increment:1 sec
end
% Convert time from seconds to days
time_days = time / (60 * 60 * 24);
```

```
% Display result
display(time_days);
% end of M - file
```

MATLAB Output:

```
time_days =
  21.2963
```

Example 1.13

Python code. This Python code, named `CH01Pcode5.py`, calculates the time it takes to fill the Craig Goch reservoir.

Python Input:

```
#This Python script, named CH01Pcode5.py, calculates
#time taken to fill Craig Goch reservoir
V = 9.2 * (10**6) #reservoir volume (m^3)
q = 5 # river discharge (m^3/s)
volume = 0 #reservoir starting volume
time = 0 #Time (seconds)

# while loop for time taken to fill reservoir
while volume < V:
    volume += q # Add discharge per sec
    time += 1 # Increment time by 1 sec
# Convert time from seconds to days
time_days = time / (60 * 60 * 24)
# Display result
print(time_days)
```

Python Output:

```
21.29629
```

1.3 FUNCTIONS AND LIBRARIES

In programming, a function is a reusable section of code designed to accomplish a particular task. Rather than rewriting the same code repeatedly, the function can be called whenever required.

In MATLAB, a function is defined as follows:

```
function [output1, output2, ...] = functionName(input1, input2, ...)
  % Function code
  output1 = ...;
  output2 = ...;
  % Additional operations can be performed here
end
```

A function in Python is structured as follows:

```
def function_name(input1, input2, ...):
  # Function code
  output1 = ...
  output2 = ...
  # Additional operations can be performed here
  return output1, output2
```

Engineering application: The application of a "function" is demonstrated through the following example. An important task in Civil Engineering is calculating the volume and weight of various structural elements, such as beams, columns, and slabs, during construction (Figure 1.8). Assuming a structural element with dimensions L (length), W (width), and H (height), the volume (V_e) and weight (Wgt) of the element are calculated as follows:

$$V_e = L \times W \times H \tag{1.4}$$

$$Wgt = \gamma \times V_e \tag{1.5}$$

where γ is the density of the element's material. For example, for concrete, $\gamma = 2400$ kg/m^3. Here, we define a MATLAB function called `CH01Mcode5.m`, which takes four input parameters—L, W, H, and γ—and calculates V_e and Wgt. Example 1.14 (script `CH01Mcode5.m`) shows this function.

It is important to note that MATLAB functions cannot be executed directly by pressing the `Run` button, as with previous MATLAB scripts. Instead, functions must be called by their names, with input and output variables, either within another MATLAB script or from the command window. For example, the function `CH01Mcode5.m` (shown in Example 1.14) can be called and executed using the following command line, assuming $L = 10\,\text{m}$, $W = 0.5\,\text{m}$, $H = 0.5\,\text{m}$, and $\gamma = 2400\,\text{kg/m}^3$:

$$[V_e, \text{Wgt}] = \text{CH01Mcode5}(10, 0.5, 0.5, 2400)$$

The outcome will be:

$$V_e = 2.5000\,\text{m}^3$$
$$Wgt = 6000\,\text{kg}$$

The Python equivalent of this function is presented in Example 1.15, named `CH01PfuncVolWgt.py`. Similar to MATLAB, Python functions cannot be executed by

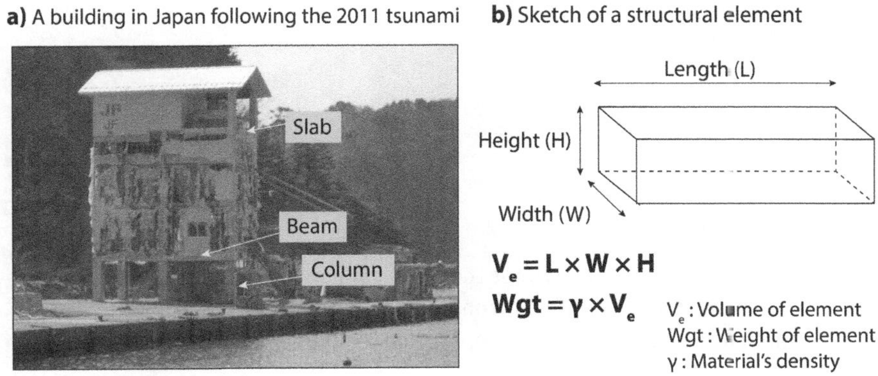

FIGURE 1.8 **a)** A damaged building in east Japan following the March 2011 Japan M 9.0 earthquake and tsunami. **b)** Sketch of a structural element and equations for calculating its volume and weight.

simply pressing the "`Run module`" button. To run a function in Python, we must call it by its name with the appropriate input parameters, either in Python IDLE or within another Python script, as shown below:

CH01PfuncVolWgt(10, 0.5, 0.5, 2400)

This produces the following output:

(2.5, 6000.0)

In the Python function of Example 1.15, these two output values correspond to the volume (V_e) and weight (Wgt) of the structural element, respectively.

Example 1.14

A MATLAB function. This MATLAB code is a function, named `CH01Mcode5.m`, that calculates volume and weight of a structural element (beam, column, slab, etc.).

MATLAB Input:

```
%Function to calculate volume and weight
function [V_e, Wgt]=CH01Mcode5(L,W,H,gamma)
clc; close all; % clear memory
%This function calculates volume and weight
%given length(L),width(W),height(H),
%and density(gamma)
%dimensions in meters and gamma in kg/m^3
```

```
%Calculate volume (V_e =L*W* H)
V_e = L * W * H; % volume in m^3
% Calculate weight (Wgt = V_e * gamma)
Wgt = V_e * gamma; % weight in kg
end
% end of M - file
```

MATLAB Output:

```
Note: This is a MATLAB function and cannot be executed by
    pressing the ``Run" button. Refer to the text for
    instructions on how to apply and execute it.
```

Example 1.15

A Python function. This Python code is a function, named `CH01PfuncVolWgt.py`, that calculates volume and weight of a structural element (beam, column, slab, etc.).

Python Input:

```
# Function to calculate volume and weight
def CH01PfuncVolWgt(L, W, H, gamma):
    """
    This function calculates volume and weight
    given length (L), width (W), height (H),
    and density (gamma). Dimensions are in
    meters and gamma in kg/m^3.
    """
    # Calculate volume (V_e = L*W*H)
    V_e = L * W * H # volume in m^3
    # Calculate weight (Wgt = V_e*gamma)
    Wgt = V_e * gamma # weight in kg
    return V_e, Wgt
```

Python Output:

```
Note: this is a Python function and cannot be executed by
    pressing the ``Run module" button. Refer to the text on how
     to apply and execute this function.
```

1.4 PLOTTING AND VISUALIZATION TECHNIQUES

MATLAB and Python are both powerful and versatile platforms for generating plots and visualizations, making them indispensable in various scientific and engineering disciplines. The most commonly used MATLAB and Python plotting tools are summarized in Tables 1.5 and 1.6, respectively. In the following sections, we explore the plotting techniques and capabilities of MATLAB and Python, utilizing the tools listed in Tables 1.5 and 1.6.

TABLE 1.5

Basic MATLAB plotting tools

Tool	Description
plot()	Creates basic 2D line plots.
title()	Adds a title to the plot.
xlabel()	Labels the x-axis.
ylabel()	Labels the y-axis.
legend()	Adds a legend to describe plotted lines.
subplot()	Creates multiple plots in a single figure window.
bar()	Creates a bar graph.
histogram()	Plots the distribution of data.
scatter()	Creates scatter plots to show data relationships.
hold on/off	Allows multiple plots in the same figure.
grid on	Turns the grid on for better visual clarity.
axis()	Manually sets axis limits.
fill()	Creates filled polygons (useful for shading areas).
pie()	Creates a pie chart.
surf()	Plots 3D surfaces.

Engineering application: We apply plotting tools in MATLAB and Python to plot the variations of the deflections of a simply-supported beam under a point load F at its midpoint along its length L (Figure 1.9). This deflection is calculated using the following equation:

$$\delta(x) = \frac{F \cdot x}{48 \cdot E \cdot I} \left(3L^2 - 4x^2\right), \quad 0 < x < \frac{L}{2} \tag{1.6}$$

where L is the length of the beam (mm), $\delta(x)$ is deflection of the beam (mm) at a distance of x from one end of the beam, F is point load (N), E is modulus of elasticity (N/mm^2), and I is moment of inertia (mm^4). For plotting purposes, the following values are considered: $L = 12000$ mm, $I = 5.0 \times 10^9$ mm^4, $E = 200$ kN/mm^2, and $F = 500$ kN.

The outcome is plotted in Figure 1.9c,d which shows that the maximum deflection is 18 mm occurring at the midpoint of the beam (L = 6 m). The MATLAB and Python codes used to plot the deflection of the beam are provided in Examples 1.16 (script `CH01Mcode6.m`) and 1.17 (script `CH01Pcode6.py`), respectively. It is evident that several of the plotting tools introduced in Tables 1.5 and 1.6 are applied in these examples.

TABLE 1.6
Basic Python plotting tools

Tool	Description
plot()	Creates basic 2D line plots (from "matplotlib").
title()	Adds a title to the plot.
xlabel()	Labels the x-axis.
ylabel()	Labels the y-axis.
legend()	Adds a legend to describe plotted lines.
bar()	Creates a bar graph (from 'matplotlib').
hist()	Plots a histogram to show data distribution.
scatter()	Creates scatter plots to show data relationships.
subplot()	Creates multiple plots in a single figure window.
grid()	Turns the grid on or off for better visual clarity.
axis()	Sets axis limits manually.
pie()	Creates a pie chart.
contour()	Plots contour lines for 3D data (from 'matplotlib').
heatmap()	Creates heatmaps (from 'seaborn').

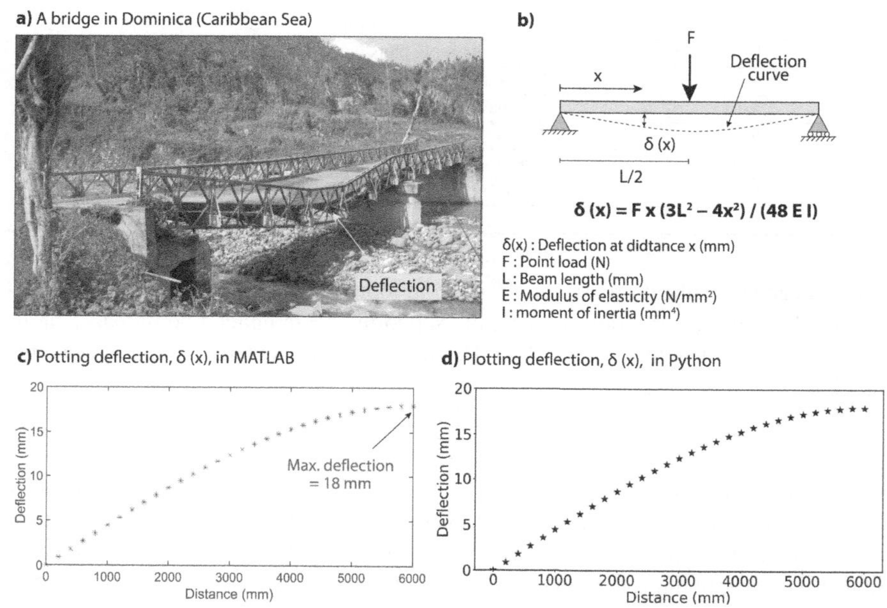

FIGURE 1.9 **a)** A damaged bridge in Dominica (the Caribbean Sea) following Hurricane Maria in 2017 [4]. **b)** Sketch showing the deflection of a simply-supported beam under a point load F at its midpoint. **c)** A MATLAB plot of the deflection of half of the beam shown in panel "b." **d)** A Python plot of the same beam.

Example 1.16

MATLAB code. This MATLAB code, named `CH01Mcode6.m`, plots the deflection of a simply-supported beam under a point load at its midpoint.

MATLAB Input:

```
%this M-file , CH01Mcode6.m,
%plots variations of deflection for
%simply-supported beam under point load
%this plot is for half of the beam
clc ; % clear screen
clear ; % clear memory
set(0,'defaultaxesfontsize',30);%fontsize
L=12000; % length in mm
I=5*(10^9);%moment of inertia (mm^4)
E=200000;%modulus of elasticity(N/mm^2)
F=500.*1000; % force (N)
% calculations of deflection(mm)
x=0:200:6000;
delta= F.*x.*(3*L.^2 -4*x.^2)./(48*E*I);
% plotting
plot(x,delta,'*k','markersize',15);
xlabel('Distance (mm)');
ylabel('Deflection (mm)');
ylim([0 20]);% limits of the y axis
set(gca,'linewidth',2.0);
% end of M - file
```

MATLAB Output:

```
Note: The output is a plot which is shown in Figure 1.9 in this
    chapter.
```

Example 1.17

Python code. This Python code, named `CH01Pcode6.py`, plots the deflection of a simply-supported beam under a point load at its midpoint.

Python Input:

```
# Python script to plot deflection of a
# simply-supported beam under point load
```

```
# This plot is for half of the beam
import numpy as np
import matplotlib.pyplot as plt
# Beam parameters
L = 12000 # length in mm
I = 5 * (10**9) #moment of inertia(mm^4)
E = 200000 #modulus of elasticity(N/mm^2)
F = 500 * 1000 # force (N)
# CalculatING deflection, delta (mm)
x =np.arange(0,6001,200,dtype=np.int64)
delta =(F*x*((3*(L**2))-(4*(x**2))))/(48*E*I)
# Plotting
plt.plot(x, delta, '*k', markersize=15)
plt.xlabel('Distance (mm)', fontsize=30)
plt.ylabel('Deflection (mm)', fontsize=30)
plt.ylim([0, 20]) # limits of the y-axis
plt.gca().tick_params(axis='both',labelsize=30,
                width=2.0)
plt.gca().spines['top'].set_linewidth(2.0)
plt.gca().spines['right'].set_linewidth(2.0)
plt.gca().spines['left'].set_linewidth(2.0)
plt.gca().spines['bottom'].set_linewidth(2.0)
plt.tight_layout()
plt.show()
# End of Python code.
```

Python Output:

```
Note: the output of this code is a plot which is shown in
    Figure 1.9 in this chapter.
```

Engineering application: As another example of plotting, we present a bar chart using MATLAB and Python for a construction activity known as "beach nourishment." In coastal engineering, beach nourishment involves adding sand, gravel, or other materials to a beach to prevent erosion and preserve its size and shape. For example, Seaford Beach in the UK (Figure 1.10) undergoes annual beach nourishment. Assume that the quantities of various materials used for beach nourishment at a coastal site are shown in Table 1.7. The MATLAB and Python codes in Examples 1.18 (script `CH01Mcode7.m`) and 1.19 (script `CH01Pcode7.py`), respectively, generate a bar chart illustrating the nourishment materials listed in Table 1.7.

Example 1.18

MATLAB code. This MATLAB code, named `CH01Mcode7.m`, makes a bar chart of materials used for beach nourishment.

TABLE 1.7
Volume of beach nourishment materials

Material	Volume (m^3)
Sand	500
Gravel	300
Shells	200
Silt	150
Rock	400
Clay	100
Geotextiles	250

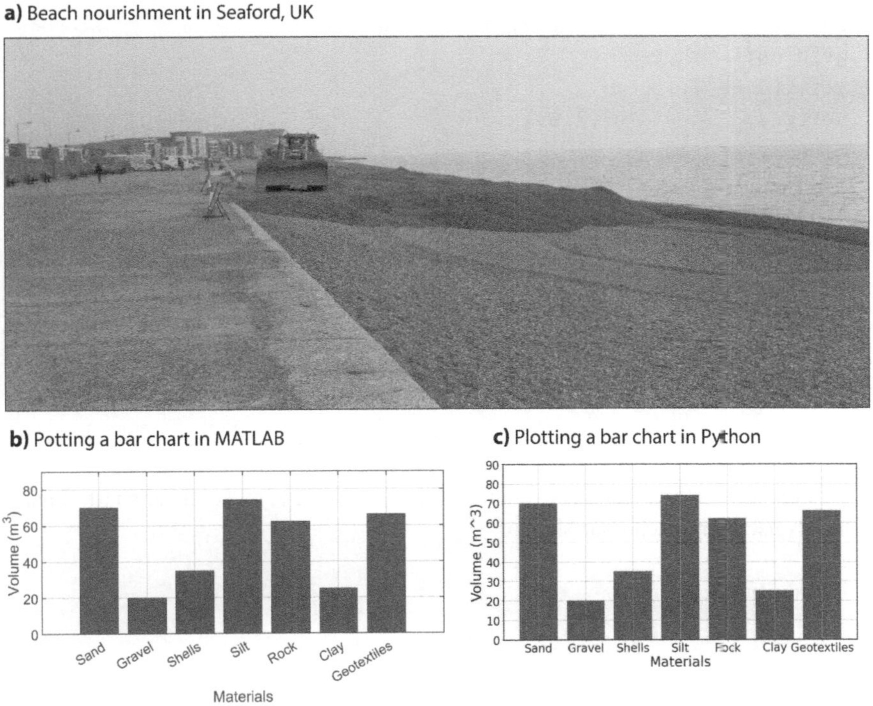

FIGURE 1.10 **a)** A photo of Seaford Beach in the UK showing beach nourishment activities. **b)** A MATLAB bar chart showing the volumes of various materials used for beach nourishment. **c)** A Python bar chart of the same.

MATLAB Input:

```
%this M-file , CH01Mcode7.m,
```

```
%This code makes a bar chart of
%materials used for beach nourishment
clc ; % clear screen
clear ; %clear memory
set(0,'defaultaxesfontsize',30);%font
% Define materials and their volumes
mat={'Sand','Gravel','Shells','Silt', ...
   'Rock', 'Clay', 'Geotextiles'};
vol=[70,20,35,74,62,25,66];%Volumes(m^3)
% Create a bar chart
figure; % Open a new figure
bar(vol,'FaceColor','b'); % blue color
xlabel('Materials');ylabel('Volume (m^3)');
set(gca,'XTickLabel',mat);%x-axis labels
grid on;%Add grid for better readability
ylim([0 90]); % limit of y axis
set(gca,'linewidth',2.0);
% end of M - file
```

MATLAB Output:

```
Note: The output is a plot which is shown in Figure 1.10 in
    this chapter.
```

Example 1.19

Python code. This Python code, named `CH01Pcode7.py`, makes a bar chart of materials used for beach nourishment.

Python Input:

```
# Python script to plot a bar chart
# of materials for beach nourishment
import matplotlib.pyplot as plt
# Define materials and volumes
mat = [
   'Sand', 'Gravel', 'Shells',
   'Silt', 'Rock', 'Clay',
   'Geotextiles'
]
vol = [70,20,35,74,62,25,66]#Volumes(m^3)
# Create a bar chart
```

```
plt.figure() #Open a new figure
plt.bar(mat, vol, color='b')#Blue color
plt.xlabel('Materials', fontsize=30)
plt.ylabel('Volume (m^3)', fontsize=30)
plt.xticks(fontsize=25) # Set fontsize
plt.yticks(fontsize=25)
plt.grid(True) # Add grids
plt.ylim([0, 90])# Limit of y-axis
plt.gca().spines['top'].set_linewidth(2.0)
plt.gca().spines['right'].set_linewidth(2.0)
plt.gca().spines['left'].set_linewidth(2.0)
plt.gca().spines['bottom'].set_linewidth(2.0)
# Show the plot
plt.tight_layout() #Adjust layout to fit
plt.show()
# End of Python code.
```

Python Output:

```
Note: the output of this code is a plot which is shown in
    Figure 1.10 in this chapter.
```

1.5 BEST PRACTICES AND DEBUGGING

1.5.1 BEST PRACTICES

Clear code structure: It is essential to write clear and well-structured code by adhering to proper indentation, spacing, and the use of meaningful comments. Code should be organized into logical sections—such as input, calculations, and output—to improve readability and maintainability.

Naming variables: Variables should be assigned meaningful and descriptive names rather than arbitrary letters. For instance, use V to represent volume and M to represent mass, ensuring clarity and consistency.

Code readability: Code should be written in a manner that is easily understood by others, or even by the programmer at a later time. Comments should be used to explain complex or non-obvious sections of the code while avoiding excessive commentary on straightforward statements.

DRY principle: The "Don't Repeat Yourself" (DRY) principle means that the same code should not be written more than once. Instead of duplicating code, loops, functions, or separate scripts should be used to manage repeated tasks. This approach not only saves time but also makes the code easier to update and maintain.

Avoid hard-coding: Avoiding hard-coding means not directly typing fixed values (such as numbers or text) into the code. Instead, variables should be used to store these values. This approach makes the code easier to modify later and more flexible. For example, rather than writing $V = 60$, it is better to use $V = V_{\text{max}}$ so that the value of V_{max} can be easily changed without altering the entire code.

Preallocate arrays: Preallocating arrays is useful because it improves the efficiency of the code, especially when working with large datasets. By allocating memory for an array in advance, the program avoids repeatedly resizing the array as data is added, which saves both time and computational resources. **In MATLAB**, preallocating arrays using commands such as `zeros( )`, `ones( )`, or `nan( )` is a common practice to improve performance, particularly when working with large arrays. Similarly, **in Python**, especially when using the `NumPy` library, arrays can be preallocated with functions like `np.zeros( )`, `np.ones( )`, or `np.empty( )` to achieve the same efficiency benefits.

Vectorization: Vectorization means performing operations on entire arrays without using loops. **In MATLAB**, vectorization is a key feature that allows users to apply operations to whole arrays at once, resulting in faster execution. Similarly, **in Python**, especially with libraries like `NumPy` or `pandas`, vectorization can be used to efficiently handle operations on arrays or Data Frames without the need for explicit loops.

1.5.2 DEBUGGING

Read error messages: Error messages in both MATLAB and Python should be read carefully, as they offer valuable insights into what went wrong and indicate the specific location of the issue within the code.

Breakpoints: Breakpoints enable the program to pause at specific points, allowing for a detailed inspection of variables and the program's behavior step by step. **In MATLAB**, breakpoints can be set directly within the editor, while **in Python**, this functionality is available through tools such as the `pdb` (Python Debugger) module or Integrated Development and Learning Environment (IDLE).

Step-by-step execution: Stepping through the code one line at a time is an effective method for identifying where errors occur. Both MATLAB and Python provide functionality that allows users to step into, over, or out of functions, enabling a detailed observation of how each line of code impacts the variables.

Inspecting variables: In MATLAB, variable values can be displayed using the workspace or functions such as `disp( )` and `fprintf( )`. In Python, similar inspection is achieved with the `print( )` function or debugging tools like `pdb`. These techniques allow users to monitor how variables change during program execution.

Commenting out code: Temporarily commenting out sections of code is an effective technique for isolating the source of an error. This approach allows for testing individual parts of the program in both MATLAB and Python, making it easier to identify and resolve issues.

Error tracing: When an error occurs, it is essential to trace it back to its origin. Understanding how the error message relates to the underlying issue in the code greatly aids in identifying and resolving the problem more effectively.

1.6 PROBLEMS

Problem 1.1: Consider the simply-supported beam of length $L = 10$ m with a point load $F = 20$ kN acting at a distance $a = 4$ m from the left support shown in Figure 1.11. Make MATLAB and Python codes that calculate and print the reactions at the supports R_A and R_B.

Answer: $R_A = 12$ kN and $R_B = 8$ kN.

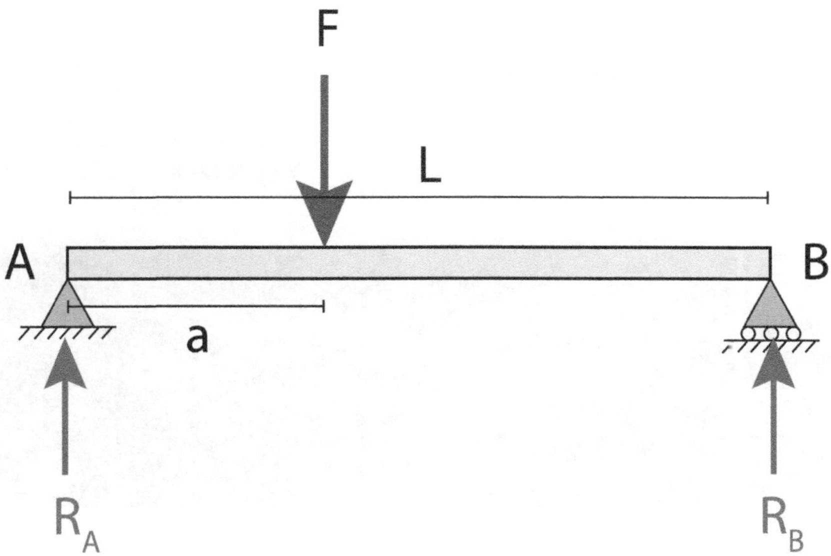

FIGURE 1.11 Sketch showing a simply-supported beam under a point load F at the distance of a from its left end.

Problem 1.2: Considering Problem 1.10, develop MATLAB and Python codes that utilize a "`for Loop`" to calculate and plot the variations of reactions R_A and R_B as the distance a varies from 0 to $L/2$ (i.e., the midpoint of the beam) in 0.5 m intervals.

Answer: For $a = 0$, the answers are: $R_A = 20$ kN and $R_B = 0$ kN. For $a = 2.5$ m, the answers are $R_A = 15$ kN and $R_B = 5$ kN. For $a = 5$ m, the answers are: $R_A = 10$ kN and $R_B = 10$ kN.

Problem 1.3: Redo Problem 1.2 by implementing a "`while loop`" to calculate and analyze the variations of reactions R_A and R_B as the distance a varies from 0 to $L/2$ (i.e., the midpoint of the beam) in 0.5 m intervals.

Problem 1.4: Figure 1.12 illustrates the Clifton suspension bridge in Bristol, UK, with a main span of approximately 214 m and tower heights of approximately 26 m above the deck. The shape of the bridge's cables can be approximated by a quadratic equation in the form of $Y = 0.00227X^2$. Assume that the origin of the coordinate system is placed at the midpoint of the bridge (see Figure 1.12). Develop MATLAB and Python scripts to plot this curve.

Clifton Suspension Bridge, Bristol (UK)

FIGURE 1.12 The Clifton suspension bridge in Bristol, UK.

Problem 1.5: Using the data provided in Problem 1.4, develop MATLAB and Python codes to calculate the height of the cable above the bridge deck at distances of 10 m, 50 m, and 85 m from the center of the bridge.

Answer: The heights of the cable are $y = 0.23$ m (for $x = 10$ m), $y = 5.68$ m (for $x = 50$ m), and $y = 16.4$ m (for $x = 85$ m).

Problem 1.6: The discharge rate of a river (Q) is calculated as the product of the cross-sectional area (A) and the water velocity (u):

$$Q = A \cdot u \tag{1.7}$$

Assuming a cross-sectional area of $A = 56\,\text{m}^2$, develop MATLAB and Python scripts to calculate the river's discharge rate for velocities in the range $u = 0.5$ to 5.5 m/s, using 0.1 m/s intervals.

Problem 1.7: The quantities of various materials used for beach nourishment at a coastal site are shown in Table 1.7. Plot these quantities using pie charts in both MATLAB and Python.

Problem 1.8: In coastal engineering, the significant wave height (H_s) is defined as the average height of the highest one-third of the waves. Using the wave records provided in Table 1.8, calculate H_s by writing MATLAB and Python scripts.

Answer: The significant wave height is $H_s = 1.3$ m.

TABLE 1.8

Wave heights for significant wave height calculation

Wave number	Wave height (m)
1	0.5
2	0.8
3	0.7
4	1.2
5	1.0
6	0.6
7	1.3
8	0.9
9	1.1
10	1.4
11	0.7
12	0.8
13	1.5
14	1.0

Problem 1.9: Using the data shown in Table 1.8, plot the wave heights using scatter plots in both MATLAB and Python.

Problem 1.10: 1.100: Referring to the wave records in Table 1.8, the goal is to count the number of waves with heights greater than or equal to 1.0 m. Use an "`if statement`" to accomplish this task in both MATLAB and Python.

Answer: The number of waves with heights greater than or equal to 1.0 m is 7.

REFERENCES

[1] David Amon et al. *Python Basics: A Practical Introduction to Python 3*. 2015.

[2] Mohammad Heidarzadeh et al. "Estimating the eruption-induced water displacement source of the 15 January 2022 Tonga volcanic tsunami from tsunami spectra and numerical modelling." In: *Ocean Engineering* 261 (2022), p. 112165.

[3] Mohammad Heidarzadeh et al. "Historical tsunami in the Makran Subduction Zone off the southern coasts of Iran and Pakistan and results of numerical modeling." In: *Ocean Engineering* 35(8-9) (2008), p. 774–786.

[4] Mohammad Heidarzadeh et al. "Storm wave runups and sea level variations for the September 2017 Hurricane Maria along the coast of Dominica, eastern Caribbean sea: Evidence from field surveys and sea-level data analysis." In: *Coastal Engineering Journal* 60(3) (2018), p. 371–384.

[5] Kenneth Lambert. *Fundamentals of Python: First Programs*. Cengage, 2019, p. 498.

[6] The MathWorks, Inc. *MATLAB R2023b*. Software available: `https://www.mathworks.com`. 2023.

2 Matrices, Vectors, and Systems of Equations

In engineering and applied sciences, understanding matrices, vectors, and systems of equations is essential, as they form the mathematical foundation for modeling and solving real-world problems. Matrices and vectors provide a structured way to handle and manipulate large sets of data, allowing for efficient computations in various fields, from structural analysis to electrical circuit design. By exploring these mathematical tools, we gain the ability to represent and solve complex systems, enabling us to model physical phenomena, optimize processes (e.g., [3]), and make informed decisions in engineering practice.

For example, in structural analysis, matrices represent the relationships between loads and forces in structures like bridges, ensuring stability and safety (Figure 2.1a). In traffic flow modeling, vectors track road volume, and systems of equations help optimize flow across networks (Figure 2.1b). Hydraulic network analysis also relies on matrices to represent flow and pressure in pipes, allowing engineers to determine efficient water distribution.

2.1 VECTORS AND MATRICES

In Chapter 1, we provided a brief introduction to creating vectors and matrices in MATLAB and Python. In this chapter, we delve deeper into operations and applications involving vectors and matrices, building on the foundational concepts introduced earlier.

A matrix is a rectangular array of numbers arranged in rows and columns. It is represented by its dimensions, $n \times m$, where n is the number of rows, and m is the number of columns. For example, a 5 × 3 matrix has 5 rows and 3 columns, as shown below:

$$B = \begin{bmatrix} 1 & 2 & 3 \\ 4 & 5 & 6 \\ 7 & 8 & 9 \\ 10 & 11 & 12 \\ 13 & 14 & 15 \end{bmatrix}$$

Each element of a matrix can be accessed using its position. For example, B(1, 3) refers to the element in the 1st row and 3rd column of the matrix, which is 3. This means you are selecting the value that is at the intersection of the first row and third column in the matrix. Similarly, B(4, 2) points to the element in the 4th row and 2nd column, which is 11. And B(5, 1) refers to the element in the 5th row and 1st column which is 13.

A vector is a special case of a matrix that has only one dimension. This means that either n=1, making it a row vector, or m=1, making it a column vector. For example, a

DOI: 10.1201/9781003511854-2

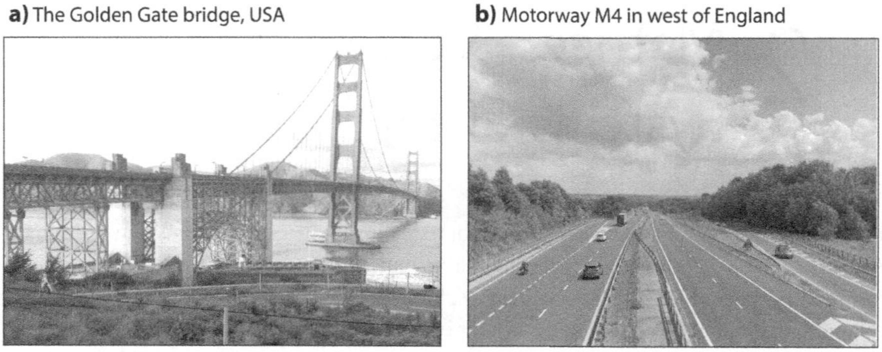

FIGURE 2.1 Examples of infrastructure systems where matrices and vectors are usually used for their design. **a)** The Golden Gate suspension bridge in San Francisco, USA. **b)** The Motorway M4 in west of England.

row vector with dimension 1 × 4 is shown below:

$$K = \begin{bmatrix} 1 & 6.5 & -3 & 0.65 \end{bmatrix}$$

In MATLAB, a matrix is created by specifying rows separated by semicolons, with elements in each row separated by spaces or commas. The matrix B above, which is a 5×3 matrix, can be written as: $B = [1, 2, 3; 4, 5, 6; 7, 8, 9; 10, 11, 12; 13, 14, 15]$ in MATLAB. And the vector K, presented earlier, is written using $K = [1, 6.5, -3, 0.65]$ in MATLAB. Considering matrix B, different elements of this matrix are accessed as: $B(1, 3) = 3$; $B(4, 2) = 11$; and $B(5, 1) = 13$.

In Python, a matrix can be defined in different ways, including using a "list of lists" or by utilizing the powerful `NumPy` library, which is specifically designed for numerical computations. Using "list of lists," the previously defined matrix B can be written as: B = [[1, 2, 3], [4, 5, 6], [7, 8, 9], [10, 11, 12], [13, 14, 15]]. By using the `NumPy` library, the same matrix is introduced as follows:

```
import numpy as np
B = np.array([[1, 2, 3], [4, 5, 6], [7, 8, 9], [10,
11, 12], [13, 14, 15]])
```

In Python, when using the `NumPy` library, the element at the first row and third column of matrix B (i.e., $B(1, 3)$ in MATLAB) is accessed as $B[0][2]$. Similarly, the element in the fourth row and second column (i.e., $B(4, 2)$ in MATLAB) is accessed as $B[3][1]$ in Python. And the element in the fifth row and first column (i.e., $B(5, 1)$ in MATLAB) is accessed as $B[4][0]$ in Python. It is important to remember that Python uses zero-based indexing, meaning it starts counting rows and columns from 0, unlike MATLAB, which starts from 1.

Engineering application: Application of vectors and matrices in engineering is discussed using an example from road construction (Figure 2.2). A road construction project involves three sections (A, B, and C), each requiring specific amounts of gravel, sand, and cement whose amounts are given in Figure 2.2b. These materials are represented in a matrix to organize the data, and vectors are used to perform calculations like total materials required for each section or cost estimation. First, the materials are represented in matrix

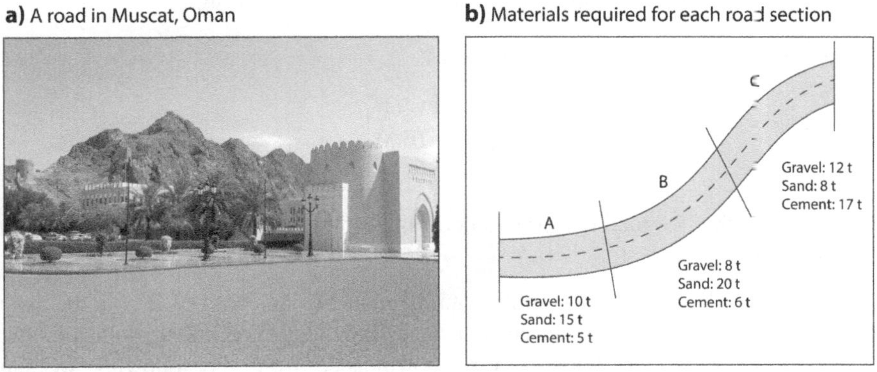

FIGURE 2.2 Application of Matrices and Vectors in Road Engineering. **a)** A road in Muscat, Oman. **b)** A three-section road and the weights of materials required for the construction of each section.

M where each row represents a section of the road, and each column represents a type of material (in tons):

$$\mathbf{M} = \begin{bmatrix} 10 & 15 & 5 \\ 8 & 20 & 6 \\ 12 & 18 & 7 \end{bmatrix}$$

The cost per ton for each material, in order of gravel, sand, and cement, is stored in a vector named **C** as follows:

$$\mathbf{C} = \begin{bmatrix} 50 \\ 30 \\ 100 \end{bmatrix}$$

where, cost of gravel = £50/ton, cost of sand = £30/ton, and cost of cement = £100/ton.

The total cost of construction ($\mathbf{C}_T$) of the road can be obtained by multiplying each row of the material matrix (**M**) by the cost vector (**C**) using the following equation:

$$\mathbf{C}_T = \mathbf{M} \cdot \mathbf{C} \tag{2.1}$$

This matrix multiplication can be done as follows:

$$\mathbf{C}_T = \begin{bmatrix} 10 & 15 & 5 \\ 8 & 20 & 6 \\ 12 & 18 & 7 \end{bmatrix} \cdot \begin{bmatrix} 50 \\ 30 \\ 100 \end{bmatrix}$$

In a step-by-step manner, the above multiplication is performed as follows:

- Cost of section 1: $(10 \times 50) + (15 \times 30) + (5 \times 100) = 1450$
- Cost of section 2: $(8 \times 50) + (20 \times 30) + (6 \times 100) = 1600$
- Cost of section 3: $(12 \times 50) + (18 \times 30) + (7 \times 100) = 1840$

Therefore, the final total cost vector becomes:

$$\mathbf{C}_T = \begin{bmatrix} 1450 \\ 1600 \\ 1840 \end{bmatrix}$$

This indicates that the total construction cost is £1450 for section A, £1600 for section B, and £1840 for section C.

This procedure and calculations can be done using MATLAB via the code presented in Example 2.1. This code, named `CH02Mcode1.m`, along with all other scripts presented in this book, can be downloaded from the following website: https://github.com/prog871/CRC-book2025

The corresponding Python script, `CH02Pcode1.py`, is provided in Example 2.2 and can be downloaded from the aforementioned website.

Example 2.1

MATLAB code. This MATLAB script, named `CH02Mcode1.m`, calculates the total cost of construction of a three-section road. This script, along with all other code presented in this book, can be downloaded from the following website: https://github.com/prog871/CRC-book2025

MATLAB Input:

```
% start of the M-file
% This code calculates the total
% construction cost for each
% section of a 3 section road
clc ; % clear screen
clear ; % clear ariables from the memory
%
M=[10,15,5;8,20,6;12,18,7];
% Cost per ton (GBP)
C = [50; 30; 100];
% Total cost calculation (C_T)
C_T = M * C;
% Display results
disp('Total cost for each section (in GBP):');
disp(C_T);
```

```
%
% end of M-file
```

MATLAB Output:

```
Total cost for each section (in GBP):
      1450
      1600
      1840
```

Example 2.2

Python code. This Python script, named `CH02Pcode1.py`, calculates the total cost of construction of a three-section road. It can be downloaded at the following website: https://github.com/prog871/CRC-book2025

Python Input:

```
# This code calculates the total
# construction cost for each
# section of a 3-section road
import numpy as np #Import NumPy
# Define matrix of quantities (M) in tons
M = np.array([
   [10, 15, 5],
   [8, 20, 6],
   [12, 18, 7]
])
# Define cost per ton (GBP) as a vector
C = np.array([50, 30, 100])
# Calculate total cost for each section (C_T)
C_T = np.dot(M, C)
# Display results
print("Total cost for each section (GBP):")
print(C_T)
# End of Python script
```

Python Output:

```
Total cost for each section (GBP):
[1450 1600 1840]
```

Engineering application: In coastal engineering, oceanic waves, such as those generated by storms, swells, and tsunamis, are typically represented as a two-column matrix. The first column denotes time, while the second column reflects the sea level oscillations caused by the oceanic phenomenon. An example of such data is presented in Table 2.1, which illustrates the case of the September 2018 tsunami in Palu, Indonesia, a disaster that claimed more than 4,000 lives [1] (Figure 2.3a). It is important to note that the time series data shown in Table 2.1 covers only the first 25 minutes of the Palu tsunami, although the tsunami generated sea level oscillations for more than six hours.

TABLE 2.1

The sea level oscillations caused by the September 2018 Palu tsunami in Indonesia [1, 2]

Time (min)	Sea level (cm)
1	1.37
2	1.67
3	1.98
4	2.29
5	2.60
6	1.92
7	3.24
8	2.57
9	2.90
10	3.24
11	3.58
12	2.93
13	3.28
14	3.63
15	2.99
16	3.35
17	2.72
18	3.09
19	3.47
20	10.86
21	-3.76
22	24.64
23	-202.97
24	-172.56
25	176.84

The data in Table 2.1 is organized in a two-column matrix, where the first column represents time in minutes, and the second column represents sea level oscillations in centimeters, caused by the 2018 Palu tsunami in Indonesia [1, 2]. This matrix is an example of a time series where each row corresponds to a specific time point and the associated sea level measurement. The matrix has dimensions of $m \times 2$, where m is the number of time points recorded. For example, if the data includes 25 time points, the matrix will be a 25×2 matrix, with each row representing one time-sea level pair.

Here, we can use MATLAB and Python to plot the time series of the Palu tsunami presented in Table 2.1. The plot of the tsunami oscillations is shown in Figure 2.3b using the MATLAB script `CH02Mcode2.m` (presented in Example 2.3). Note that this MATLAB script requires an input file named `CH02-data-tsunami-Palu.txt` which can be downloaded from the book website at: https://github.com/prog871/CRC-book2025.

The corresponding Python script is provided in Example 2.4 with the script named as `CH02Pcode2.py`, which plots the sea level oscillations caused by the 2018 Palu tsunami in Indonesia (see Figure 2.3b).

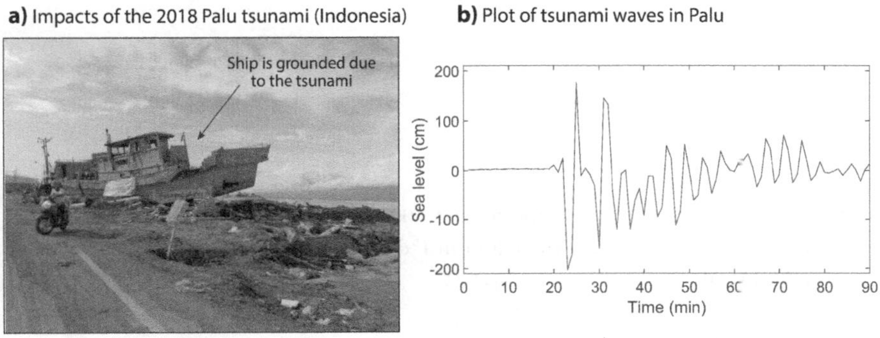

FIGURE 2.3 **a)** An example of damage made by the September 2018 Palu tsunami in Indonesia where a ship is moved inland by the tsunami [1, 2]. **b)** Plot of the tsunami oscillations.

Example 2.3

MATLAB code. This MATLAB script, named `CH02Mcode2.m`, plots the sea level oscillations caused by the 2018 tsunami in Palu, Indonesia. It requires an input file named `CH02-data-tsunami-Palu.txt`.

MATLAB Input:

```
% This code plots the oscillatins
% of the 2018 Palu tsunami
% note that this codes needs
% an input file named
% 'CH02-data-tsunami-Palu.txt'.
clear; close all; clc
set(0,'defaultaxesfontsize',28);
% Raed input data (tsunami time series)
data=load('CH02-data-tsunami-Palu.txt');
% plot time series
plot(data(:,1),data(:,2),'b','linewidth',1.5)
```

```
% add axis lables
xlabel('Time (min)'); ylabel('Sea level (cm)');
% limit of x-axis & y-axis
xlim([0 90]); ylim([-210 210]);
set(gca,'linewidth', 1.5);
% end of M - file
```

MATLAB Output:

```
The outcome of this code is a figure which is shown in Figure
    2.3b.
```

Example 2.4

Python code. This Python script, named `CH02Pcode2.py`, plots the sea level oscillations caused by the 2018 tsunami in Palu, Indonesia. It requires an input file named `CH02-data-tsunami-Palu.txt`.

Python Input:

```
import numpy as np
import matplotlib.pyplot as plt
# Read input data (tsunami time series)
data=np.loadtxt('CH02-data-tsunami-Palu.txt')
# Plot time series
plt.plot(data[:,0],data[:,1],'b',linewidth=1.5)
# Add axis labels
plt.xlabel('Time (min)')
plt.ylabel('Sea level (cm)')
# Set limits for x-axis and y-axis
plt.xlim([0, 90])
plt.ylim([-210, 210])
# Set axis line width
plt.gca().tick_params(width=1.5)
# Show the plot
plt.show()
# End of Python code
```

Python Output:

```
The outcome of this code is a figure which is shown in Figure
    2.3b.
```

2.2 SYSTEMS OF LINEAR EQUATIONS

In many engineering applications, systems of linear equations arise, which need to be solved simultaneously to determine key parameters for the associated engineering problem. These equations often represent physical laws, constraints, or relationships between variables, such as forces in structures, currents in electrical circuits, or flow rates in hydraulic networks. Understanding how to solve such systems is a fundamental skill for engineers and provides a foundation for more advanced concepts.

In the simplest case, consider the following linear system comprising two equations:

$$\begin{cases} a_1x + b_1y = c_1, \\ a_2x + b_2y = c_2. \end{cases} \tag{2.2}$$

This system represents two straight lines in a two-dimensional plane, where x and y are the unknowns, and $a_1, a_2, b_1, b_2, c_1, c_2$ are known constants. The solution to this system is the point of intersection of these two lines, provided they intersect. An example of a linear system consisting of two equations is given below:

$$\begin{cases} 2x + 3y = 6, \\ x - y = 1. \end{cases} \tag{2.3}$$

The solution to this linear system corresponds to the point where the two lines intersect, which occurs at $x = 1.8$ and $y = 0.8$, as illustrated in Figure 2.4.

Systems of linear equations, such as the one presented in Equations 2.2, can be solved using a variety of methods, including substitution, elimination, and matrix-based techniques. These methods are commonly introduced in high school and further explored in college-level mathematics and engineering courses. In this context, we will focus on matrix-based methods to solve the system, using both MATLAB and Python for implementation.

To solve a system of linear equations in MATLAB, we can utilize matrix operations, specifically matrix inversion or matrix division. Below is a step-by-step guide on how to approach this using MATLAB.

Step 1: Represent the system of equations in matrix form.

The system of equation shown in Equation 2.2 can be represented in matrix form as:

$$A \cdot x = b \tag{2.4}$$

where,

$$A = \begin{bmatrix} a_1 & a_2 \\ b_1 & b_2 \end{bmatrix} \tag{2.5}$$

$$x = \begin{bmatrix} x \\ y \end{bmatrix} \tag{2.6}$$

$$b = \begin{bmatrix} c_1 \\ c_2 \end{bmatrix} \tag{2.7}$$

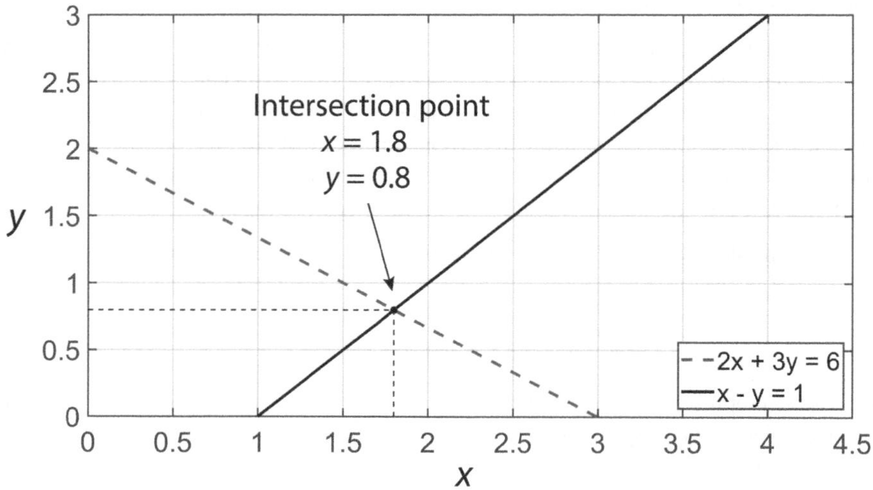

FIGURE 2.4 An example of a linear system consisting of two equations, $2x + 3y = 6$ and $x - y = 1$, whose solution is the point of intersection of the two lines, located at $x = 1.8$ and $y = 0.8$.

Step 2: Define the matrices in MATLAB.

In this step, the matrices shown in Equations 2.5, and 2.7 are represented in MATLAB as follows:

```
A = [a1, b1; a2, b2]; % Coefficient matrix
b = [c1; c2]; % Constant matrix
```

Note that in MATLAB, the comma (,) and semicolon (;) are used to structure matrices appropriately. The comma is used to separate elements within the same row, while the semicolon is used to indicate the end of a row and the start of a new row.

Step 3: Solve the system using matrix division.

MATLAB provides a convenient way to solve systems of linear equations using the backslash operator (\). This operator performs matrix division, which is equivalent to multiplying by the inverse of the coefficient matrix. The solution can be found as:

$$x = A\backslash b; \tag{2.8}$$

Engineering application: Here, we apply the concept of solving systems of linear equations to address an engineering problem. Consider a simply supported beam, as shown in Figure 2.5, which is 20 m long and subjected to a point load of 10 kN located 7 m from its left-hand end. The two endpoints of the beam are designated as points A and B (Figure 2.5). The objective is to calculate the reaction forces (R_A and R_B) at the two supports of the beam.

According to the principles of statics, we know that the sum of the vertical forces (F_y) must be zero, which gives the following equation:

$$\sum F_y = 0 \tag{2.9}$$

which results in the following equation referring to Figure 2.5b:

$$R_A + R_B = 10 \tag{2.10}$$

Based on the principles of statics, the total moment (M) about point A must be zero,

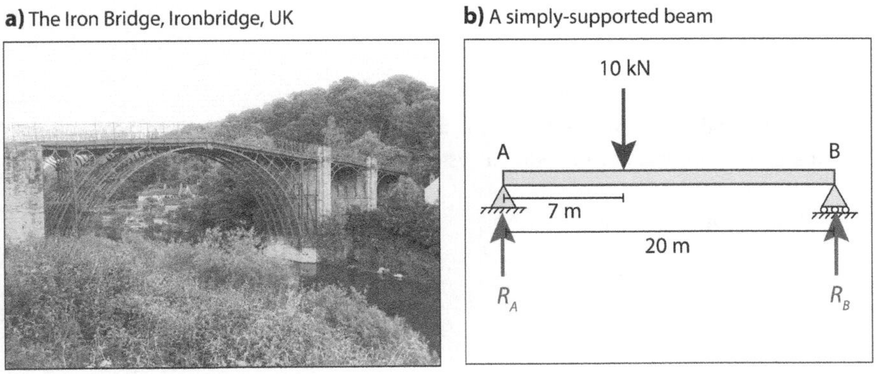

FIGURE 2.5 a) The Iron Bridge located in Ironbridge, UK. b) A simply-supported beam.

leading to the following equation:

$$\sum M_A = 0 \tag{2.11}$$

Since the reaction at point A does not produce any moment about point A, Equation 2.11, assuming clockwise moments are negative, leads to the following:

$$(20 \times R_B) - (7 \times 10) = 0 \tag{2.12}$$

which is rearranged to:

$$(20 \times R_B) = 70 \tag{2.13}$$

which can be rewritten as:

$$(0 \times R_A) + (20 \times R_B) = 70 \tag{2.14}$$

Here, Equations 2.10 and 2.14 form a system of two linear equations as follows:

$$\begin{cases} R_A + R_B = 10, \\ (0 \times R_A) + (20 \times R_B) = 70. \end{cases} \tag{2.15}$$

In matrix form, Equation 2.15 is represented as follows:

$$\begin{bmatrix} 1 & 1 \\ 0 & 20 \end{bmatrix} \begin{bmatrix} R_A \\ R_B \end{bmatrix} = \begin{bmatrix} 10 \\ 70 \end{bmatrix} \tag{2.16}$$

We apply Equation 2.8 to solve this linear system. This can be achieved using MATLAB and Python, as demonstrated in Example 2.5 and Example 2.6, respectively. The corresponding scripts are named `CH02Mcode3.m` and `CH02Pcode3.py`. The outputs of these scripts yield the following vector, representing R_A and R_B:

$$\mathbf{x} = \begin{bmatrix} R_A \\ R_B \end{bmatrix} = \begin{bmatrix} 6.5 \\ 3.5 \end{bmatrix}$$

Example 2.5

MATLAB code. This MATLAB script, named `CH02Mcode3.m`, solves the reactions of a simply-supported beam using a system of two linear equations.

MATLAB Input:

```
% This code solves the reactions
% of a simply-supported beam
% using a system of two linear
% equations
clear; close all; clc
set(0,'defaultaxesfontsize',28);
% Define the coefficient matrix A
A = [1, 1; 0, 20];
% Dfine the right-hand side vector b
b = [10; 70];
% Solve the system Ax = b
x = A \ b;
% Display the results
R_A = x(1);
R_B = x(2);
% prinr the results
fprintf('R_A is %.2f kN.\n', R_A);
fprintf('R_B is %.2f kN.\n', R_B);
% end of M - file
```

MATLAB Output:

```
R_A is 6.50 kN.
R_B is 3.50 kN.
```

Example 2.6

Python code. This Python code, named `CH02Pcode3.py`, solves the reactions of a simply-supported beam using a system of two linear equations.

Python Input:

```
#This code solves the reactions
#of a simply-supported beam
#using a system of 2 linear equations
import numpy as np
# Define the coefficient matrix A
A = np.array([[1, 1], [0, 20]])
# Define right-hand side vector b
b = np.array([10, 70])
# Solve the system Ax = b
x = np.linalg.solve(A, b)
# Extract the results
R_A = x[0]
R_B = x[1]
# Print the results
print(f'R_A is {R_A:.2f} kN.')
print(f'R_B is {R_B:.2f} kN.')
# End of Python script
```

Python Output:

```
R_A is 6.50 kN.
R_B is 3.50 kN.
```

Engineering application: As another engineering application, we use a system of linear equations to calculate the forces in a triangular truss of three elements subjected to an external load of 10 kN, as shown in Figure 2.6b. A steel truss bridge located in the Caribbean island nation of Dominica is shown in Figure 2.6a. The purpose of this exercise is to determine the internal forces in the truss elements (F_{AB}, F_{AC}, and F_{BC}). The truss has a pinned support at joint B and a roller support at joint C (see Figure 2.6b). Based on the principles of statics, two reactions must be considered at joint B, denoted as R_{Bx} and R_{By}, since it is a pinned joint. In contrast, only one reaction, R_{Cy}, exists at joint C because it is a roller support (see Figure 2.6c).

A widely used method for solving trusses is **the method of joints**, a static analysis technique that examines each joint to ensure equilibrium. This involves ensuring that the sum of all forces acting on the members connected to each joint is zero, maintaining the overall equilibrium of the truss. Based on the free-body diagrams shown in Figure 2.6c,

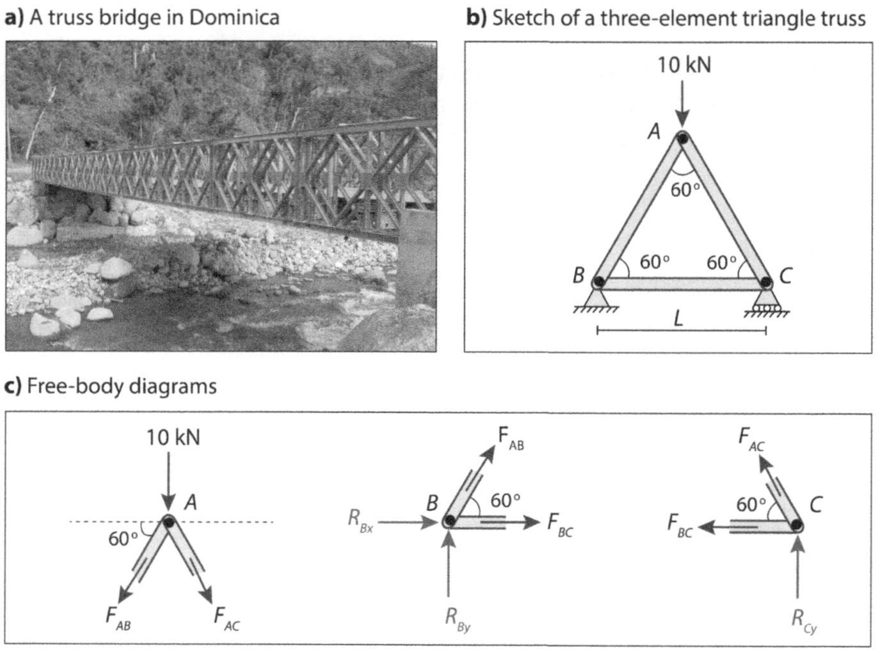

FIGURE 2.6 **a)** A truss bridge in Dominica. **b)** Sketch showing a three-element triangle truss under an external load of 10 kN. **c)** The free-body diagrams at the joints A, B, and C.

the following two equations apply at each joint:

$$\sum F_x = 0 \tag{2.17}$$

$$\sum F_y = 0 \tag{2.18}$$

Here, F_x and F_y represent the components of the forces acting in the horizontal and vertical directions, respectively. These equations ensure equilibrium in both directions for each joint in the truss. Applying Equations 2.17 and 2.18 to joint A, the following relationships are obtained, assuming that forces directed to the right and upward are considered positive:

$$F_{AC}\cos(60°) - F_{AB}\cos(60°) = 0 \tag{2.19}$$

$$-10 - F_{AC}\sin(60°) - F_{AB}\sin(60°) = 0 \tag{2.20}$$

At joint C (see Figure 2.6c), we apply only Equation 2.17 which results in the following relationship:

$$-F_{BC} - F_{AC}\cos(60°) = 0 \tag{2.21}$$

Examining Equations 2.19, 2.20, and 2.21, it can be observed that a system of three linear equations has been formed, with three unknowns (F_{AB}, F_{AC}, and F_{BC}). This linear system is represented in matrix form as follows:

$$\begin{bmatrix} \frac{1}{2} & -\frac{1}{2} & 0 \\ -\frac{\sqrt{3}}{2} & -\frac{\sqrt{3}}{2} & 0 \\ -\frac{1}{2} & 0 & -1 \end{bmatrix} \begin{bmatrix} F_{AC} \\ F_{AB} \\ F_{BC} \end{bmatrix} = \begin{bmatrix} 0 \\ 10 \\ 0 \end{bmatrix} \tag{2.22}$$

The linear system presented in Equation 2.22 can be solved by applying Equation 2.8 for linear systems, yielding the following outcome:

$$\begin{bmatrix} F_{AC} \\ F_{AB} \\ F_{BC} \end{bmatrix} = \begin{bmatrix} -5.77 \\ -5.77 \\ 2.89 \end{bmatrix}$$

Therefore, $F_{AC} = -5.77$ kN, $F_{AB} = -5.77$ kN, and $F_{BC} = 2.89$ kN. The negative force values indicate that the actual direction of these forces is opposite to those assumed in Figure 2.6c.

The MATLAB and Python scripts, named `CH02Mcode4.m` and `CH02Pcode4.py`, presented in Examples 2.7 and 2.8, respectively, can be used to solve the system of linear equations presented in Equation 2.22.

Example 2.7

MATLAB code. This MATLAB script, named `CH02Mcode4.m`, solves the internal forces of a triangular truss using a system of three linear equations.

MATLAB Input:

```
% This code solves the forces
% of a triangular truss
% using a system of three linear
% equations
clear; close all; clc
set(0,'defaultaxesfontsize',28);
% Define the coefficient matrix A
A = [1/2, -1/2, 0;
    -sqrt(3)/2, -sqrt(3)/2, 0;
    -1/2, 0, -1];
% Define the constants vector b
b = [0; 10; 0];
% Solve for unknowns (F_AC,F_AB,F_BC)
x = A \ b;
% Display the solution
disp('Solution:');
disp(['F_AC = ',num2str(x(1))]);
disp(['F_AB = ',num2str(x(2))]);
```

```
disp(['F_BC = ',num2str(x(3))]);
% end of M - file
```

MATLAB Output:

```
Solution:
F_AC = -5.7735
F_AB = -5.7735
F_BC = 2.8868
```

Example 2.8

Python code. This Python code, named `CH02Pcode4.py`, solves the internal forces of a triangular truss using a system of three linear equations.

Python Input:

```
# This code solves the forces
# of a triangular truss
# using a system of three
# linear equations
import numpy as np
# Define coefficient matrix A
A = np.array([[1/2, -1/2, 0],
  [-np.sqrt(3)/2,-np.sqrt(3)/2,0],
  [-1/2, 0, -1]])
# Define constants vector b
b = np.array([0, 10, 0])
# Solve for F_AC,F_AB,F_BC
x = np.linalg.solve(A, b)
# Display the solution
print("Solution:")
print(f"F_AC = {x[0]}")
print(f"F_AB = {x[1]}")
print(f"F_BC = {x[2]}")
```

Python Output:

```
Solution:
F_AC = -5.773502691896258
F_AB = -5.773502691896258
F_BC = 2.886751345948129
```

2.3 DETERMINANTS AND MATRIX INVERSION

2.3.1 DETERMINANTS

The **determinant** of a matrix is a special number that provides useful information about the matrix. For a 2×2 matrix A, the determinant, $det(A)$, is calculated as:

$$A = \begin{bmatrix} a & b \\ c & d \end{bmatrix} \tag{2.23}$$

$$\det(A) = ad - bc \tag{2.24}$$

For example, given the matrix:

$$A = \begin{bmatrix} 1 & 2 \\ 3 & 4 \end{bmatrix} \tag{2.25}$$

The determinant is:

$$\det(A) = (1 \times 4) - (2 \times 3) = 4 - 6 = -2 \tag{2.26}$$

In mathematical notation, the determinant of matrix A is represented by the following symbol:

$$\det(A) = \begin{vmatrix} 1 & 2 \\ 3 & 4 \end{vmatrix} = -2 \tag{2.27}$$

For a 3×3 matrix named B as follows:

$$B = \begin{bmatrix} a & b & c \\ d & e & f \\ g & h & i \end{bmatrix} \tag{2.28}$$

The determinant of B, denoted $\det(B)$, is calculated as:

$$\det(B) = \begin{vmatrix} a & b & c \\ d & e & f \\ g & h & i \end{vmatrix} = a(ei - fh) - b(di - fg) + c(dh - eg) \tag{2.29}$$

The determinant helps us understand the properties of the matrix. If the determinant is **non-zero**, the matrix has an inverse. If the determinant is **zero**, the matrix does not have an inverse. Manual calculation of determinants for large matrices is both cumbersome and time-consuming. Therefore, using computational tools such as MATLAB or Python is highly recommended. In **MATLAB**, the determinant of a matrix, named A, can be calculated using the following command:

$$\text{determinant} = \det(A) \tag{2.30}$$

In **Python**, calculation of the determinant of matrix A is performed using the `numpy` library, specifically the `numpy.linalg.det` function, as follows:

$$\text{determinant} = \text{np.linalg.det(A)} \tag{2.31}$$

Engineering application: The concept of the determinant is highly valuable in engineering, as it can be applied to solve various engineering problems. In surveying and structural engineering, determinants are used to calculate the area of a triangle based on the coordinates of its corners; i.e., $(x_1, y_1), (x_2, y_2), (x_3, y_3)$. This method is particularly useful when analyzing triangular land plots or structural components. Referring to the triangular land plot marked by the triangle ABC in Figure 2.7b, the area of the land can be calculated using the following equation:

$$Area = \frac{1}{2} \begin{vmatrix} x_1 & y_1 & 1 \\ x_2 & y_2 & 1 \\ x_3 & y_3 & 1 \end{vmatrix} \tag{2.32}$$

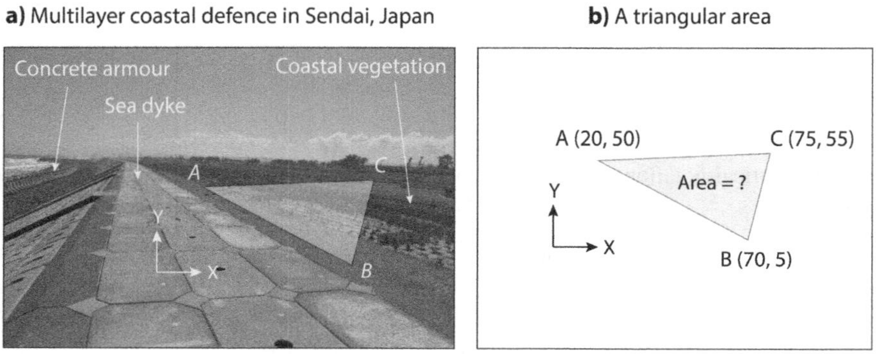

FIGURE 2.7 **a)** The multilayer coastal defense system in Sendai, Japan. **b)** A triangular area whose area is to be calculated.

Given the coordinates of points A, B and C in Figure 2.7b, and applying Equation 2.32, the area of the triangle ABC is calculated as follows:

$$Area = \frac{1}{2} \begin{vmatrix} 20 & 50 & 1 \\ 70 & 5 & 1 \\ 75 & 55 & 1 \end{vmatrix} = 1362.5 \tag{2.33}$$

The area bounded by the triangle ABC in Figure 2.7b is 1362.5 m^2. These calculations can be performed using script `CH02Mcode5.m` (see Example 2.9) in MATLAB, and script `CH02Pcode5.py` (Example 2.10) in Python.

Example 2.9

MATLAB code. This MATLAB script, named `CH02Mcode5.m`, calculates the area of a triangle ABC by knowing the coordinates of A, B, and C.

MATLAB Input:

```
% This code calculates the
% area of a triangle ABC
% by knowing the coordinates
% of A, B and C
clear; close all; clc
set(0,'defaultaxesfontsize',28);
% Give coordinates of A,B,C
A = [20, 50];
B = [70, 5];
C = [75, 55];
% Construct the matrix
matrix = [
   A(1), A(2), 1;
   B(1), B(2), 1;
   C(1), C(2), 1
];
% Calculate the determinant
determinant = det(matrix);
% Calculate the area
Area = 0.5*abs(determinant);
% Display the result
fprintf('Area = %.2f m^2\n',Area);
% end of M - file
```

MATLAB Output:

```
Area = 1362.50 m^2
```

Example 2.10

Python code. This Python code, named `CH02Pcode5.py`, calculates the area of a triangle ABC by knowing the coordinates of A, B, and C.

Python Input:

```
# This code calculates the
# area of a triangle ABC
# by knowing the
# coordinates of A,B,and C
import numpy as np
# Give coordinates A,B,C
A = [20, 50]
```

```
B = [70, 5]
C = [75, 55]
# Construct the matrix
matrix = np.array([
    [A[0], A[1], 1],
    [B[0], B[1], 1],
    [C[0], C[1], 1]
])
# Calculate the determinant
deter = np.linalg.det(matrix)
# Calculate the area
Area = 0.5 * abs(deter)
# Display the result
print(f'Area={Area:.2f} m^2')
# End of Python script
```

Python Output:

```
Area=1362.50 m^2
```

2.3.2 MATRIX INVERSION

The **inverse** of a matrix is a special matrix that, when multiplied by the original matrix, results in the **identity matrix**. The identity matrix acts similarly to the number 1 in regular multiplication. For a matrix A, its inverse A^{-1} satisfies:

$$A \times A^{-1} = I \tag{2.34}$$

where I is the identity matrix. For example, the identity matrix of order 2×2 is given by:

$$I = \begin{bmatrix} 1 & 0 \\ 0 & 1 \end{bmatrix} \tag{2.35}$$

In simple terms, matrix inversion "reverses" the effect of multiplying by the matrix. Not every matrix has an inverse, and a matrix will only have an inverse if its determinant is **non-zero**. The determinant is crucial when determining if a matrix is invertible. If the determinant of a matrix is **zero**, the matrix does not have an inverse. If the determinant is non-zero, the matrix has an inverse.

If the inverse of a matrix exists, then for a 2×2 matrix given by $A = \begin{bmatrix} a & b \\ c & d \end{bmatrix}$, the inverse can be computed using the following formula:

$$A^{-1} = \frac{1}{\det(A)} \begin{bmatrix} d & -b \\ -c & a \end{bmatrix} \tag{2.36}$$

where $\det(A) = ad - bc$ is the determinant of matrix A. The inverse exists only if $\det(A) \neq 0$; otherwise, the matrix is singular and does not have an inverse.

Example: For the following matrix A, determine whether an inverse exists. If it does, compute the inverse matrix A^{-1}.

$$A = \begin{bmatrix} 2 & 1 \\ 3 & 4 \end{bmatrix} \tag{2.37}$$

Solution:

To compute the inverse A^{-1}, we first calculate the determinant of A. For matrix A, the determinant is calculated as:

$$\det(A) = (2)(4) - (1)(3) = 8 - 3 = 5 \tag{2.38}$$

Since the determinant is non-zero, the matrix is invertible.

For a 2×2 matrix $A = \begin{bmatrix} a & b \\ c & d \end{bmatrix}$, the formula for the inverse is:

$$A^{-1} = \frac{1}{\det(A)} \begin{bmatrix} d & -b \\ -c & a \end{bmatrix} \tag{2.39}$$

Substituting the values for matrix A:

$$A^{-1} = \frac{1}{5} \begin{bmatrix} 4 & -1 \\ -3 & 2 \end{bmatrix} \tag{2.40}$$

Thus, the inverse of matrix A is:

$$A^{-1} = \begin{bmatrix} \frac{4}{5} & -\frac{1}{5} \\ -\frac{3}{5} & \frac{2}{5} \end{bmatrix} \tag{2.41}$$

To verify, one can multiply A and A^{-1} to confirm that the result is the identity matrix I.

For large matrices, manually calculating the inverse becomes increasingly complex. Therefore, computational tools such as MATLAB or Python are often used. In **MATLAB**, the inverse of a matrix A can be determined using the following command:

$$A^{-1} = \text{inv(A);} \tag{2.42}$$

And in **Python**, the inverse of matrix A is made using following command from the `numpy` library:

$$A^{-1} = \text{np.linalg.inv(A)} \tag{2.43}$$

Matrix inversion is a fundamental concept in linear algebra, widely used in engineering to solve systems of linear equations. Consider the equation:

$$A\mathbf{x} = \mathbf{B} \tag{2.44}$$

where A is an $n \times n$ matrix of coefficients, $\mathbf{x}$ is an $n \times 1$ vector of unknowns, and $\mathbf{B}$ is an $n \times 1$ vector of constants. The goal is to determine the vector $\mathbf{x}$. If the matrix A is invertible, the solution can be expressed as:

$$\mathbf{x} = A^{-1}\mathbf{B} \tag{2.45}$$

where A^{-1} is the inverse of matrix A.

Example: As an example, here we solve the following system of equations:

$$\begin{cases} 2x_1 + 3x_2 = 8 \\ x_1 - 4x_2 = -2 \end{cases} \tag{2.46}$$

Solution:

This system can be written in matrix form as:

$$\begin{bmatrix} 2 & 3 \\ 1 & -4 \end{bmatrix} \begin{bmatrix} x_1 \\ x_2 \end{bmatrix} = \begin{bmatrix} 8 \\ -2 \end{bmatrix} \tag{2.47}$$

Here, $A = \begin{bmatrix} 2 & 3 \\ 1 & -4 \end{bmatrix}$, $\mathbf{x} = \begin{bmatrix} x_1 \\ x_2 \end{bmatrix}$, and $\mathbf{B} = \begin{bmatrix} 8 \\ -2 \end{bmatrix}$. To solve for $\mathbf{x}$, compute A^{-1} and multiply by $\mathbf{B}$:

$$A^{-1} = \frac{1}{\det(A)} \begin{bmatrix} -4 & -3 \\ -1 & 2 \end{bmatrix} = \frac{1}{-11} \begin{bmatrix} -4 & -3 \\ -1 & 2 \end{bmatrix} \tag{2.48}$$

$$\mathbf{x} = A^{-1}\mathbf{B} = \frac{1}{-11} \begin{bmatrix} -4 & -3 \\ -1 & 2 \end{bmatrix} \begin{bmatrix} 8 \\ -2 \end{bmatrix} = \begin{bmatrix} 2.36 \\ 1.09 \end{bmatrix} \tag{2.49}$$

Thus, the solution is:

$$x_1 = 2.36, \quad x_2 = 1.09$$

This example can be solved using MATLAB and Python with the scripts provided in Examples 2.11 and 2.12, respectively. The scripts are named `CH02Mcode6.m` (MATLAB) and `CH02Pcode6.py` (Python) and are available for download from the book's website: https://github.com/prog871/CRC-book2025

Example 2.11

MATLAB code. This MATLAB script, named `CH02Mcode6.m`, solves a system of two linear equations using matrix inversion.

MATLAB Input:

```
% This code olves a system
% of two linear equations
% using matrix inversion.
% 2x1 + 3x2 = 8
% x1 - 4x2 = -2
clear; close all; clc
set(0,'defaultaxesfontsize',28);
% Define coefficient matrix A
A = [2, 3; 1, -4];
% Define constant matrix B
```

```
B = [8; -2];
% Compute the inverse of A
A_inv = inv(A);
% Calculate solution x
x = A_inv * B;
% Display the values of x1 and x2
x1 = x(1);
x2 = x(2);
fprintf('x1 = %.2f\n', x1);
fprintf('x2 = %.2f\n', x2);
% end of M - file
```

MATLAB Output:

```
x1 = 2.36
x2 = 1.09
```

Example 2.12

Python code. This Python code, named CH02Pcode6.py, solves a system of two linear equations using matrix inversion.

Python Input:

```
# This code olves a system
# of two linear equations
# using matrix inversion.
# 2x1 + 3x2 = 8
# x1 - 4x2 = -2
import numpy as np
# Define coefficient matrix A
A = np.array([[2, 3], [1, -4]])
# Define the constant matrix B
B = np.array([8, -2])
# Compute the inverse of A
A_inv = np.linalg.inv(A)
# Calculate solution vector x
x = np.dot(A_inv, B)
# Extract values of x1 and x2
x1 = x[0]
x2 = x[1]
# Display values of x1 and x2
```

```
print(f"x1 = {x1:.2f}")
print(f"x2 = {x2:.2f}")
# end of Python code.
```

Python Output:

```
x1 = 2.36
x2 = 1.09
```

Engineering application: In structural engineering, the response of a beam is typically described in terms of its *degrees of freedom*, which include both the vertical displacement (d) and the rotation (or slope) angle (θ) (Figure 2.8b). These quantities are grouped into a displacement vector, denoted by **u**, and related to the applied loading (**P**) and the stiffness matrix of the beam (K) through the following general matrix equation:

$$\mathbf{P} = K \cdot \mathbf{u} \tag{2.50}$$

Here, the displacement vector **u** and the loading vector **P** are defined as:

$$\mathbf{u} = \begin{bmatrix} d \\ \theta \end{bmatrix}, \quad \mathbf{P} = \begin{bmatrix} F \\ M \end{bmatrix} \tag{2.51}$$

where d is the vertical displacement, θ is the rotation at the point of interest, F is the applied vertical force, and M is the applied moment (rotational force).

By applying matrix inversion, the displacement vector can be computed from:

$$\mathbf{u} = K^{-1}\mathbf{P} \tag{2.52}$$

Referring to the cantilever beam shown in Figure 2.8b, the stiffness matrix (K) for this configuration is:

$$K = \frac{EI}{L^3}\begin{bmatrix} 12 & -6L \\ -6L & 4L^2 \end{bmatrix} \tag{2.53}$$

where E is Young's modulus (modulus of elasticity), I is the second moment of area (moment of inertia), and L is the length of the beam.

In the case where only a vertical force F is applied at the free end and no external moment is applied, the loading vector simplifies to:

$$\mathbf{P} = \begin{bmatrix} F \\ 0 \end{bmatrix} \tag{2.54}$$

Therefore, the governing equation becomes:

$$\begin{bmatrix} F \\ 0 \end{bmatrix} = \frac{EI}{L^3}\begin{bmatrix} 12 & -6L \\ -6L & 4L^2 \end{bmatrix}\begin{bmatrix} d \\ \theta \end{bmatrix} \tag{2.55}$$

This form clearly indicates that the applied load consists only of a vertical force F, while the moment component is zero, as no rotational force is imposed at the free end.

a) Aftermath of the 2024 Noto earthquake (Japan) **b)** A cantilever beam

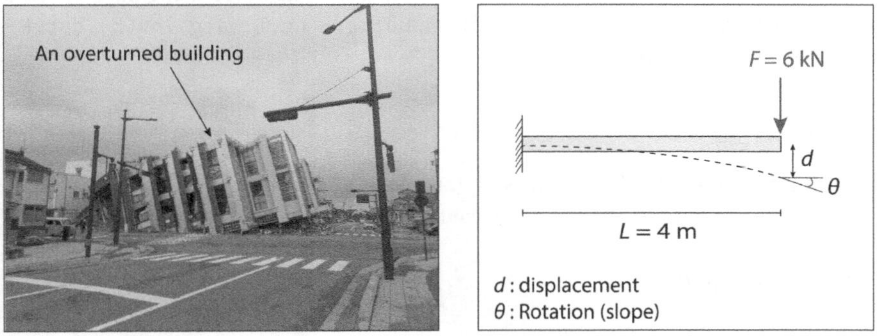

FIGURE 2.8 **a)** An overturned building due to the January 2024 Noto earthquake in Japan. **b)** A cantilever beam.

For this analysis, we consider a concrete cantilever beam with a length $L = 4000\,\text{mm}$, a square cross section of 300 mm × 300 mm, a Young's Modulus of $E = 30,000\,\text{N/mm}^2$, and a second moment of area $I = 6.75 \times 10^8\,\text{mm}^4$. When a force of $F = 6\,\text{kN}$ is applied at the free end, the displacement (d) and rotation (θ) at the cantilever's free end are calculated using the beam's stiffness matrix. The equation that relates the force, displacement, and rotation is Equation 2.55. Substituting the given values of $L = 4000\,\text{mm}$, $E = 30,000\,\text{N/mm}^2$, $I = 6.75 \times 10^8\,\text{mm}^4$, and $F = 6\,\text{kN}$, we get the following equation:

$$\begin{bmatrix} -6000 \\ 0 \end{bmatrix} = \frac{(30,000)(6.75 \times 10^8)}{(4000)^3} \begin{bmatrix} 12 & -6 \times 4000 \\ -6 \times 4000 & 4 \times 4000^2 \end{bmatrix} \begin{bmatrix} d \\ \theta \end{bmatrix} \tag{2.56}$$

This simplifies to:

$$\begin{bmatrix} -6000 \\ 0 \end{bmatrix} = 316.40625 \begin{bmatrix} 12 & -24000 \\ -24000 & 64000000 \end{bmatrix} \begin{bmatrix} d \\ \theta \end{bmatrix} \tag{2.57}$$

Now, to solve for the displacement and rotation, we apply matrix inversion:

$$\begin{bmatrix} d \\ \theta \end{bmatrix} = \left(316.40625 \begin{bmatrix} 12 & -24000 \\ -24000 & 64000000 \end{bmatrix}\right)^{-1} \begin{bmatrix} -6000 \\ 0 \end{bmatrix} \tag{2.58}$$

$$\begin{bmatrix} d \\ \theta \end{bmatrix} = \begin{bmatrix} -6.32\,\text{mm} \\ -0.00237\,\text{radians} \end{bmatrix} \tag{2.59}$$

The results show that the vertical displacement at the free end is $d = -6.32\,\text{mm}$, indicating a downward deflection, and the rotation at the free end is $\theta = -0.0024$ radians, indicating a clockwise rotation (see Figure 2.8b).

MATLAB and Python can be used to perform the calculations outlined above, with the corresponding scripts provided in Examples 2.13 (MATLAB script `CH02Mcode7.m`) and 2.14 (Python script `CH02Pcode7.py`). Both MATLAB and Python scripts compute these displacements by applying matrix inversion. This is achieved using the commands `Delta = inv(K) * F_vec` in MATLAB and `Delta = np.linalg.inv(K).dot(F_vec)` in Python, respectively.

Example 2.13

MATLAB code. This MATLAB script, named `CH02Mcode7.m`, calculates deflection and rotation of the free end of a cantilever beam.

MATLAB Input:

```
%This MATLAB code calculates
%deflection and rotation of
%the free end of a cantilever
%beam using marix inversion
clear; close all; clc
set(0,'defaultaxesfontsize',28);
% Given values
L =4000;%Length of beam(mm)
E =30000;%Young Modulus(N/mm^2)
I =6.75e8;%2nd area moment (mm^4)
F =6e3; % Force (N)
% Stiffness matrix calculation
k1 = (E * I) / L^3;% Stiffness
% Stiffness matrix
K = k1 * [12,-6*L;-6*L,4*L^2];
% Force vector
F_vec = [-F; 0];
% Matrix inversion
Delta = inv(K) * F_vec;
% Displacement(d), rotation(theta)
d = Delta(1); %displacement(mm)
theta = Delta(2);%rotation(radians)
% Display results
fprintf('d = %.2f mm\n', d);
fprintf('theta = %.4f radians\n',theta);
% end of M - file
```

MATLAB Output:

```
d = -6.32 mm
theta = -0.0024 radians
```

Example 2.14

Python code. This Python code, named `CH02Pcode7.py`, calculates deflection and rotation of the free end of a cantilever beam.

Python Input:

```
#This Python code calculates
#deflection and rotation of
#the free end of a cantilever
#beam using marix inversion
import numpy as np
# Given values
L = 4000 #Length of beam (mm)
E = 30000 #Young Modulus(N/mm^2)
I = 6.75e8 #2nd area moment(mm^4)
F = 6e3 # Force (N)
# Stiffness matrix calculation
k1 = (E * I)/L**3 #Stiffness
# Stiffness matrix
K = k1 * np.array([[12, -6 * L],
        [-6 * L, 4 * L**2]])
# Force vector
F_vec = np.array([-F, 0])
# Matrix inversion
Delta = np.linalg.inv(K).dot(F_vec)
# Displacement (d),rotation(theta)
d = Delta[0] # Displacement(mm)
theta = Delta[1] #Rotation(radians)
# Display results
print(f'd = {d:.2f} mm')
print(f'theta = {theta:.4f} radians')
# end of Python code.
```

Python Output:

```
d = -6.32 mm
theta = -0.0024 radians
```

2.4 EIGENVALUES AND EIGENVECTORS

In engineering, many systems are represented and analyzed using matrices, especially in areas such as structural analysis, vibration analysis, and control systems. Eigenvalues and eigenvectors are fundamental in understanding how these systems behave, helping to identify stable and unstable modes of operation.

An eigenvector is a special vector that, when a matrix is applied to it, only changes in size (either stretched or shrunk), while its direction remains unchanged. In other words, an eigenvector points in the same direction, and the matrix transformation results in a scaled version of the vector.

Mathematically, an eigenvector **v** of a matrix **A** is a non-zero vector that, when multiplied by **A**, results in a scaled version of itself. The scaling factor is called the eigenvalue λ. The equation is:

$$\mathbf{A}\mathbf{v} = \lambda\mathbf{v} \tag{2.60}$$

where **A** is a square matrix, **v** is the eigenvector, and λ is the eigenvalue.

To solve for the eigenvector **v** and eigenvalue λ, rewrite the Equation 2.60 into a form that can be solved:

$$\mathbf{A}\mathbf{v} - \lambda\mathbf{v} = 0 \tag{2.61}$$

Factor out the eigenvector **v** from the right-hand side, which results in the following equation:

$$(\mathbf{A} - \lambda\mathbf{I})\mathbf{v} = 0 \tag{2.62}$$

Here, **I** represents the identity matrix of the same dimension as **A**.

In Equation 2.62, the values of λ are eigenvalues, which satisfy the following characteristic equation:

$$\det(\mathbf{A} - \lambda\mathbf{I}) = 0 \tag{2.63}$$

It is noted that `det(x)` represents the determinant operator. This equation is referred to as the **characteristic equation**. Solving it will yield the eigenvalues λ.

By substituting the eigenvalues into Equation 2.62, a system of linear equations is produced which can be solved for the eigenvector **v**.

Example: Find the eigenvalues and eigenvectors of the following matrix:

$$\mathbf{A} = \begin{bmatrix} 4 & 1 \\ 2 & 3 \end{bmatrix}$$

Solution:

Step 1: Finding the Eigenvalues. To determine the eigenvalues, we begin by solving the characteristic equation, which is given by:

$$\det(\mathbf{A} - \lambda\mathbf{I}) = 0 \tag{2.64}$$

where **I** is the identity matrix and λ is the eigenvalue. For our matrix **A**, we subtract $\lambda\mathbf{I}$ from **A** to obtain:

$$\mathbf{A} - \lambda\mathbf{I} = \begin{bmatrix} 4-\lambda & 1 \\ 2 & 3-\lambda \end{bmatrix}$$

Next, we compute the determinant of this matrix:

$$\det(\mathbf{A} - \lambda\mathbf{I}) = (4-\lambda)(3-\lambda) - (1)(2) \tag{2.65}$$

Simplifying:

$$(4-\lambda)(3-\lambda)-2=0$$

Expanding the terms:

$$\lambda^2-7\lambda+10=0$$

Now, we solve this quadratic equation using the quadratic formula:

$$\lambda=\frac{-(-7)\pm\sqrt{(-7)^2-4(1)(10)}}{2(1)} \tag{2.66}$$

This simplifies to:

$$\lambda=\frac{7\pm\sqrt{49-40}}{2}=\frac{7\pm\sqrt{9}}{2}=\frac{7\pm 3}{2}$$

Thus, the two eigenvalues are:

$$\lambda_1=5 \quad \text{and} \quad \lambda_2=2 \tag{2.67}$$

Step 2: Finding the Eigenvectors. We find the eigenvectors corresponding to each eigenvalue. For each eigenvalue, Equation 2.62 is solved as follows.
Eigenvector for $\lambda=5$:

Substitute $\lambda=5$ into $\mathbf{A}-\lambda\mathbf{I}$:

$$\mathbf{A}-5\mathbf{I}=\begin{bmatrix}4-5 & 1\\ 2 & 3-5\end{bmatrix}=\begin{bmatrix}-1 & 1\\ 2 & -2\end{bmatrix}$$

Now, solve the system of equations:

$$\begin{bmatrix}-1 & 1\\ 2 & -2\end{bmatrix}\begin{bmatrix}v_1\\ v_2\end{bmatrix}=\begin{bmatrix}0\\ 0\end{bmatrix} \tag{2.68}$$

From the first row of the matrix, we get:

$$-v_1+v_2=0 \quad \Rightarrow \quad v_1=v_2$$

Thus, the eigenvector corresponding to $\lambda=5$ is:

$$\mathbf{v}_1=\begin{bmatrix}1\\ 1\end{bmatrix} \tag{2.69}$$

Eigenvector for $\lambda=2$:

Now, substitute $\lambda=2$ into $\mathbf{A}-\lambda\mathbf{I}$:

$$\mathbf{A}-2\mathbf{I}=\begin{bmatrix}4-2 & 1\\ 2 & 3-2\end{bmatrix}=\begin{bmatrix}2 & 1\\ 2 & 1\end{bmatrix}$$

Solve the system of equations:

$$\begin{bmatrix} 2 & 1 \\ 2 & 1 \end{bmatrix} \begin{bmatrix} v_1 \\ v_2 \end{bmatrix} = \begin{bmatrix} 0 \\ 0 \end{bmatrix} \tag{2.70}$$

From the first row of the matrix, we get:

$$2v_1 + v_2 = 0 \quad \Rightarrow \quad v_2 = -2v_1$$

Thus, the eigenvector corresponding to $\lambda = 2$ is:

$$\mathbf{v}_2 = \begin{bmatrix} 1 \\ -2 \end{bmatrix} \tag{2.71}$$

In summary, for matrix $\mathbf{A} = \begin{bmatrix} 4 & 1 \\ 2 & 3 \end{bmatrix}$, the eigenvalues are:

$$\lambda_1 = 5 \quad \text{and} \quad \lambda_2 = 2$$

with corresponding eigenvectors:

$$\mathbf{v}_1 = \begin{bmatrix} 1 \\ 1 \end{bmatrix}, \quad \mathbf{v}_2 = \begin{bmatrix} 1 \\ -2 \end{bmatrix}$$

To calculate the eigenvalues and eigenvectors of a matrix **A** in **MATLAB**, use the 'eig()' function. For example, if **A** is your matrix, simply run the following command in MATLAB:

$$\boxed{[\mathbf{v}, D] = \text{eig}(\mathbf{A})}$$

The diagonal elements of matrix D are the eigenvalues, and the columns of matrix **v** are the corresponding eigenvectors.

In Python, use the following command:

```
eigenvalues, eigenvectors = np.linalg.eig(A)
```

Here, we present **MATLAB** and **Python** scripts that can be used to calculate the eigenvalues and eigenvectors of the matrix $\mathbf{A} = \begin{bmatrix} 4 & 1 \\ 2 & 3 \end{bmatrix}$ through Examples 2.15 (script `CH02Mcode8.m`) and 2.16 (script `CH02Pcode8.py`), respectively.

Example 2.15

MATLAB code. This MATLAB script, named `CH02Mcode8.m`, calculates the eigenvalues and eigenvectors of a matrix.

MATLAB Input:

```
%This MATLAB code calculates
%the eigenvalues and eigenvectors
```

```
%of a matrix A=[4, 1; 2, 3]
clear; close all; clc
set(0,'defaultaxesfontsize',28);
% Given matrix A
A = [4, 1; 2, 3];
% Use the eig() for
[eig_vec, eig_val] = eig(A);
%The diagonal of eigenvalue matrix
%contains the eigenvalues
lam1 = eig_val(1,1);%1st eigenvalue
lam2 = eig_val(2,2);%2nd eigenvalue
% Display eigenvalues
fprintf('Eigenvalue 1: %.2f\n',lam1);
fprintf('Eigenvalue 2: %.2f\n',lam2);
% The eigenvectors are stored in the
% columns of the eigenvectors matrix
v1= eig_vec(:,1);%1st eigenvector
v2= eig_vec(:,2);%2nd eigenvector
% Display eigenvectors- note: usually
% the first element of eigenvectors
% is 1; that is why below the elements
% are divided to the first element
fprintf(['Eigenvector 1: ' ...
 '[%.2f, %.2f]\n'], v1(1)/v1(1), ...
   v1(2)/v1(1));
fprintf(['Eigenvector 2: ' ...
 '[%.2f, %.2f]\n'], v2(1)/v2(1), ...
   v2(2)/v2(1));
% end of M - file
```

MATLAB Output:

```
Eigenvalue 1: 5.00
Eigenvalue 2: 2.00
Eigenvector 1: [1.00, 1.00]
Eigenvector 2: [1.00, -2.00]
```

Example 2.16

Python code. This Python code, named `CH02Pcode8.py`, calculates the eigenvalues and eigenvectors of a matrix.

Python Input:

```
#This Python code calculates
#the eigenvalues & eigenvectors
#of a matrix A=[4, 1; 2, 3]
import numpy as np
#Given matrix A
A = np.array([[4, 1], [2, 3]])
#Use numpy.linalg.eig() for
#eigenvalues & eigenvectors
eig_val,eig_vec=np.linalg.eig(A)
#The diagonal of eig_val
#contains the eigenvalues
lam1=eig_val[0] #1st eigenvalue
lam2=eig_val[1] #2nd eigenvalue
#Display eigenvalues
print(f'Eigenvalue 1:{lam1:.2f}')
print(f'Eigenvalue 2:{lam2:.2f}')
#The eigenvectors are stored in
#the columns of eig_vec
v1=eig_vec[:, 0] #1st eigenvector
v2=eig_vec[:, 1] #2nd eigenvector
#Display eigenvectors (usually
#normalized by dividing by first element)
print(f'Eigenvec 1:[{v1[0]/v1[0]:.2f},',end='')
print(f'{v1[1]/v1[0]:.2f}]')
print(f'Eigenvec 2:[{v2[0]/v2[0]:.2f},',end='')
print(f'{v2[1]/v2[0]:.2f}]')
# end of Python code.
```

Python Output:

```
Eigenvalue 1:5.00
Eigenvalue 2:2.00
Eigenvec 1:[1.00,1.00]
Eigenvec 2:[1.00,-2.00]
```

Engineering application: We demonstrate the application of eigenvalues and eigenvectors in engineering through an example in earthquake engineering. Buildings experience severe shaking and oscillation during earthquakes (Figure 2.9a). To design earthquake-resilient structures, it is essential to model these oscillations and evaluate the earthquake-induced drifts.

The first step in designing buildings that can resist earthquakes is to calculate how they naturally vibrate when no outside forces, like earthquakes, are acting on them. Every

building has a natural frequency of oscillations, which depends on how tall it is and how strong its structure is. Understanding this helps engineers predict how the building will respond during an earthquake. In this example, we calculate the natural oscillations of a two-story building.

Consider the two-story building illustrated in Figure 2.9b with story masses denoted as m_1 and m_2 and stiffness factors k_1 and k_2. Based on principles of structural engineering, the equations of motion for the structural system shown in Figure 2.9b lead to the following matrix equation:

$$\left([\mathbf{k}] - \omega^2 [\mathbf{M}]\right) [\mathbf{v}] = 0 \tag{2.72}$$

where $[\mathbf{k}]$ is the stiffness matrix, ω^2 is the square of the angular frequency, $[\mathbf{M}]$ is the mass matrix, and $[\mathbf{v}]$ is the eigenvector, also referred to as the mode shape, which describes the deformation pattern the structure undergoes during vibration at its natural frequency. The eigenvalues ω (angular frequency of building oscillations) are obtained by calculating the determinant of the following matrix equation:

$$\det\left([\mathbf{k}] - \omega^2 [\mathbf{M}]\right) = 0 \tag{2.73}$$

The mass matrix $[\mathbf{M}]$ for a two-story building with story masses m_1 and m_2 (Figure 2.9b) is given by:

$$[\mathbf{M}] = \begin{bmatrix} m_1 & 0 \\ 0 & m_2 \end{bmatrix} \tag{2.74}$$

The stiffness matrix $[\mathbf{k}]$ for a two-story building, where k_1 and k_2 are the stiffness values for the first and second stories respectively (Figure 2.9b), is given by:

$$[\mathbf{k}] = \begin{bmatrix} k_1 + k_2 & -k_2 \\ -k_2 & k_2 \end{bmatrix} \tag{2.75}$$

Therefore, Equation 2.72 becomes:

$$\left(\begin{bmatrix} k_1 + k_2 & -k_2 \\ -k_2 & k_2 \end{bmatrix} - \omega^2 \begin{bmatrix} m_1 & 0 \\ 0 & m_2 \end{bmatrix}\right) \begin{bmatrix} v_1 \\ v_2 \end{bmatrix} = 0 \tag{2.76}$$

Which is further rearranged to:

$$\begin{bmatrix} k_1 + k_2 - \omega^2 m_1 & -k_2 \\ -k_2 & k_2 - \omega^2 m_2 \end{bmatrix} \begin{bmatrix} v_1 \\ v_2 \end{bmatrix} = 0 \tag{2.77}$$

Referring to Figure 2.9b and by substituting the values of mass and stiffness in the equation, we have:

$$\begin{bmatrix} 12 - 0.01\omega^2 & -6 \\ -6 & 6 - 0.008\omega^2 \end{bmatrix} \begin{bmatrix} v_1 \\ v_2 \end{bmatrix} = 0 \tag{2.78}$$

Note that story masses m_1 and m_2 were divided by gravitational acceleration g = 9810 mm/s^2 before entering them into the above equation.

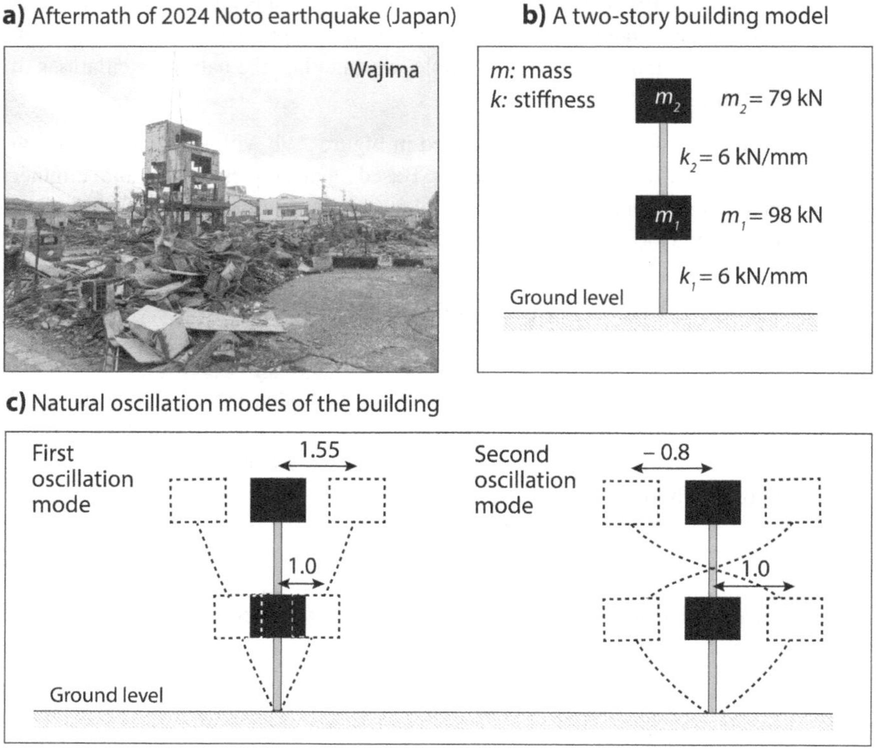

FIGURE 2.9 **a)** A destroyed community in Wajima due to the January 2024 Noto earthquake in Japan. **b)** A two-story model of a building. **c)** The two natural oscillation modes of the two-story building.

The eigenvalues ω (angular frequency of building oscillations) are obtained by calculating the determinant of the following matrix equation:

$$\det\left(\begin{bmatrix} 12-0.01\omega^2 & -6 \\ -6 & 6-0.008\omega^2 \end{bmatrix}\right) = 0 \tag{2.79}$$

which is calculated as below:

$$36 - 0.156\omega^2 + 0.00008\omega^4 = 0 \tag{2.80}$$

By solving the above equation, the four roots of the equation are: $\omega = \pm 16.35$ and $\omega = \pm 41.01$. For this problem, only positive values are accepted. Therefore, the eigenvalues are:

$$\omega_1 = 16.35\,, \quad \omega_2 = 41.01$$

The natural oscillation periods are calculated by using the equation $T = \frac{2\pi}{\omega}$, which gives the following values:

$$T_1 = \frac{2\pi}{\omega_1} = 0.38\,\text{s}, \quad T_2 = \frac{2\pi}{\omega_2} = 0.15\,\text{s}.$$

By having natural frequency modes (ω_1 and ω_2), we can now proceec to calculate oscillation modes (or eigenvectors) using Equations 2.72 and 2.78.

For $\omega_1 = 16.35\,\text{Hz}$, the corresponding eigenvector is:

$$\mathbf{v}_1 = \begin{bmatrix} 1.0 \\ 1.55 \end{bmatrix}$$

And the second eigenvector corresponding to $\omega_2 = 41.01\,\text{Hz}$ becomes:

$$\mathbf{v}_2 = \begin{bmatrix} 1.0 \\ -0.8 \end{bmatrix}$$

These two eigenvectors represent the two natural oscillation modes of the building as shown in Figure 2.9c.

The above calculations of natural oscillation periods and modes for a two-story building (Figure 2.9b) can be done using MATLAB and Python through scripts `CH02Mcode9.m` (Example 2.17) and `CH02Pcode9.py` (Example 2.18), respectively.

Example 2.17

MATLAB code. This MATLAB script, named `CH02Mcode9.m`, calculates the natural periods and modes of free oscillations of a two-story building.

MATLAB Input:

```
%This MATLAB code calculates
%the natural periods and modes
%of a two-story building.
clear; close all; clc
set(0,'defaultaxesfontsize',28);
m1=98/9810;%Mass, 1st story(kN.s^2/mm)
m2=79/9810;%Mass, 2nd story(kN.s^2/mm)
k1= 6;%Stiffness, 1st story(kN/mm)
k2= 6;%Stiffness, 2nd story(kN/mm)
%[M]:mass matrix,[K]:stiffness matrix
M = [m1, 0; 0, m2];
K = [k1 + k2, -k2; -k2, k2];
%Solve the eigenvalue problem
% [K - w^2 * M] * v = 0
% |K - w^2 * M|=0
%solve: 36-0.156w^2+0.00008w^4=0
```

```
% using function 'roots' in MATLAB
p=[0.00008 0 -0.156 0 36];
r=roots(p);
% find acceptable roots (eigenvalues)
k=1;
for i=1:4
   if r(i)≥0
      x(k)=r(i); k=k+1;
   end
end
% Sort 1st and 2nd mode
w1=x(1); w2=x(2);
if x(1)>x(2)
   w1=x(2); w2=x(1);
end
% Eigenvalues (angular frequencies)
fprintf('Eigenval 1 (w_1):%.2f\n',w1);
fprintf('Eigenval 2 (w_2):%.2f\n',w2);
% Calculate the natural periods
T1= 2*pi/w1; %1st natural period
T2= 2*pi/w2; %2nd natural period
% Display the natural periods
fprintf('Period 1 (T1):%.2f s\n',T1);
fprintf('Period 2 (T2):%.2f s\n',T2);
% 1st eigenvector (1st mode shape)
MK1=[k1 + k2-(m1*w1^2), -k2; ...
   -k2, k2-0.008*w1^2];
[eig_vec1,D]=eig(MK1);
v1= eig_vec1(:, 1); %1st eigenvector
v1=v1/v1(1);%Normalize to first element
% 2nd eigenvector (2nd mode shape)
MK2=[k1 + k2-(m1*w2^2), -k2; ...
   -k2, k2-0.008*w2^2];
[eig_vec2,D]=eig(MK2);
v2= eig_vec2(:, 2); %1st eigenvector
v2=v2/v2(1);%Normalize to first element
fprintf('Eigenvec 1:[%.2f,%.2f]\n',v1(1),v1(2));
fprintf('Eigenvec 2:[%.2f,%.2f]\n',v2(1),v2(2));
% end of M - file
```

MATLAB Output:

```
Eigenval 1 (w_1):16.35
Eigenval 2 (w_2):41.02
```

```
Period 1 (T1):0.38 s
Period 2 (T2):0.15 s
Eigenvec 1:[1.00,1.55]
Eigenvec 2:[1.00,-0.80]
```

Example 2.18

Python code. This Python script, named CH02Pcode9.py, calculates the natural periods and modes of free oscillations of a two-story building.

Python Input:

```
#This Python code calculates
#the natural periods and modes
#of a two-story building.
import numpy as np
# Mass and stiffness values
m1=98/ 9810 #Mass,1st story(kN.s^2/mm)
m2=79/ 9810 #Mass,2nd story(kN.s^2/mm)
k1=6 #Stiffness,1st story (kN/mm)
k2=6 #Stiffness,2nd story (kN/mm)
#Mass matrix[M], stiffness matrix[K]
M=np.array([[m1,0],[0,m2]])
K=np.array([[k1+k2,-k2],[-k2,k2]])
# Solve:36-0.156w^2+0.00008w^4=0
p = [0.00008, 0, -0.156, 0, 36]
r = np.roots(p)
# Find acceptable roots(eigenvalues)
x = [root for root in r
    if np.isreal(root) and root >= 0]
x = sorted(np.real(x)) #Sort roots
w1, w2 = x[0], x[1]
# Eigenvalues (angular frequencies)
print(f"Eigenval 1 (w_1):{w1:.2f}")
print(f"Eigenval 2 (w_2):{w2:.2f}")
# Calculate natural periods
T1 = 2*np.pi/w1 #1st period
T2 = 2*np.pi/w2 #2nd period
print(f"Period 1 (T1):{T1:.2f}s")
print(f"Period 2 (T2):{T2:.2f}s")
# 1st eigenvector(1st mode shape)
```

```
MK1 = K - (w1**2) * M
eig_vals1,eig_vecs1=np.linalg.eig(MK1)
idx1 = np.argmin(np.abs(eig_vals1))
v1=eig_vecs1[:,idx1] #Select eigenvector
v1=v1/v1[0] #Normalize to first element
# 2nd eigenvector (2nd mode shape)
MK2 = K - (w2**2) * M
eig_vals2, eig_vecs2 = np.linalg.eig(MK2)
idx2 = np.argmin(np.abs(eig_vals2))
v2=eig_vecs2[:,idx2] #Select eigenvector
v2=v2/v2[0] #Normalize to first element
print(f"Eigenvec 1:[{v1[0]:.2f},{v1[1]:.2f}]")
print(f"Eigenvec 2:[{v2[0]:.2f},{v2[1]:.2f}]")
# end of Python code.
```

Python Output:

```
Eigenval 1 (w_1):16.35
Eigenval 2 (w_2):41.02
Period 1 (T1):0.38s
Period 2 (T2):0.15s
Eigenvec 1:[1.00,1.56]
Eigenvec 2:[1.00,-0.80]
```

2.5 PROBLEMS

Problem 2.1: In coastal engineering, the squared amplitude of waves during tsunamis and storms is often used as a surrogate measure for wave energy and duration. Using the tsunami data provided in the input file named `CH02-data-tsunami-Palu.txt` (available at the book's website: https://github.com/prog871/CRC-book2025), plot the time history of the squared amplitude of the tsunami waves using both MATLAB and Python.

Problem 2.2: Referring to the road construction example illustrated in Figure 2.2, and assuming that Section B requires the same quantity of materials as Section A, reconstruct the materials matrix ($\mathbf{M}$), the cost vector ($\mathbf{C}$), and compute the total construction cost ($\mathbf{C}_T$) for the road.

Answer: Total construction cost for the entire road is £4740.00.

$$\mathbf{M} = \begin{bmatrix} 10 & 15 & 5 \\ 10 & 15 & 5 \\ 12 & 18 & 7 \end{bmatrix} \quad \mathbf{C}_T = \begin{bmatrix} 1450 \\ 1450 \\ 1840 \end{bmatrix}$$

Problem 2.3: Use both MATLAB and Python to solve the following system of equations:

$$\begin{cases} 1.0x - 5.0y = 0.0, \\ 3.0x + 9.5y = 1.3. \end{cases} \tag{2.81}$$

Answer: $x = 0.266$, $y = 0.0531$.

Problem 2.4: For the simply-supported beam shown in Figure 2.5b, reposition the point load to a distance of 9 m from the left support. Formulate a system of two linear equations to determine the reactions R_A and R_B, and solve them using MATLAB and Python.

Answer: R_A = 5.5 kN, R_B = 4.5 kN.

Problem 2.5: The truss illustrated in Figure 2.10 has six unknowns: three support reactions (R_{AX}, R_{AY}, and R_{CY}) and three axial forces in the members (F_{AB}, F_{AC}, and F_{BC}). Using both MATLAB and Python, formulate a system of six linear equations and solve for the aforementioned unknowns.

Answer: $F_{AB} = 24.17$ kN, $F_{AC} = -7.34$ kN, $F_{BC} = 5.42$ kN, $R_{AX} = 12$ kN, $R_{AY} = -14.5$ kN, $R_{CY} = -5.5$ kN.

Problem 2.6: Using the concept of determinants, calculate the area of a triangular plot of land with corner coordinates at $(25, 2)$, $(25, 111)$, and $(78, 2)$. All units are in meters.

Answer: Area = 2888.50 m^2.

Problem 2.7: Apply the concepts of eigenvalues and eigenvectors to calculate the natural periods and modes of oscillation for the three-story building shown in Figure 2.11. For this building, the mass matrix **M** and stiffness matrix **K** are given in the following forms:

$$[\mathbf{M}] = \begin{bmatrix} m_1 & 0 & 0 \\ 0 & m_2 & 0 \\ 0 & 0 & m_3 \end{bmatrix} \tag{2.82}$$

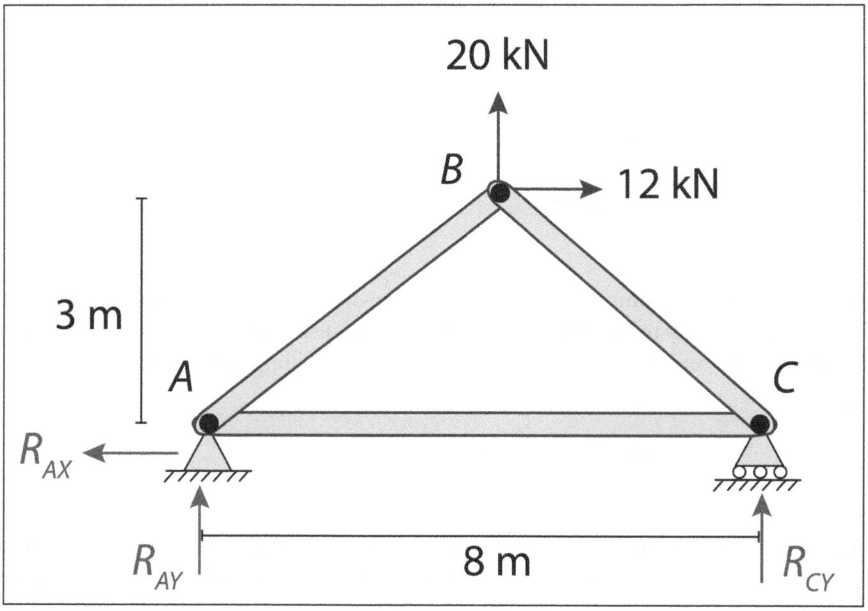

FIGURE 2.10 A truss with three members and two simple supports.

$$[\mathbf{K}] = \begin{bmatrix} k_1 + k_2 & -k_2 & 0 \\ -k_2 & k_2 + k_3 & -k_3 \\ 0 & -k_3 & k_3 \end{bmatrix} \tag{2.83}$$

Answer: $w_1 = 11.43, w_2 = 31.5, w_3 = 44.36, T_1 = 0.55\,\text{s}, T_2 = 0.2\,\text{s}, T_3 = 0.14\,\text{s}$. The eigenvectors are:

$$v_1 = \begin{bmatrix} 1 \\ 1.78 \\ 2.17 \end{bmatrix}, v_2 = \begin{bmatrix} 1 \\ 0.32 \\ -0.9 \end{bmatrix}, v_3 = \begin{bmatrix} 1 \\ -1.34 \\ -0.8 \end{bmatrix}.$$

Problem 2.8: Using MATLAB and Python, compute the eigenvalues and eigenvectors of the following matrix:

$$[\mathbf{X}] = \begin{bmatrix} -2 & 3.5 & 1 \\ 0 & 1.3 & 2 \\ 0.5 & 4.1 & 9 \end{bmatrix} \tag{2.84}$$

Answer: The eigenvalues are: 10.0158, -1.9305, and 0.2146. And the corresponding normalized eigenvectors are:

$$v_1 = \begin{bmatrix} 1 \\ 1.53 \\ 6.66 \end{bmatrix}, v_2 = \begin{bmatrix} 1 \\ 0.04 \\ -0.06 \end{bmatrix}, v_3 = \begin{bmatrix} 1 \\ 0.75 \\ -0.41 \end{bmatrix}.$$

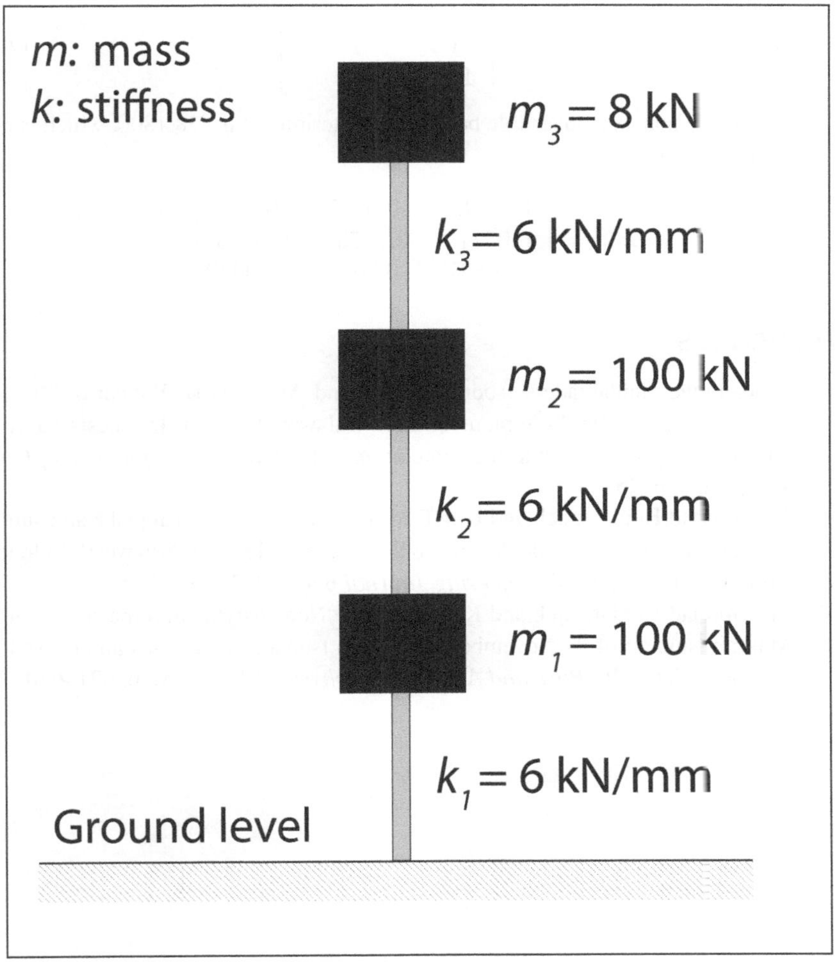

FIGURE 2.11 A three-story building model.

Problem 2.9: Referring to the cantilever beam shown in Figure 2.8, calculate the displacement at the free end of the beam under a point load of 9 kN, for a beam length of 6 m.

Answer: $d = -32.00$ mm, $\theta = -0.0080$ radians.

Problem 2.10: First, determine whether the following matrix is invertible. If it is, compute its inverse:

$$\mathbf{Z} = \begin{bmatrix} 11 & 2 & 0 \\ -4.5 & -1.1 & -5 \\ 1.5 & 14.1 & 7.2 \end{bmatrix} \tag{2.85}$$

Answer: Yes, the matrix is invertible because its determinant is 738.1800, which is non-zero. The inverse of **Z** is:

$$\mathbf{Z}^{-1} = \begin{bmatrix} 0.0848 & -0.0195 & -0.0135 \\ 0.0337 & 0.1073 & 0.0745 \\ -0.0837 & -0.2060 & -0.0042 \end{bmatrix}$$

REFERENCES

[1] Mohammad Heidarzadeh, Abdul Muhari, and Antonius B Wijanarto. "Insights on the source of the 28 September 2018 Sulawesi tsunami, Indonesia based on spectral analyses and numerical simulations." In: *Pure and Applied Geophysics* 176 (2019), p. 25–43.

[2] Mohammad Heidarzadeh and Iyan E Mulia. "A new dual earthquake and submarine landslide source model for the 28 September 2018 Palu (Sulawesi), Indonesia tsunami." In: *Coastal Engineering Journal* 65(1) (2023), p. 97–109.

[3] Mohammad Heidarzadeh and Kenji Satake. "New insights into the source of the Makran tsunami of 27 November 1945 from tsunami waveforms and coastal deformation data." In: *Pure and Applied Geophysics* 172 (2015), p. 621–640.

3 Limits, Differentiation, and Integration

This chapter introduces the fundamental concepts of limits, differentiation, and integration, presenting their foundational principles, common methodologies, and a range of practical applications. By examining these core components of mathematical analysis, readers will develop a deeper understanding of the processes that govern both natural phenomena and engineered systems. Whether pursued for advanced academic study or applied in real-world contexts, a thorough understanding of these topics is essential for harnessing the full potential of calculus.

3.1 LIMITS

Consider a scenario in which you are walking toward a specific point along a path. As you approach the point, the remaining distance decreases progressively. Similarly, imagine filling a glass with water from a tap that is only partially open. The glass does not fill instantaneously; rather, the water level rises gradually over time. As time progresses, the water level approaches the brim but does not overflow immediately. These situations are analogous to the charging process of a capacitor in an electrical circuit containing a resistor. Over time, the voltage across the capacitor gradually approaches the maximum voltage supplied by the battery, yet it does not reach this value instantaneously.

Limits play an important role in various engineering projects, such as calculating the altitude of an airplane or designing railway transportation systems (see Figure 3.1). Understanding the limit of an airplane's altitude helps in designing ascent profiles, fuel management, and ensuring that the aircraft remains within safe operating conditions during its climb. Similar principles apply in railway transportation, where the performance of a train may approach a steady speed or position over time, which is essential for safety and efficiency [1]. Using limits, engineers can analyze the behavior of these systems as they approach critical thresholds, allowing for precise calculations and improved designs. This concept also extends to structural and thermomechanical systems, where limits help evaluate stress, deformation, and thermal behavior in components such as beams, plates, and shells [3].

For example, in the design of bridges, limits are used to calculate the deflection of the bridge deck under various loads by solving differential equations that approximate the behavior of the structure as it approaches equilibrium [2, 4]. Limits are also critical in assessing stress and strain in materials near failure points to ensure the safety and longevity of the bridge. Using the railway train analogy, a limit is like the station toward which the train is heading. No matter how close the train gets, the limit tells us the destination point (zero distance, in this case) that the train is approaching. This helps us understand behaviors in calculus, such as how functions behave as their input values get very close to a particular point.

DOI: 10.1201/9781003511854-3

FIGURE 3.1 Examples of applications of mathematical limits in engineering. **a)** An airplane steadily climbs to its cruising altitude. **b)** A train on Great Western Railway in the UK. These examples mirror the idea of a limit, where a function's value approaches a particular number as the input gets close to some point.

In addition, in dynamic load analysis, such as wind or traffic forces, limits help engineers model oscillatory behaviors and damping effects over time. Similarly, in the thermal analysis of internal combustion engines, limits are essential in optimizing thermal efficiency. For instance, they are used in thermodynamic models to study system behavior under extreme conditions, such as approaching isothermal or adiabatic states. The Carnot efficiency, which defines the theoretical maximum efficiency of a heat engine, relies on limits to analyze the effects of temperature differences approaching zero or infinity. The concept of a limit helps us understand how a quantity behaves as it gets closer to a specific value. In engineering, limits are often used to describe what happens under extreme conditions. For instance, on a busy motorway (see Figure 3.2), as the number of cars increases

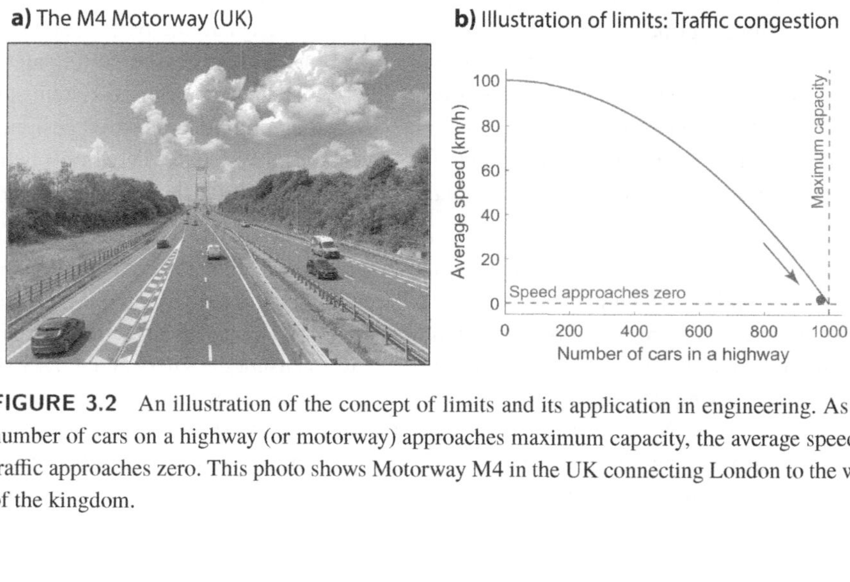

FIGURE 3.2 An illustration of the concept of limits and its application in engineering. As the number of cars on a highway (or motorway) approaches maximum capacity, the average speed of traffic approaches zero. This photo shows Motorway M4 in the UK connecting London to the west of the kingdom.

and approaches the road's full capacity, the average speed of traffic slows down and gets closer to zero. Although the speed never instantly becomes zero, the idea of a limit allows us to analyze this gradual change and predict system performance as conditions become critical.

3.1.1 MATHEMATICAL DEFINITION OF A LIMIT

In mathematics, a limit describes how a function (like the voltage over time) gets closer to a particular value as the input (like time) increases. The capacitor's charging process is a physical example of this idea. The limit of a function is an output value that the function approaches as the input approaches a particular value. It is a way to describe the behavior of a function near a point, even if the function is undefined at that point. Mathematically, we can write the limit as follows:

$$\lim_{x \to a} f(x) = L \tag{3.1}$$

Here, x is the independent variable, a is the value approached by x, and L is the value approached by $f(x)$. Thus, the limit of $f(x)$ as $x \to a$ is L.

To find the limit of a function as the independent variable approaches a specific value, you can start by directly substituting that value into the function. If this gives you a valid result, then you have found the limit. However, if direct substitution is not possible or leads to an indeterminate form such as $\frac{0}{0}$, you can try simplifying the function using algebraic techniques.

Engineering application: An RC circuit is an electric circuit that contains a resistor (R) and a capacitor (C). The voltage across the capacitor as a function of time is given by:

$$V(t) = V_0 \left(1 - \mathrm{e}^{-\frac{t}{RC}}\right) \tag{3.2}$$

where $V(t)$ is the voltage at time t, V_0 is the source voltage, R is the resistance, C is the capacitance, and e is the base of the natural logarithm. This equation describes the charging process of a capacitor in an RC circuit (see Figure 3.3). Initially, when $t = 0$, the exponential term $\mathrm{e}^{-\frac{t}{RC}}$ is equal to 1, so:

$$V(0) = V_0 \left(1 - \mathrm{e}^{-\frac{0}{RC}}\right) = V_0(1 - 1) = 0 \tag{3.3}$$

This indicates that the voltage across the capacitor is zero at the initial time ($t = 0$). As time progresses, the exponential term $\mathrm{e}^{-\frac{t}{RC}}$ decreases. The fact that $\mathrm{e}^{-\frac{t}{RC}}$ approaches zero as $t \to \infty$ is fundamental to understanding the long-term behavior of $V(t)$:

$$\lim_{t \to \infty} \mathrm{e}^{-\frac{t}{RC}} = \mathrm{e}^{-\infty} = \frac{1}{\mathrm{e}^{\infty}} = \frac{1}{\infty} = 0 \tag{3.4}$$

This implies that the voltage across the capacitor approaches the supply voltage V_0 as time tends to approach infinity:

$$\lim_{t \to \infty} V(t) = V_0(1 - 0) = V_0 \tag{3.5}$$

Physically, this means that the capacitor gradually charges up to the source voltage V_0, and the circuit eventually reaches a steady state where the voltage across the capacitor remains constant.

Let us explore this concept with specific values. Suppose $V_0 = 10\,\text{V}$ (volts), $R = 1\,\text{k}\Omega$ (kilo-ohms), and $C = 1\,\mu\text{F}$ (microfarads). The time constant τ of the circuit is:

$$\tau = RC = (1 \times 10^3) \times (1 \times 10^{-6}) = 1\,\text{ms} \tag{3.6}$$

where ms stands for "milliseconds," which is equal to one thousandth of a second ($1\,\text{ms} = 10^{-3}\,\text{s}$). Using the expression for $V(t)$, we can calculate the voltage at various points in time:

At $t = 0$, as calculated above:

$$V(0) = 0\,\text{V} \tag{3.7}$$

At $t = \tau = 1\,\text{ms}$ (see Figure 3.3):

$$V(\tau) = 10\left(1 - \text{e}^{-1}\right) = 10\,(1 - 0.3679) = 6.321\,\text{V} \tag{3.8}$$

At $t = 2\tau = 2\,\text{ms}$:

$$V(2\tau) = 10\left(1 - \text{e}^{-2}\right) = 10\,(1 - 0.1353) = 8.647\,\text{V} \tag{3.9}$$

At $t = 3\tau = 3\,\text{ms}$:

$$V(3\tau) = 10\left(1 - \text{e}^{-3}\right) = 10\,(1 - 0.0498) = 9.502\,\text{V} \tag{3.10}$$

At $t = 4\tau = 4\,\text{ms}$:

$$V(4\tau) = 10\left(1 - \text{e}^{-4}\right) = 10\,(1 - 0.0183) = 9.817\,\text{V} \tag{3.11}$$

At $t = 5\tau = 5\,\text{ms}$:

$$V(5\tau) = 10\left(1 - \text{e}^{-5}\right) = 10\,(1 - 0.0067) = 9.933\,\text{V} \tag{3.12}$$

At $t = 10\tau = 10\,\text{ms}$:

$$V(10\tau) = 10\left(1 - \text{e}^{-10}\right) = 10\,(1 - 0.0000454) = 9.9995\,\text{V} \tag{3.13}$$

As $t \to \infty$, $\text{e}^{-\frac{t}{RC}} \to 0$, hence:

$$\lim_{t \to \infty} V(t) = 10\,\text{V} \tag{3.14}$$

This shows that the voltage across the capacitor approaches $V_0 = 10\,\text{V}$ as $t \to \infty$, illustrating the concept of a limit in mathematics (see Figure 3.3).

Example: For the function $f(x) = x + 1$, evaluate the limit of $f(x)$ as x approaches 0. In mathematical terms, compute the following:

$$\lim_{x \to 0} (x + 1) \tag{3.15}$$

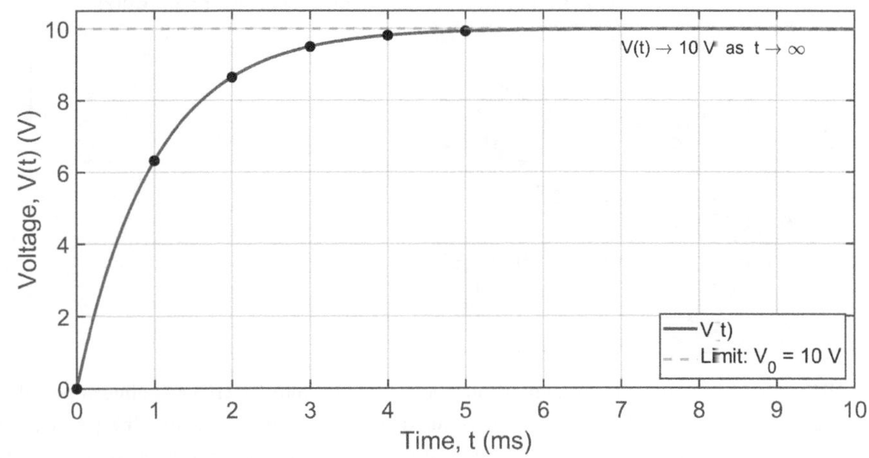

FIGURE 3.3 Voltage across the capacitor $V(t)$ as a function of time in an RC circuit. The curve shows how the voltage gradually increases and approaches the supply voltage $V_0 = 10$ V as time progresses. Key time points, $t = n\tau$ (where τ is the time constant), are marked; the pink dashed line represents the asymptotic value of the voltage as $t \to \infty$.

Solution:

The given function, $f(x) = x + 1$, is a polynomial function. Polynomial functions are continuous everywhere, which means that their value at a given point is the same as their limit at that point. This property simplifies the calculation of limits for polynomials because we can directly substitute the point into the function.

To evaluate the limit as x approaches 0, we substitute $x = 0$ into the expression for $f(x) = x + 1$:

$$f(0) = 0 + 1 = 1 \tag{3.16}$$

Thus, the limit is:

$$\lim_{x \to 0} (x + 1) = 1 \tag{3.17}$$

Since $f(x) = x + 1$ is continuous everywhere, there are no abrupt jumps, holes, or asymptotes in the function. Therefore, the value of the function at $x = 0$ coincides with the limit as $x \to 0$.

Figure 3.4 illustrates the connection between a mathematical concept and a real-world phenomenon. On the left, a photograph of a river serves as a metaphor for continuity in nature, representing the uninterrupted flow of water. This smooth, continuous movement of the river parallels the concept of continuity in mathematical functions. On the right, a MATLAB plot of the function $\sin(x)$ over the interval $[0, 2\pi]$ provides an example of a continuous mathematical function, demonstrating a curve without any breaks, gaps, or discontinuities.

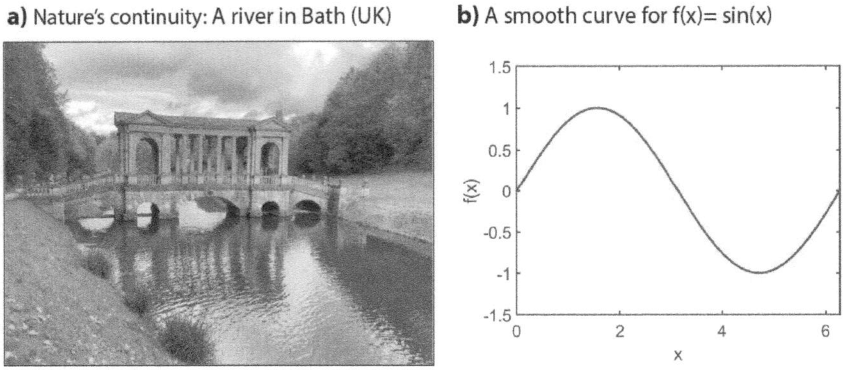

FIGURE 3.4 Illustration of continuity in mathematics and nature. The river's uninterrupted flow (**a**) visually reinforces the idea of a continuous function such as a $\sin(x)$ function (**b**), where every point on the graph is smoothly connected, much like the steady, ongoing movement of water in the river.

3.1.2 CALCULATING LIMITS USING MATLAB AND PYTHON

MATLAB's built-in **Symbolic Math Toolbox** streamlines complex symbolic computations, such as calculating limits, derivatives, and integrals, with concise and user-friendly syntax. This toolbox simplifies the computational process and integrates seamlessly with MATLAB's visualization capabilities, enabling users to graphically explore mathematical functions and their behaviors. MATLAB's comprehensive environment allows these symbolic tools to be used in conjunction with other features, such as numerical methods and data analysis, making it a versatile platform for both theoretical exploration and practical application.

Python's `SymPy` library is a powerful tool for symbolic mathematics that offers a wide range of functionalities, including the calculation of limits, derivatives, integrals, and solutions to algebraic equations. Its flexibility makes it particularly useful for integrating mathematical computations into larger, more complex programming tasks. For example, `SymPy` can be used in conjunction with libraries like `NumPy` or `Matplotlib` to perform symbolic computations and visualize results, or as part of a web-based application to dynamically calculate and display mathematical solutions. The library's symbolic capabilities allow for exact calculations, as opposed to numerical approximations, which is advantageous for tasks requiring precise solutions.

Example: Evaluate the limit of function $f(x) = x+1$ as x approaches 0 using MATLAB and Python:

$$\lim_{x \to 0}(x+1) \tag{3.18}$$

Solution:

Example 3.1 presents a MATLAB script for calculating the limit of the function $f(x) = x + 1$ as x approaches 0. The MATLAB script, named `CH03Mcode1.m`, is available through the supplementary website (https://github.com/prog871/CRC-book2025), which provides access to all the codes presented in this book.

The Python script, named `CH03Pcode1.py`, provided in Example 3.2 can be used to calculate the limit of the function $f(x) = x + 1$ as x approaches 0.

Example 3.1

MATLAB code. This MATLAB script, named `CH03Mcode1.m`, calculates the limit of the function $f(x) = x + 1$ as x approaches 0. This MATLAB script, along with all other scripts presented in this book, can be downloaded from the following website: https://github.com/prog871/CRC-book2025

MATLAB Input:

```
% Define the function
f = @(x) x + 1;
%
% Define the point of interest
x0 = 0;
%
% Calculate the limit
limit_value = f(x0);
%
% Display the result
fprintf( ['The limit of f(x) = x + 1 as x ' ...
   'approaches %d is %.2f\n'], x0, limit_value );
```

MATLAB Output:

```
The limit of f(x) = x + 1 as x approaches 0 is 1.00
```

Example 3.2

Python code. This Python code, named `CH03Pcode1.py`, calculates the limit of the function $f(x) = x + 1$ as x approaches 0.

Python Input:

```
# Define the function
def f(x):
   return x + 1
# Define the point of interest
x0 = 0
# Calculate the limit
```

```
limit_value = f(x0)
# Display the result
print("The limit of f(x) = x + 1 as x \
approaches 0 is %.2f" % limit_value)
```

Python Output:

```
The limit of f(x) = x + 1 as x approaches 0 is 1.00
```

Example: Evaluate the limit of the function $f(x) = \frac{1}{x}$ as x approaches 2 using MATLAB and Python:

$$\lim_{x \to 2} \frac{1}{x}$$

Solution:

To understand the behavior of $f(x)$ as x approaches 2, we can plug in values of x that are close to 2. Let us compute some values of $f(x)$ for points near $x = 2$. Approaching from the left ($x < 2$):

- If $x = 1.9 \mapsto f(1.9) = \frac{1}{1.9} = 0.5263$
- If $x = 1.99 \mapsto f(1.99) = \frac{1}{1.99} = 0.5025$

Approaching from the right ($x > 2$)

- If $x = 2.01 \mapsto f(2.01) = \frac{1}{2.01} = 0.4975$
- If $x = 2.1 \mapsto f(2.1) = \frac{1}{2.1} = 0.4762$

As x gets closer to 2 from either side, the value of $f(x)$ gets closer to 0.5. From the left ($x < 2$), the function values $f(x)$ decrease and get closer to 0.5, while from the right ($x > 2$), the function values $f(x)$ increase and get closer to 0.5. Since the function values converge to the same number, 0.5, from both sides of $x = 2$, we conclude that the limit exists and is equal to 0.5. Therefore:

$$\lim_{x \to 2} \frac{1}{x} = 0.5 \tag{3.19}$$

which means that as x approaches 2, the function $f(x) = \frac{1}{x}$ approaches the value 0.5. In the following, we solve this problem using MATLAB with the script `CH03Mcode2.m`, provided in Example 3.3.

To solve the limit of $f(x) = \frac{1}{x}$ as x approaches 2, the Python script `CH03Pcode2.py` provided in Example 3.4 can be used.

Example 3.3

MATLAB code. This MATLAB script, named `CH03Mcode2.m`, calculates the limit of the function $f(x) = \frac{1}{x}$ as x approaches 2.

MATLAB Input:

```
% Use symbolic math
syms x;
%
% Define the function
f = 1 / x;
%
% Calculate the limit as x approaches 2
limit_value = limit(f, x, 2);
%
% Display the result
fprintf(['The limit of f(x) = 1/x as x ' ...
   'approaches 2 is %.2f\n'], double(limit_value));
```

MATLAB Output:

```
The limit of f(x) = 1/x as x approaches 2 is 0.50
```

Example 3.4

Python code. This Python code, named `CH03Pcode2.py`, calculates the limit of the function $f(x) = \frac{1}{x}$ as x approaches 2.

Python Input:

```
# Import sympy
from sympy import symbols, limit
#
# Define the symbolic variable
x = symbols('x')
#
# Define the function
f = 1 / x
#
# Calculate the limit as x approaches 2
limit_value = limit(f, x, 2)
#
# Display the result
print("The limit of f(x) = 1/x as x \
approaches 2 is %.2f" % float(limit_value))
```

Python Output:

```
The limit of f(x) = 1/x as x approaches 2 is 0.50
```

Example: Evaluate the limit of rational functions $f(x) = \frac{\sin x}{x^2+1}$ as x approaches 0 using MATLAB and Python:

$$\lim_{x \to 0} f(x) = \lim_{x \to 0} \frac{\sin x}{x^2 + 1}$$

Solution:

To evaluate the limit, consider the given function. When $x = 0$, the numerator $\sin(0) = 0$, and the denominator, $0^2 + 1 = 1$, result in:

$$f(0) = \frac{0}{1} = 0 \tag{3.20}$$

Since the function is continuous at $x = 0$, the limit exists and is equal to $f(0)$. Thus, we have

$$\lim_{x \to 0} \frac{\sin x}{x^2 + 1} = 0 \tag{3.21}$$

MATLAB and Python can be used to solve this problem as shown in Examples 3.5 (script `CH03Mcode3.m`) and 3.6 (script `CH03Pcode3.py`), respectively.

Example 3.5

MATLAB code. This MATLAB script, named `CH03Mcode3.m`, calculates the limit of the function $f(x) = \frac{\sin x}{x^2+1}$ as x approaches 0.

MATLAB Input:

```
% Use symbolic math
syms x;
%
% Define the function
f = sin(x) / (x^2 + 1);
%
% Calculate the limit as x approaches 0
limit_value = limit(f, x, 0);
%
% Display the result
fprintf(['The limit of f(x) = sin(x) / (x^2 + 1) ' ...
    'as x approaches 0 is %.2f\n'], double(limit_value));
```

MATLAB Output:

```
The limit of f(x) = sin(x) / (x^2 + 1) as x approaches 0 is
    0.00
```

Example 3.6

Python code. This Python code, named `CH03Pcode3.py`, calculates the limit of the function $f(x) = \frac{\sin x}{x^2+1}$ as x approaches 0.

Python Input:

```
# Import sympy
from sympy import symbols, sin, limit
#
# Define the symbolic variable
x = symbols('x')
#
# Define the function
f = sin(x) / (x**2 + 1)
#
# Calculate the limit as x approaches 0
limit_value = limit(f, x, 0)
#
# Display the result
print(f"The limit of f(x) = sin(x) / (x^2 + 1) as x \
approaches 0 is {float(limit_value):.2f}")
```

Python Output:

```
The limit of f(x) = sin(x) / (x^2 + 1) as x approaches 0 is
    0.00
```

3.2 DIFFERENTIATION

Differentiation is a fundamental concept in mathematics and engineering concerned with determining the rate of change of a function with respect to its independent variable. In practical terms, it allows us to understand how a quantity varies as its input changes. This is particularly valuable in applications where the rate of change at a specific point is of interest. In bridge engineering, differentiation is crucial for analyzing how structural quantities—such as deflection, internal force, and stress—change along the length of the bridge. For instance, in suspension bridges, the main cable typically forms a catenary curve, and differentiation is used to calculate the forces in the suspension cables at various locations along the span (Figure 3.5).

The result of differentiation is called a "derivative," which provides a mathematical way to describe the slope of the tangent line at any point on the graph of the function. This slope gives us valuable insights into the behavior of the function, such as where it is increasing, decreasing, or staying constant.

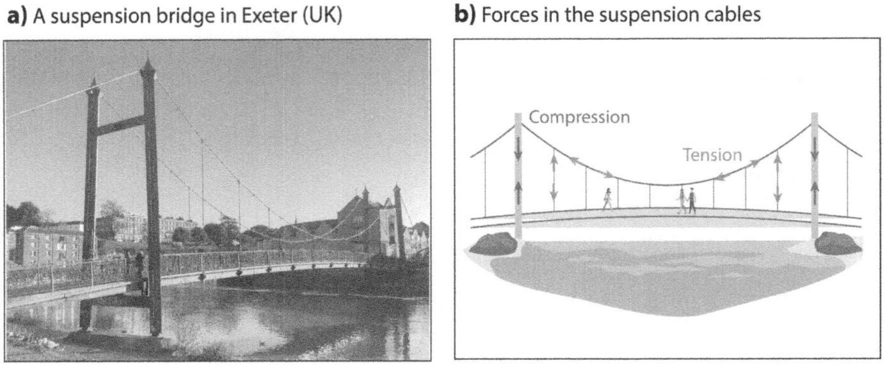

FIGURE 3.5 Application of differentiation in engineering. **a)** Differentiation is used to analyze the rate of change of various quantities such as deflection, force, and stress along the bridge. **b)** The shape of the main cable follows a catenary curve, and differentiation is used to calculate forces in the suspension cables at different locations. Compressive forces are shown in blue, while tensile forces are depicted in red.

3.2.1 MATHEMATICAL DEFINITION OF DIFFERENTIATION

In mathematical terms, the derivative of function $f(t)$ with respect to t is denoted as either $\frac{df}{dt}$ or $f'(t)$. Imagine driving a car along a highway, with your speed recorded at various moments. The speed of the car is determined by the rate at which the distance traveled changes over time—this is precisely where the concept of differentiation becomes relevant.

To differentiate a single term of the form ax^n, where a is a constant, x is the variable, and n is a non-negative integer, we apply the power rule of differentiation. This rule is expressed as:

$$\frac{d}{dx}(ax^n) = anx^{n-1} \tag{3.22}$$

The power rule simplifies the process of differentiation by providing a systematic approach: for each term of the form ax^n, multiply the exponent n by the coefficient a, then reduce the exponent by one. The derivative is the sum of the resulting terms. Note that the derivative of a constant term (i.e., a term with no x) is always zero.

For a general polynomial:

$$f(x) = a_n x^n + a_{n-1}x^{n-1} + ... + a_1 x + a_0 \tag{3.23}$$

We should look at each term one at a time and apply the rule to each term. Its derivative is then:

$$f'(x) = na_n x^{n-1} + (n-1)a_{n-1}x^{n-2} + ... + a_1 \tag{3.24}$$

Engineering application: Now, let us look into a few examples to explore how differentiation works for functions. Suppose the distance $u(t)$ (in km) that the car has traveled at time t (in hours) is given by the following parabola function:

$$u(t) = t^2 + t \tag{3.25}$$

This means that the distance traveled depends on time t, and as t increases, the distance $u(t)$ increases. To find the speed of the car at any given time t, we need to determine the rate of change of $u(t)$ with respect to t. In mathematical terms, the speed is the derivative of $u(t)$ with respect to t, which is denoted as either $\frac{du}{dt}$ or $u'(t)$.

Differentiating $u(t) = t^2 + t$, gives:

$$\frac{du}{dt} = u'(t) = 2t + 1 \tag{3.26}$$

which represents the car's speed (in km/h) at any time t.

3.2.2 DIFFERENTIATION USING MATLAB AND PYTHON

Differentiation can be performed in both MATLAB and Python using built-in functions. In MATLAB, the `diff()` function is commonly used to compute derivatives. Similarly, Python provides the `diff()` function for differentiation. In both environments, symbolic differentiation requires defining the symbolic variables first. This is done using the `syms` command in MATLAB and the `symbols()` function from the `sympy` library in Python.

Example: Apply MATLAB and Python to calculate the derivative of $u(t) = t^2 + t$ considering symbolic differentiation in both environments.

Solution:

MATLAB is used to calculate the derivative of $u(t) = t^2+t$, as demonstrated in Example 3.7 using the script `CH03Mcode4.m`, which employs the `diff()` function. In Python, the `diff()` function is used, as illustrated in Example 3.8 (script `CH03Pcode4.py`).

Example 3.7

MATLAB code. This MATLAB script, named `CH03Mcode4.m`, calculates the derivative of function $f(x) = t^2 + t$ using the "diff" and "simplify" functions in MATLAB's Symbolic Math Toolbox:

MATLAB Input:

```
% Define the symbolic variable
syms t
% Define the function f(t)
f = t^2 + t;
% Calculate the derivative of f(t)
f_derivative = diff(f, t);
% Simplify the derivative (if applicable)
f_derivative_simplified = simplify(f_derivative);
```

```
% Display the results
fprintf('The derivative of f(t) is %s\n', char(f_derivative));
```

MATLAB Output:

```
The derivative of f(t) is 2*t + 1
```

Example 3.8

Python code. This Python code, named CH03Pcode4.py, calculates the derivative of function $f(x) = t^2 + t$.

Python Input:

```
from sympy import symbols, diff
# Define the symbolic variable
t = symbols('t')
# Define the function f(t)
f = t**2 + t
# Calculate the derivative of f(t)
f_derivative = diff(f, t)
# Display the result
print(f"The derivative of f(t) is {f_derivative}")
```

Python Output:

```
The derivative of f(t) is 2*t + 1
```

3.2.3 THE PRODUCT RULE OF DIFFERENTIATION

To differentiate functions of the form $f(x) = (x^2 + 1)(x - 1)$, we apply the **product rule**. The product rule for differentiation is given by:

$$\frac{d}{dx}[u(x)v(x)] = u'(x)v(x) + u(x)v'(x) \tag{3.27}$$

Example: Use the product rule of differentiation to compute the derivative of the following function:

$$f(x) = (x^2 + 1)(x - 1)$$

Solution:

Let: $u(x) = x^2 + 1$ and $v(x) = x - 1$. First, we compute $u'(x)$ and $v'(x)$:

$$u'(x) = \frac{d}{dx}(x^2 + 1) = 2x \tag{3.28}$$

$$v'(x) = \frac{d}{dx}(x - 1) = 1 \tag{3.29}$$

Now, we need to apply the product rule:

$$f'(x) = u'(x)v(x) + u(x)v'(x) \tag{3.30}$$

Substitute the values of $u(x)$, $u'(x)$, $v(x)$, and $v'(x)$:

$$f'(x) = (2x)(x - 1) + (x^2 + 1)(1) \tag{3.31}$$

Simplify the expression:

$$f'(x) = 2x(x - 1) + (x^2 + 1) \tag{3.32}$$

$$f'(x) = 2x^2 - 2x + x^2 + 1 \tag{3.33}$$

$$f'(x) = 3x^2 - 2x + 1 \tag{3.34}$$

Thus, the derivative is:

$$f'(x) = 3x^2 - 2x + 1 \tag{3.35}$$

Examples 3.9 and 3.10 provide scripts for solving this problem using MATLAB and Python, respectively, through the files `CH03Mcode5.m` and `CH03Pcode5.py`. Both scripts utilize the `diff()` function to perform symbolic differentiation.

Example 3.9

MATLAB code. This MATLAB script, named `CH03Mcode5.m`, calculates the derivative of function $f(x) = (x^2 + 1)(x - 1)$.

MATLAB Input:

```
% Define the symbolic variable
syms x;
% Define the function
f = (x^2 + 1) * (x - 1);
% Calculate the derivative
f_prime = diff(f, x);
% Display the result
fprintf('The derivative of f(x) is %s\n', char(f_prime));
```

MATLAB Output:

```
The derivative of f(x) is 2*x*(x - 1) + x^2 + 1
```

Example 3.10

Python code. This Python code, named CH03Pcode5.py, calculates the derivative of function $f(x) = (x^2 + 1)(x - 1)$ using:

Python Input:

```
from sympy import symbols, diff
# Define the symbolic variable
x = symbols('x')
# Define the function
f = (x**2 + 1) * (x - 1)
# Calculate the derivative
f_prime = diff(f, x)
# Display the result
print(f"The derivative of f(x) is {f_prime}")
```

Python Output:

```
The derivative of f(x) is x**2 + 2*x*(x - 1) + 1
```

3.2.4 THE QUOTIENT RULE OF DIFFERENTIATION

For rational functions, differentiation requires the use of the **quotient rule**. Consider the function $f(x) = \frac{x^3}{x^2+1}$ as an example. To differentiate functions of this form, we apply the quotient rule as outlined below:

$$\frac{d}{dx}\left[\frac{u(x)}{v(x)}\right] = \frac{u'(x)v(x) - u(x)v'(x)}{[v(x)]^2} \tag{3.36}$$

Example: Apply the quotient rule of differentiation to find the derivative of the following function:

$$f(x) = \frac{x^3}{x^2 + 1}$$

Solution:

Let: $u(x) = x^3$ and $v(x) = x^2 + 1$. First, compute $u'(x)$ and $v'(x)$:

$$u'(x) = \frac{d}{dx}(x^3) = 3x^2 \tag{3.37}$$

$$v'(x) = \frac{d}{dx}(x^2 + 1) = 2x \tag{3.38}$$

Now, we can apply the quotient rule:

$$f'(x) = \frac{u'(x)v(x) - u(x)v'(x)}{[v(x)]^2} \tag{3.39}$$

Substitute the values of $u(x)$, $u'(x)$, $v(x)$, and $v'(x)$:

$$f'(x) = \frac{(3x^2)(x^2+1) - (x^3)(2x)}{(x^2+1)^2} \tag{3.40}$$

We can simplify the numerator:

$$f'(x) = \frac{3x^2(x^2+1) - 2x^4}{(x^2+1)^2} \tag{3.41}$$

$$f'(x) = \frac{3x^4 + 3x^2 - 2x^4}{(x^2+1)^2} \tag{3.42}$$

Combine like terms in the numerator:

$$f'(x) = \frac{x^4 + 3x^2}{(x^2+1)^2} \tag{3.43}$$

Thus, the derivative is:

$$f'(x) = \frac{x^4 + 3x^2}{(x^2+1)^2} \tag{3.44}$$

The same function, `diff()`, can be used in both MATLAB and Python to calculate the derivative of function $f(x) = \frac{x^3}{x^2+1}$, as illustrated in Examples 3.11 and 3.12. These examples make use of the scripts `CH03Mcode6.m` and `CH03Pcode6.py`, respectively.

Example 3.11

MATLAB code. This MATLAB script, named `CH03Mcode6.m`, calculates the derivative of function $f(x) = \frac{x^3}{x^2+1}$.

MATLAB Input:

```
% Define the symbolic variable
syms x;
% Define the function
f = x^3 / (x^2 + 1);
% Calculate the derivative
f_prime = diff(f, x);
% Display the result
fprintf('The derivative of f(x) is %s\n', char(f_prime));
```

MATLAB Output:

```
The derivative of f(x) is (3*x^2)/(x^2 + 1) - (2*x^4)/(x^2 + 1)
    ^2
```

Example 3.12

Python code. This Python code, named CH03Pcode6.py, calculates the derivative of function $f(x) = \frac{x^3}{x^2+1}$ using:

Python Input:

```
from sympy import symbols, diff
# Define the symbolic variable
x = symbols('x')
# Define the function
f = x**3 / (x**2 + 1)
# Calculate the derivative
f_prime = diff(f, x)
# Display the result
print(f"The derivative of f(x) is {f_prime}")
```

Python Output:

```
The derivative of f(x) is -2*x**4/(x**2 + 1)**2 + 3*x**2/(x**2
    + 1)
```

3.3 INTEGRATION

In mathematics, integration is the process of finding the total accumulation of a quantity. It helps us calculate the area under a curve, which is known as definite integration, or to reconstruct a function from its rate of change, which is called indefinite integration. In bridge engineering, integration plays a key role by allowing us to calculate important quantities like the total load on the bridge, stress distribution, displacement, and the overall weight, including distributed loads (Figure 3.6). In simple terms, integration helps us understand the overall effect of different forces across the whole structure. It is important to note that integration and differentiation are related processes – they are essentially inverse operations.

3.3.1 MATHEMATICAL DEFINITION OF INTEGRATION

The integral of the derivative of a function $f'(x)$ returns the original function $f(x)$, up to an additive constant:

$$\int f'(x)\,dx = f(x) + C \tag{3.45}$$

where C is an arbitrary constant of integration. It is also worth noting that, under certain conditions, the derivative of the integral of a function returns the original function:

$$\frac{d}{dx}\left(\int f(x)\,dx\right) = f(x) \tag{3.46}$$

a) The Hammersmith bridge (UK)

b) Distributed loads over span of the bridge

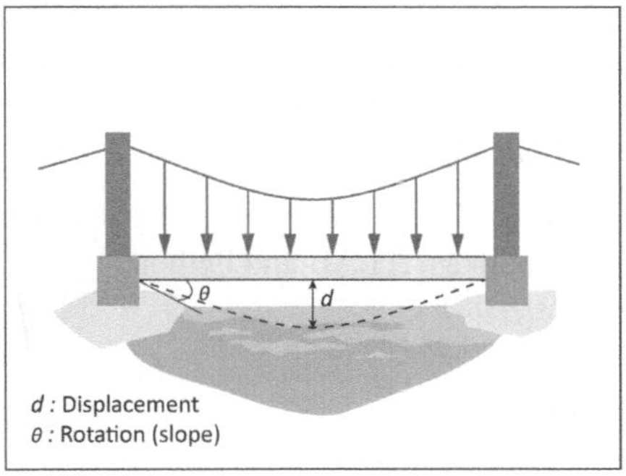

FIGURE 3.6 An illustration of the application of integration in engineering. **a)** Integration is essential in calculating the total effects over the span of the bridge, such as total load, stress distribution, and displacement. **b)** Using integration to compute the total weight of the bridge including the distributed loads across its length.

These relationships highlight the fact that differentiation and integration are inverse operations.

Integration can be challenging, especially when a function does not have a closed-form anti-derivative, for example $\int \mathrm{e}^{-x^3}\,dx$. Let us evaluate the integral of a few functions.

3.3.2 THE POWER RULE FOR INTEGRATION

Polynomial functions are often seen in basic calculus when determining areas under a curve or solving problems in physics where motion follows simple polynomial laws such as distance as a function of time. These functions are among the simplest types of integral, for example, $\int x^2 dx$. The **power rule** states:

$$\int x^n\,dx = \frac{x^{n+1}}{n+1} + C \quad \text{for } n \neq -1 \tag{3.47}$$

In this case, for the integral $\int x^2\,dx$, we have $n = 2$. Therefore, the integral can be evaluated using the power rule as follows:

$$\int x^2\,dx = \frac{x^{2+1}}{2+1} + C = \frac{x^3}{3} + C \tag{3.48}$$

The final answer is:

$$\int x^2\,dx = \frac{x^3}{3} + C \tag{3.49}$$

3.3.3 INTEGRATION IN MATLAB AND PYTHON

In MATLAB, symbolic integration is performed using the `int()` function. In Python, the equivalent operation can be carried out using the `integrate()` function from the `sympy` library. Before performing symbolic integration in either environment, the variables involved must first be declared as symbolic using the `syms` command in MATLAB or the `symbols()` function in Python. Both tools are capable of computing indefinite and definite integrals, and they provide useful options for simplifying or evaluating expressions symbolically.

Example: Evaluate the following integral by applying symbolic integration techniques using both MATLAB and Python:

$$\int x^2\,dx$$

Solution:

In MATLAB, symbolic integration is performed using the `int()` function. Example 3.13 (script `CH03Mcode7.m`) demonstrates how to compute the integral $\int x^2\,dx$ using this function. In Python, the equivalent operation can be carried out using the `integrate()` function from the `sympy` library, as illustrated in Example 3.14 (script `CH03Pcode7.py`).

Example 3.13

MATLAB code. This MATLAB script, named `CH03Mcode7.m`, calculates the integral of function $f(x) = x^2$.

MATLAB Input:

```
% Define the symbolic variable
syms x
% Define the function
f = x^2;
% Calculate the indefinite integral
indefinite_integral = int(f, x);
% Display the result
disp('The indefinite integral of x^2 is');
disp(indefinite_integral);
% Calculate the definite integral (example: from 0 to 1)
a = 0; % lower limit
b = 1; % upper limit
definite_integral = int(f, x, a, b);
% Display the result
fprintf( ...
  ['The indefinite integral of x^2 is %s, ' ...
   'and the definite integral from %d to %d is %s\n'], ...
  char(indefinite_integral), a, b, char(definite_integral) ...
);
```

MATLAB Output:

```
The indefinite integral of x^2 is x^3/3, and the definite
    integral from 0 to 1 is 1/3
```

Example 3.14

Python code. This Python code, named `CH03Pcode7.py`, calculates the integral of function $f(x) = x^2$.

Python Input:

```
from sympy import symbols, integrate
# Define the variable
x = symbols('x')
# Define the function
f = x**2
# Calculate the indefinite integral
indefinite_integral = integrate(f, x)
print("The indefinite integral of x^2 is:")
print(indefinite_integral)
# Calculate the definite integral
a = 0 # lower limit
b = 1 # upper limit
definite_integral = integrate(f, (x, a, b))
print(f"The definite integral of x^2 from {a} to \
{b} is {definite_integral}")
```

Python Output:

```
The indefinite integral of x^2 is:
x**3/3
The definite integral of x^2 from 0 to 1 is 1/3
```

3.3.4 THE SUBSTITUTION RULE FOR INTEGRATION

Now let us evaluate functions where a substitution simplifies the expression. **Substitution** is a technique used to transform the integral into a more manageable form. Problems like $\int 2x\sqrt{x^2+1}\,dx$ often arise in physics (e.g., in mechanics when dealing with energy expressions involving $\sqrt{x^2+1}$) and engineering calculations where root functions appear due to geometry or design constraints.

Example: Evaluate the following integral using the substitution rule for integration:

$$\int 2x\sqrt{x^2+1}\,dx$$

Solution:

Let:

$$u = x^2 + 1 \quad \text{so that} \quad du = 2x\,dx \tag{3.50}$$

We can now rewrite the integral substitute u and du:

$$\int 2x\sqrt{x^2+1}\,dx = \int \sqrt{u}\,du \tag{3.51}$$

By applying the power rule, we rewrite $\sqrt{u}$ as $u^{\frac{1}{2}}$:

$$\int u^{\frac{1}{2}}\,du = \frac{u^{\frac{3}{2}}}{\frac{3}{2}} + C = \frac{2}{3}u^{\frac{3}{2}} + C \tag{3.52}$$

By substituting back $u = x^2 + 1$:

$$\frac{2}{3}(x^2+1)^{\frac{3}{2}} + C \tag{3.53}$$

Thus, the final answer is:

$$\int 2x\sqrt{x^2+1}\,dx = \frac{2}{3}(x^2+1)^{\frac{3}{2}} + C \tag{3.54}$$

MATLAB and Python can be used to perform the above calculations, as demonstrated in Examples 3.15 and 3.16 through the scripts `CH03Mcode8.m` and `CH03Pcode8.py`, respectively. It can be seen that the `int()` function in MATLAB and the `integrate()` function in Python are used to perform the integration.

Example 3.15

MATLAB code. This MATLAB script, named `CH03Mcode8.m`, calculates the integral of function $f(x) = 2x\sqrt{x^2+1}$.

MATLAB Input:

```
% Define the symbolic variable and function
syms x;
% Define the integrand
f = 2 * x * sqrt(x^2 + 1);
% Compute the integral
integral_result = int(f, x);
% Display the result
fprintf('The integral of the function is %s\n', ...
    char(integral_result));
```

MATLAB Output:

```
The integral of the function is (2*(x^2 + 1)^(3/2))/3
```

Example 3.16

Python code. This Python code, named `CH03Pcode8.py`, calculates the integral of function $f(x) = 2x\sqrt{x^2+1}$.

Python Input:

```
from sympy import symbols, integrate, sqrt
# Define the symbolic variable
x = symbols('x')
# Define the integrand
f = 2 * x * sqrt(x**2 + 1)
# Compute the integral
integral_result = integrate(f, x)
# Display the result
print(f"The integral of the function is \
{integral_result}")
```

Python Output:

```
The integral of the function is 2*x**2*sqrt(x**2 + 1)/3 + 2*
    sqrt(x**2 + 1)/3
```

3.3.5 INTEGRATION BY PARTS

Another type of integration is **integration by parts**. It is used in a scenario where two functions are multiplied, and their integration requires the technique of integration by parts. This situation is common in applied mathematics, such as when solving problems involving exponential growth/decay combined with linear changes, or when analyzing financial models and engineering systems where products of functions appear.

Example: Evaluate the following integral using the method "integration by parts:"

$$\int xe^x \, dx$$

Solution:

The formula for integration by parts is:

$$\int u \, dv = uv - \int v \, du \tag{3.55}$$

By choosing u and dv, we have

$$u = x \quad \text{and} \quad dv = \mathrm{e}^x \, dx \tag{3.56}$$

Then:

$$du = dx \quad \text{and} \quad v = \mathrm{e}^x \tag{3.57}$$

By applying the formula and substituting into it:

$$\int x\mathrm{e}^x \, dx = x\mathrm{e}^x - \int \mathrm{e}^x \, dx \tag{3.58}$$

After simplification, the integral of e^x is e^x:

$$\int x\mathrm{e}^x \, dx = x\mathrm{e}^x - \mathrm{e}^x + C \tag{3.59}$$

Then, the final answer is

$$\int x\mathrm{e}^x \, dx = \mathrm{e}^x(x - 1) + C \tag{3.60}$$

The MATLAB and Python codes provided in Examples 3.17 (script `CH03Mcode9.m`) and 3.18 (script `CH03Pcode9.py`) can be used to perform the integral $\int x\mathrm{e}^x \, dx$.

Example 3.17

MATLAB code. This MATLAB script, named `CH03Mcode9.m`, calculates the integral of function $f(x) = x\mathrm{e}^x \, dx$.

MATLAB Input:

```
% Define the symbolic variable
syms x
% Define the function
f = x * exp(x);
% Calculate the integral
integral_result = int(f, x);
% Display the result
fprintf('The integral of x*e^x is %s\n', ...
   char(integral_result));
```

MATLAB Output:

```
The integral of x*e^x is exp(x)*(x - 1)
```

Example 3.18

Python code. This Python code, named CH03Pcode9.py, calculates the integral of function $f(x) = xe^x\, dx$.

Python Input:

```
from sympy import symbols, exp, integrate
# Define the symbolic variable
x = symbols('x')
# Define the function
f = x * exp(x)
# Calculate the integral
integral_result = integrate(f, x)
# Display the result
print(f"The integral of the function x*e^x \
is {integral_result}")
```

Python Output:

```
The integral of the function x*e^x is (x - 1)*exp(x)
```

3.4 PROBLEMS

Problem 3.1: A city's population $P(t)$ (in thousands) is modeled by the following rational function:

$$P(t) = \frac{100t}{t^2+1}$$

where t is the number of years since 2022. Determine the limit of $P(t)$ as $t \to \infty$, which represents the long-term population trend.

Answer: $\lim_{t\to\infty} P(t) = 0$

Problem 3.2: In an electrical circuit, the voltage $V(R)$ across a resistor is given by the following equation:

$$V(R) = \frac{\cos(R)}{R^2+R+1}$$

where R (in ohms) is the resistance. Evaluate the limit of $V(R)$ as $R \to 0$, which approximates the voltage behavior for very small resistances.

Answer: $\lim_{R\to 0} V(R) = 1$

Problem 3.3: The concentration $C(t)$ of a drug in the bloodstream (in mg/L) is modeled by:

$$C(t) = \frac{50}{t^2 + 5t + 6}$$

where t is is the time (in hours) after the drug is administered. Find the limit of $C(t)$ as $t \to \infty$, representing the concentration of the drug after a long period.

Answer: $\lim_{t\to\infty} C(t) = 0$

Problem 3.4: A factory produces widgets, and the production rate depends on the number of hours worked, h. The production function is given by:

$$P(h) = \frac{h^3}{h^2 + 4}$$

Determine the rate of change of production $P'(h)$ with respect to hours worked h. Interpret $P'(h)$ when $h = 5$.

Answer: $P'(h)\big|_{h=5} = 1.1$

Problem 3.5: The temperature t (in degrees Celsius) of a beverage is modeled by the following equation:

$$T(t) = 75\mathrm{e}^{-0.05t} + 25$$

where t is the time in minutes. Find the rate of change of the temperature $T'(t)$ with respect to time t. Evaluate $T'(t)$ when $t = 10$, and explain what it means in the context of cooling.

Answer: $T'(t)\big|_{t=10} = -2.27$

Problem 3.6: A company spends x dollars on online advertising, and the revenue generated is modeled by:

$$R(x) = \frac{500x}{x + 100}$$

Find the derivative $R'(x)$, which represents the marginal revenue (additional revenue generated per additional dollar spent). Evaluate $R'(x)$ for $x = 200$ and interpret the result.

Answer: $R'(x)\big|_{x=200} = 0.56$

Problem 3.7: The position of a car (in meters) moving along a straight road is given by:

$$s(t) = 5t^2 + \frac{200}{t+1}$$

where t is the time in seconds. Compute the velocity $v(t) = s'(t)$ of the car. Find $v(t)$ at $t = 4$ seconds and interpret the result.

Answer: $v(t)\big|_{t=4} = 32 \text{ m/s}$

Problem 3.8: Water flows through a pipe at a velocity given by $v(t) = 5t^2 + 3$ (in m/s), where t is the time in seconds. Find the total volume of water that flows through the pipe during the first 4 s.

Answer: 118.7 m^3

Problem 3.9: The population of a certain species in a protected area grows at a rate given by:

$$r(t) = 100\mathrm{e}^{-0.2t}$$

where t is the time in years and $r(t)$ is the individual per year. Determine the total increase in the population over the first 10 years.

Answer: 432.3 individuals

Problem 3.10: A car consumes fuel at a rate of $C(v) = 0.05v^2 + 0.8$ (liters per hour), where v is the speed of the car in km/h. If the car maintains a constant speed of 60 km/h for 3 h, calculate the total fuel consumed during the trip.

Answer: 542.4 liters

REFERENCES

[1] Mohammad Heidarzadeh et al. *A Practical Approach to Advanced Mathematical Modelling in Civil Engineering*. Oxford University Press, 2025.

[2] Mohammad Heidarzadeh et al. “Field surveys of tsunami runup and damage following the January 2024 Mw 7.5 Noto (Japan sea) tsunamigenic earthquake.” In: *Ocean Engineering* 307 (2024), p. 118140.

[3] Payam Khazaeinejad. “Fundamental Solutions for Beams, Plates, and Shells Under Thermomechanical Actions”. PhD thesis. The University of Edinburgh, 2016.

[4] Seyed Mahdi Zahraei and Mohammad Heidarzadeh. “Destructive effects of the 2003 Bam Earthquake on structures.” In: *Asian Journal of Civil Engineering* 8(3) (2007), p. 329–342.

4 Derivatives and Differential Equations

In this chapter, we delve into the calculus of functions involving more than one variable. We begin by examining gradients and directional derivatives, which describe how a function changes with respect to direction. The concept of tangent planes, which represent the best linear approximation to a surface at a given point, will also be introduced.

We then explore single, double, and triple integrals Which are powerful tools for calculating areas, volumes, and other quantities in higher-dimensional spaces. To simplify complex integrals, we will discuss the method of changing variables using Jacobians. The chapter concludes with an introduction to first-order and second-order differential equations, which describe rates of change and are fundamental to modeling physical, engineering, and natural systems. Mastering these topics will enhance your ability to solve a wide range of real-world problems across various scientific and engineering disciplines.

4.1 FUNCTIONS OF SEVERAL VARIABLES AND DERIVATIVES

Functions of several variables and their derivatives are central topics in multivariable calculus and can help describe phenomena such as material bending and stretching in high-rise buildings and dams [1, 2] (see Figure 4.1). These physical behaviors, especially under thermal and mechanical loads, are often modeled using functions involving multiple spatial and material parameters [3–5]. Consider a recipe that requires multiple ingredients—such as flour, sugar, and eggs—to produce a final dish. Similarly, a function of several variables takes multiple inputs, such as (x, y, z), and combines them accord-

FIGURE 4.1 Multivariable functions and their derivatives help describe phenomena such as material bending and stretching in high-rise buildings (**a**) and dams (**b**).

DOI: 10.1201/9781003511854-4

ing to a specific rule to produce an output. Just as the quality of the dish depends on the interaction of its ingredients, the behavior of a multivariable function depends on the relationship between its inputs.

These functions take multiple inputs—such as coordinates on a map, where x represents the east-west direction and y represents the north-south direction—and return a single output, such as the temperature at that location. For example, the following function:

$$f(x, y) = x^2 + y^3 \tag{4.1}$$

combines two inputs, x and y, to produce a single output value. Visualizing the output of such a function can be thought of as forming a three-dimensional surface. Each point on this surface corresponds to the value of the function at a particular combination of x and y. Imagine walking over a hilly terrain (Figure 4.2); the height of the hill at any given point represents the function's value at that location.

If the hill is sliced horizontally at a fixed height, the resulting intersection forms a curve known as a *level curve*. These curves are analogous to contour lines on a topographic map, indicating all the points that share the same elevation—or in this case, the same function value.

Now consider the hill shown in Figure 4.2 once again. At any given point on the surface, you may wish to determine how steep the hill is in a specific direction. *Partial derivatives* measure how a function changes as you vary one variable while keeping the others constant. For instance, if you walk purely in the east-west direction (changing x), the partial derivative with respect to x indicates the rate of change—or steepness—in that direction. When a quantity, such as elevation, depends on multiple variables, partial derivatives allow us to assess how the quantity changes with respect to each individual variable independently. This concept is fundamental in various applications, including topographic mapping, structural analysis on uneven terrain, and more (see, for example, Figure 4.2).

4.2 CALCULATING PARTIAL DERIVATIVE

To calculate a partial derivative, we follow a process similar to ordinary differentiation, but with one key difference: when taking the derivative with respect to one variable, we treat all other variables as constants. Partial derivatives are denoted using the special symbol ∂, which looks like a rounded version of the letter d, to distinguish them from ordinary derivatives.

Example: Calculate partial derivatives respect to x and y for the following function:

$$f(x, y) = x^2 + y^3 \tag{4.2}$$

Solution:

Let us find the partial derivative of f with respect to x. Since we are only interested in how f changes as x changes, we treat y as a constant (like a number). The derivative of

a) A hill in Dorchester (UK) **b)** Illustration of partial derivatives

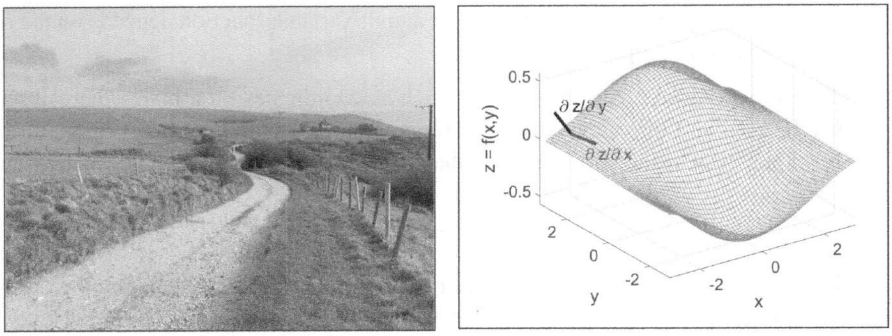

FIGURE 4.2 Multivariable functions and their derivatives help describe phenomena such as material bending and stretching in high-rise buildings and dams. **a)** On a hill, the elevation at any point depends on its spatial coordinates. **b)** A schematic illustration of $z = f(x, y)$, showing how the partial derivatives $\frac{\partial z}{\partial x}$ and $\frac{\partial z}{\partial y}$ quantify the rate of change in elevation as the variables x and y vary.

x^2 with respect to x is $2x$, and since y^3 is treated as a constant, its derivative is zero. Therefore:

$$\frac{\partial f}{\partial x} = 2x \tag{4.3}$$

Now, let us find the partial derivative with respect to y. This time, we treat x as a constant. The derivative of y^3 with respect to y is $3y^2$, and x^2 is treated as constant, and thus its derivative is zero. Therefore:

$$\frac{\partial f}{\partial y} = 3y^2 \tag{4.4}$$

In summary:

- When calculating $\frac{\partial f}{\partial x}$, treat y as a constant.
- When calculating $\frac{\partial f}{\partial y}$, treat x as a constant.

Example: Assume that you are standing in a field and the temperature around you depends on your position. We can describe the temperature (T) as a function of two variables as follows:

$$T(x, y) = 20 - x^3 - y^2 \tag{4.5}$$

where x denotes how far east or west you are, y shows how far north or south you are, and $T(x, y)$ is the temperature at that point in the field. The farther you move from the origin $(0, 0)$ the cooler it gets because the temperature decreases as x^3 or y increases. The highest temperature (20°C) is at the origin. If we plot this function, as shown in Figure 4.3, it looks like an upside-down bowl.

Now, let us figure out how temperature changes if you move in a specific direction. The partial derivative with respect to x is:

$$\frac{\partial T}{\partial x} = -3x^2 \tag{4.6}$$

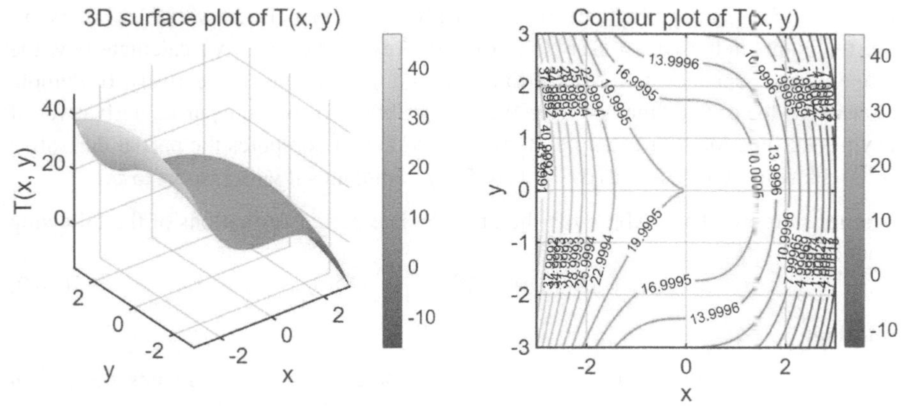

FIGURE 4.3 Graph of a function of several variables showing temperature, $T(x, y)$, as a function of x and y. The temperature is high at the center and gradually sloping downward. **a)** The 3D surface plot illustrates the function $T(x, y) = 20 - x^3 - y^2$, showing how the temperature decreases as the x and y values deviate from the origin. **b)** The contour plot provides a top-down view of the same function, with each contour line representing a level of constant temperature. The spacing between the lines indicates how rapidly the temperature changes in different regions, with denser lines showing steeper gradients.

It shows how fast the temperature changes if you move east $(+x)$ or west $(-x)$, while staying at the same y position. For example, if you are at position $(1, 0)$, the rate of temperature change with respect to x is:

$$\frac{\partial T}{\partial x} = -3(1)^2 = -3 \tag{4.7}$$

which means the temperature decreases by 3° C, for every unit you move east at this point. The partial derivative with respect to y is:

$$\frac{\partial T}{\partial y} = -2y \tag{4.8}$$

which shows how fast the temperature changes if you move north $(+y)$ to south $(-y)$. If you are at position $(0, 3)$, the rate of temperature change with respect to y is:

$$\frac{\partial T}{\partial y} = -2(3) = -6 \tag{4.9}$$

Here, the temperature decreases by 6°C for every unit you move north at this point.

4.2.1 USING MATLAB AND PYTHON TO CALCULATE PARTIAL DERIVATIVE

Partial derivatives can be computed symbolically in both MATLAB and Python using built-in tools. In MATLAB, the `diff` command from the Symbolic Math Toolbox is used to differentiate symbolic expressions with respect to a chosen variable. For example, the commands `dTdx = diff(T, x)` and `dTdy = diff(T, y)` calculate how the temperature T changes with respect to the x- and y-directions, respectively. In Python, you can use the `diff()` method from the `sympy` library to compute partial derivatives of a symbolic function. For example, `dTdx = T.diff(x)` computes the partial derivative of T with respect to x, and `dTdy = T.diff(y)` computes it with respect to y.

Example: Apply MATLAB and Python to calculate partial derivations of the following function:

$$T(x, y) = 18 - \frac{3}{2}x^2 - y^2 \tag{4.10}$$

Solution:

Example 4.1 presents a MATLAB script, `CH04Mcode1.m`, that evaluates the partial derivatives of $T(x, y) = 18 - \frac{3}{2}x^2 - y^2$ at the points $(1, 0)$ and $(0, 3)$.

In MATLAB, the `diff` function is used to compute symbolic partial derivatives. In Example 4.1, the commands `dTdx = diff(T, x)` and `dTdy = diff(T, y)` calculate how the temperature T changes with respect to x and y, respectively.

Example 4.2 gives the corresponding Python implementation, `CH04Pcode1.py`. In Python, you can use the `diff()` method to find partial derivatives of a symbolic function. In this script, the command `dTdx = T.diff(x)` is used to compute the partial derivative of T with respect to x. The command `dTdy = T.diff(y)` computes partial derivative with respect to y.

Example 4.1

MATLAB code. This MATLAB script, named `CH04Mcode1.m`, evaluates partial derivative of the function $T(x, y) = 18 - \frac{3}{2}x^2 - y^2$ with respect to x at point (1,0) and with respect to y at point (0,3). This MATLAB script is available through the companion website (`https://github.com/prog871/CRC-book2025`), which provides access to all the code presented in this book.

MATLAB Input:

```
% Define symbolic variables
syms x y
%
% Define the temperature function T(x,y)
T = 18 - (3/2)*x^2 - y^2;
%
% Compute partial derivatives
dTdx = diff(T, x);
dTdy = diff(T, y);
%
```

```
% Evaluate partial derivative w.r.t. x at (1,0)
val_dTdx_1_0 = subs(dTdx, [x, y], [1, 0]);
%
% Evaluate partial derivative w.r.t. y at (0,3)
val_dTdy_0_3 = subs(dTdy, [x, y], [0, 3]);
%
% Display results
disp(['dT/dx at (1,0) = ', char(val_dTdx_1_0)]);
disp(['dT/dy at (0,3) = ', char(val_dTdy_0_3)]);
```

MATLAB Output:

```
dT/dx at (1,0) = -3
dT/dy at (0,3) = -6
```

Example 4.2

Python code. This Python script, named CH04Pcode1.py, evaluates partial derivative of the function $T(x, y) = 18 - \frac{3}{2}x^2 - y^2$ with respect to x at point (1,0) and with respect to y at point (0,3).

Python Input:

```
import sympy
#
# Define symbols
x, y = sympy.symbols('x y', real=True)
#
# Define the temperature function T(x,y)
T = 18 - (3*x**2)/2 - y**2
#
# Compute partial derivatives
dTdx = T.diff(x)
dTdy = T.diff(y)
#
# Evaluate partial derivative w.r.t. x at (1,0)
val_dTdx_1_0 = dTdx.subs({x: 1, y: 0})
#
# Evaluate partial derivative w.r.t. y at (0,3)
val_dTdy_0_3 = dTdy.subs({x: 0, y: 3})
#
# Print results
```

```
print(f"dT/dx at (1,0) = {val_dTdx_1_0}")
print(f"dT/dy at (0,3) = {val_dTdy_0_3}")
```

Python Output:

```
dT/dx at (1,0) = -3
dT/dy at (0,3) = -6
```

Example: Calculate the partial derivatives of the following function with respect to x and y using MATLAB and Python:

$$T(x,y) = xe^y + y\sin x \tag{4.11}$$

Solution:

Example 4.3 shows a MATLAB script, `CH04Mcode2.m`, which computes the partial derivatives of $T(x,y) = xe^y + y\sin x$ at the points $(2,1)$ and $(1,2)$ using the commands `dTdx = diff(T, x)` and `dTdy = diff(T, y)`. Note that the exact values may differ slightly due to numerical precision.

Example 4.4 presents the corresponding Python script, `CH04Pcode2.py`, which uses the commands `dTdx = T.diff(x)` and `dTdy = T.diff(y)` to compute the partial derivatives.

Example 4.3

MATLAB code. This MATLAB script, named `CH04Mcode2.m`, evaluates the partial derivative of the function $T(x,y) = x\,e^y + y\sin(x)$ with respect to x at point (2,1) and with respect to y at point (1,2).

MATLAB Input:

```
% Define symbolic variables
syms x y
%
% Define the function T(x,y)
T = x*exp(y) + y*sin(x);
%
% Compute partial derivatives
dTdx = diff(T, x); % Partial derivative w.r.t. x
dTdy = diff(T, y); % Partial derivative w.r.t. y
%
% Evaluate partial derivative w.r.t. x at (2,1)
val_dTdx_2_1 = subs(dTdx, [x, y], [2, 1]);
%
```

```
% Evaluate partial derivative w.r.t. y at (1,2)
val_dTdy_1_2 = subs(dTdy, [x, y], [1, 2]);
%
% Convert to double (numerical) for display
val_dTdx_2_1_num = double(val_dTdx_2_1);
val_dTdy_1_2_num = double(val_dTdy_1_2);
%
% Display results
disp(['dT/dx at (2,1) = ', num2str(val_dTdx_2_1_num)]);
disp(['dT/dy at (1,2) = ', num2str(val_dTdy_1_2_num)]);
```

MATLAB Output:

```
dT/dx at (2,1) = 2.3021
dT/dy at (1,2) = 8.2305
```

Example 4.4

Python code. This Python code, named `CH04Pcode2.py`, evaluates partial derivative of the function $T(x, y) = x\, e^y + y \sin(x)$ with respect to x at point (2,1) and with respect to y at point (1,2).

Python Input:

```
import sympy
#
# Define symbols
x, y = sympy.symbols('x y', real=True)
#
# Define the function T(x,y)
T = x*sympy.exp(y) + y*sympy.sin(x)
#
# Compute partial derivatives
dTdx = T.diff(x) # Partial derivative w.r.t. x
dTdy = T.diff(y) # Partial derivative w.r.t. y
#
# Evaluate partial derivative w.r.t. x at (2,1)
val_dTdx_2_1 = dTdx.subs({x: 2, y: 1})
#
# Evaluate partial derivative w.r.t. y at (1,2)
val_dTdy_1_2 = dTdy.subs({x: 1, y: 2})
#
```

```
# Convert to a numerical approximation
val_dTdx_2_1_num = val_dTdx_2_1.evalf()
val_dTdy_1_2_num = val_dTdy_1_2.evalf()
#
# Print results
print(f"dT/dx at (2,1) = {val_dTdx_2_1_num}")
print(f"dT/dy at (1,2) = {val_dTdy_1_2_num}")
```

Python Output:

```
dT/dx at (2,1) = 2.30213499191190
dT/dy at (1,2) = 8.23052708373855
```

4.3 GRADIENT, DIRECTIONAL DERIVATIVE, AND TANGENT PLANES

4.3.1 GRADIENT

The gradient functions as a compass indicating the direction of the steepest ascent. It points toward the direction where the slope is greatest, and the length of this "compass arrow" represents the steepness of that slope. When movement occurs in a combined direction—such as northeast on a hill—the change in the function depends on both the gradient and the direction of movement. This concept is known as the directional derivative, which quantifies how steep the slope feels in the specific direction of travel.

The gradient encodes the direction and rate of steepest increase of a multivariable function. Tangent planes linearize the function near a specific point, aiding in calculations such as optimizing slope usage in architecture, civil engineering, and landscape design (see Figure 4.4).

Mathematically, the gradient of a function $f(x, y)$ is a vector composed of its partial derivatives with respect to each variable. It is written as:

$$\nabla f(x, y) = \left(\frac{\partial f}{\partial x}, \frac{\partial f}{\partial y}\right)$$

This vector points in the direction of the steepest increase in the value of the function, and its magnitude indicates how rapidly the function increases in that direction.

Example: Calculate the gradient for the following function:

$$T(x, y) = 20 - x^3 - y^2 \tag{4.12}$$

Solution:

Gradient combines the two partial derivatives as follows:

$$\Delta T = \left(\frac{\partial T}{\partial x}, \frac{\partial T}{\partial y}\right) = (-3x^2, -2y) \tag{4.13}$$

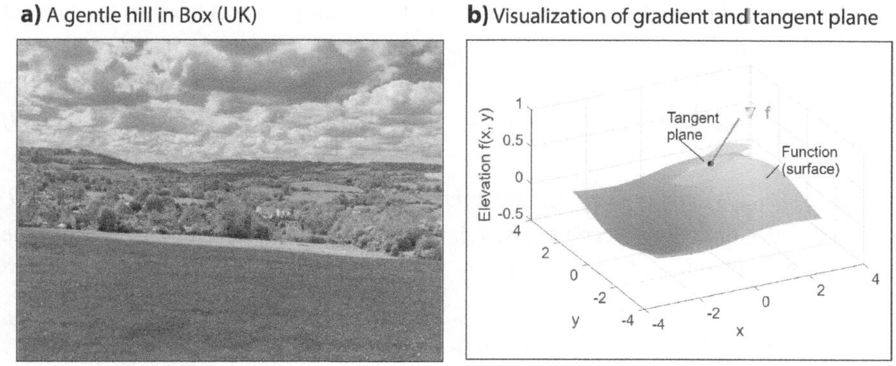

FIGURE 4.4 Visualizing the gradient on sloping land. **a)** A photo of a green land area with gentle hills. **b)** A diagram showing that the gradient ∇f points toward the direction of steepest ascent, and the tangent plane approximates the surface at a given point.

For the case of this function, the gradient is like an arrow pointing in the direction where the temperature drops the fastest. For example, at point $(1, 2)$, the gradient is:

$$\Delta T(1,2) = \left(-3(1)^2), -2(2)\right) = (-3, -4) \tag{4.14}$$

This means that, at the point $(1, 2)$, the temperature increases fastest in the direction $(-3, -4)$. The length of the gradient is $\sqrt{(-3)^2 + (-4)^2} = \sqrt{25} = 5$, which tells us how steep the change is.

A **critical point** is where the gradient is zero, meaning there is no change in any direction. For the function $T(x, y) = 20 - x^3 - y^2$, the gradient is given by:

$$\Delta T = \left(\frac{\partial T}{\partial x}, \frac{\partial T}{\partial y}\right) = (-3x^2, -2y) \tag{4.15}$$

To find the critical points, we solve:

$$-3x^2 = 0 \quad \text{and} \quad -2y = 0 \tag{4.16}$$

Solving these equations:

$$x = 0 \tag{4.17}$$

$$y = 0 \tag{4.18}$$

So the gradient is zero at the point $(0, 0)$. This is a critical point. For this function, it represents the maximum temperature of 20°C, which occurs at the origin

4.3.2 TANGENT PLANES

For a function like $z = T(x, y)$, the tangent plane at a point (x_0, y_0) is the "best linear approximation" of function T near that point. In other words, near point (x_0, y_0), the surface can be approximated by the plane. Mathematically, the equation of the tangent plane at (x_0, y_0) is:

$$z = T(x_0, y_0) + \frac{\partial T}{\partial x}(x_0, y_0)\ (x - x_0) + \frac{\partial T}{\partial y}(x_0, y_0)\ (y - y_0) \tag{4.19}$$

where $\frac{\partial T}{\partial x}(x_0, y_0)$ and $\frac{\partial T}{\partial y}(x_0, y_0)$ represents the partial derivatives (the slopes) of T in the x and y directions at point (x_0, y_0), respectively. The tangent plane touches the surface at exactly one point, $(x_0, y_0, T(x_0, y_0))$, and its slope in each direction matches that of the function at that point.

Example: Find the tangent plane at point $(x_0, y_0) = (1, 2)$ for the following function:

$$T(x, y) = 20 - x^3 - y^2 \tag{4.20}$$

Solution:

The general form of the tangent plane to a function $T(x, y)$ at the point (x_0, y_0) is given by:

$$z = T(x_0, y_0) + \frac{\partial T}{\partial x}(x_0, y_0)\ (x - x_0) + \frac{\partial T}{\partial y}(x_0, y_0)\ (y - y_0) \tag{4.21}$$

First, we compute each term at the point $(1, 2)$ as follows.

The partial derivatives of $T(x, y)$ are:

$$\frac{\partial T}{\partial x} = -3x^2, \qquad \frac{\partial T}{\partial y} = -2y \tag{4.22}$$

Now we evaluate them at the point $(1, 2)$:

$$\begin{aligned} T(1, 2) &= 20 - (1)^3 - (2)^2 = 20 - 1 - 4 = 15 \\ \frac{\partial T}{\partial x}(1, 2) &= -3(1)^2 = -3 \\ \frac{\partial T}{\partial y}(1, 2) &= -2(2) = -4 \end{aligned} \tag{4.23}$$

Substituting these values into Equation (4.21), we obtain the equation of the tangent plane:

$$\begin{aligned} z &= 15 + (-3)(x - 1) + (-4)(y - 2) \\ &= 15 - 3x + 3 - 4y + 8 \\ &= 26 - 3x - 4y \end{aligned} \tag{4.24}$$

This plane approximates the surface near point $(1, 2)$. As shown in Figure 4.5, if you zoom in close enough to $(1, 2)$, on the 3D surface, the surface looks almost “flat” and coincides with this plane.

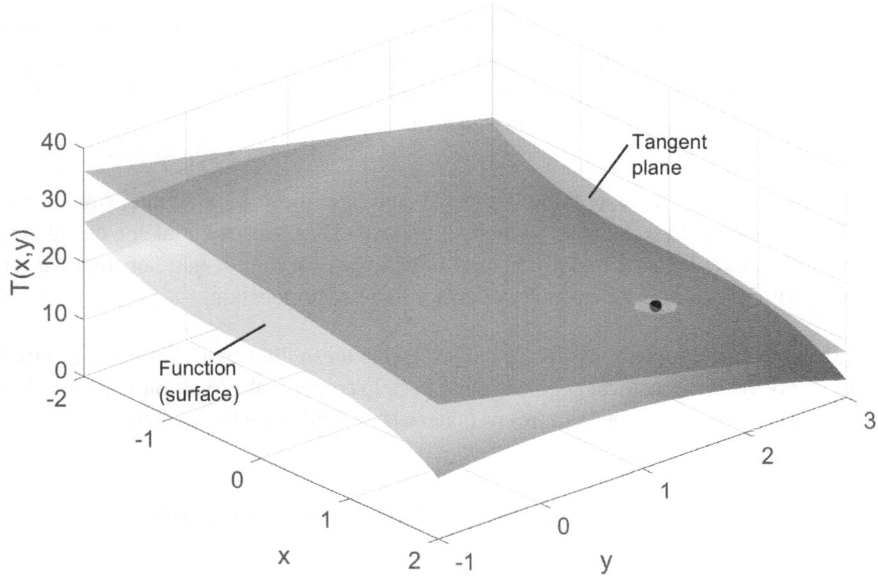

FIGURE 4.5 Surface (colored object) with its tangent plane (brown plate) for function $T(x, y) = 20 - x^3 - y^2$ at $(x_0, y_0) = (1, 2)$. The black dot $(1, 2, T(1, 2))$ marks the point of tangency, illustrating how the plane locally approximates the surface.

4.4 SINGLE, DOUBLE, AND TRIPLE INTEGRALS

There are two types of integrals:

- **Indefinite integrals**: focus on finding the family of all anti-derivatives of a function, and therefore include an arbitrary constant.
- **Definite integrals**: measure accumulated quantities such as areas, volumes, displacements, or probabilities (in applied contexts).

An indefinite integral yields a family of functions and is useful for solving differential equations or determining velocity from acceleration. In contrast, a definite integral produces a single numerical value by evaluating the function at the endpoints, and is essential for tasks such as calculating exact areas under curves, total work done by a variable force, average values of functions, and more.

Types of integrals based on dimensions

- **Single integrals** involve one variable and are used to integrate over a one-dimensional region, such as a line or an interval on the x-axis. To understand this, imagine walking along a road where, at each point, you measure a certain value—such as the height of a signpost. A single integral sums these values continuously along the road, giving the total accumulated quantity over the path.

- **Double integrals** involve two variables and are used to compute quantities over a two-dimensional region, such as an area in the xy-plane. For example, a double integral can be used to find the total mass of a flat plate with varying density across its surface, or the total elevation over a geographic area.

- **Triple integrals** involve three variables and are used to integrate over three-dimensional regions, such as volumes in space. They are essential for calculating quantities like the total mass of a solid object with varying density, or the total energy within a volume where energy density varies with location.

Single, double, and triple integrals are indispensable for finding areas, volumes, and total mass. In civil and mechanical engineering, they help compute the loads on structures like piers or dams, as well as the volumes of materials or fluids (see Figure 4.6).

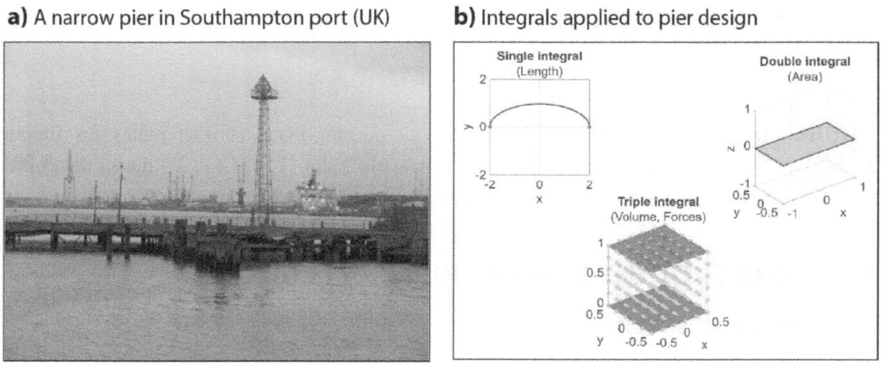

FIGURE 4.6 Integrals are applied to design of ports and harbors. **a)** A photo showing a pier in Southampton port, UK. **b)** A sketch of how single, double, or triple integrals can be used to calculate lengths, areas, volumes, and fluid forces acting on the pier structure.

4.4.1 SINGLE INTEGRALS

Mathematically, for function $f(x)$, a single integral can be written as:

$$\begin{aligned} &\text{Indefinite integral}: \int f(x)\,dx \\ &\text{Definite integral}: \int_a^b f(x)\,dx \end{aligned} \tag{4.25}$$

where a and b are the endpoints with a being the lower limit of the integral and b being the upper limit of the integral. Evaluation of these integrals results in:

$$\begin{aligned}
&\text{Indefinite integral}: \int f(x)\,dx = F(x) + C \\
&\text{Definite integral}: \int_a^b f(x)\,dx = F(x)|_a^b = F(b) - F(a)
\end{aligned} \tag{4.26}$$

where C is the constant of integration and the first derivation of function $F(x)$ is $f(x)$; i.e., $F(x)' = f(x)$. Note that any constant disappears upon differentiation.

Example: For the polynomial function $f(x) = x^n$, calculate its indefinite integral. Then, evaluate the definite integral of $f(x) = x^2$ over the interval $(0, 1)$.

Solution:

The indefinite integral is evaluated as:

$$\int x^n\,dx = \frac{x^{n+1}}{n+1} + C \tag{4.27}$$

For the function $f(x) = x^2$, the definite integral over the interval $(0, 1)$ is calculated as follows:

$$\begin{aligned}
\int_0^1 x^2\,dx &= \left[\frac{x^{(2+1)}}{(2+1)}\right]_0^1 \\
&= \left[\frac{x^3}{3}\right]_0^1 \\
&= \frac{1^3}{3} - \frac{0^3}{3} \\
&= \frac{1}{3} - 0 \\
&= \frac{1}{3}
\end{aligned} \tag{4.28}$$

The shaded area in Figure 4.7 corresponds to this definite integral which calculates the total accumulation of the function's values over the interval, bounded by the x-axis, from $x = 0$ to $x = 1$. This demonstrates how integration provides a means to determine the precise area under a curve, even when it takes a nonlinear form like x^2.

Later in this chapter, we demonstrate how integration can be efficiently performed using computational tools in MATLAB and Python.

4.4.2 DOUBLE INTEGRALS

The concept of integration can be extended from a one-dimensional line to a two-dimensional region, such as a field. A double integral involves integrating a function of two variables over a specified area in the plane. Consider, for example, a large rectangular pasture where each point is associated with a measurable quantity—such as grass

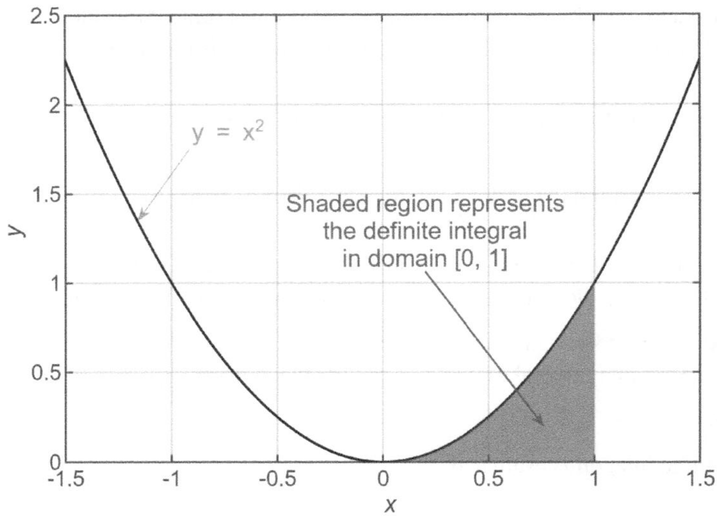

FIGURE 4.7 Graph illustrating the area under the curve $y = x^2$ over the interval $[0, 1]$. The shaded region represents the definite integral $\int_0^1 x^2 dx$ which quantifies the total area under the curve within the specified bounds.

height or the amount of sunlight received. A double integral systematically accumulates these values across the entire region in a continuous manner. Mathematically, a double integral is expressed as:

$$\iint_R f(x, y)\, dA = F(x, y) + g(y) \tag{4.29}$$

where $g(y)$ is an arbitrary function of y (it acts like the "constant of integration," but can depend on y), R is the two-dimensional region of the plane and dA represents an infinitesimal area element and is typically written as $dA = dx\, dy$ or $dA = dy\, dx$.

Example: Evaluate the double integral of the function $f(x, y) = x + y$ over the two-dimensional region R, where $x \in (0, 1)$ and $y \in (0, 1)$:

$$\iint_R (x + y)\, dA = \int_0^1 \int_0^1 (x + y)\, dx\, dy \tag{4.30}$$

Solution:

Since this is a double integral, we need to perform two successive integrations. We can begin with either x or y; here, we start by integrating with respect to x, followed by integration with respect to y, labeling them as **Step 1** and **Step 2**, respectively. Thus, we have:

$$\iint_R (x+y)\,dA = \underbrace{\int_0^1 \underbrace{\int_0^1 (x+y)\,dx}_{\text{Step 1}}\,dy}_{\text{Step2}} \tag{4.31}$$

Step 1: Treat y as a constant while integrating with respect to x. First break the integral into separate terms:

$$\int_0^1 (x+y)\,dx = \int_0^1 x\,dx + \int_0^1 y\,dx \tag{4.32}$$

Integrate each term separately:

$$\begin{aligned}\int_0^1 (x+y)\,dx &= \int_0^1 x\,dx + \int_0^1 y\,dx \\ &= \left[\frac{x^2}{2}\right]_{x=0}^{x=1} + [xy]_{x=0}^{x=1} \\ &= \left(\frac{1^2}{2} - \frac{0^2}{2}\right) + (y-0) \\ &= \left(\frac{1}{2} - 0\right) + y \\ &= \frac{1}{2} + y\end{aligned} \tag{4.33}$$

By substituting the result of step 1 integration into equation (4.31), we have:

$$\begin{aligned}\iint_R (x+y)\,dA &= \int_0^1 \underbrace{\int_0^1 (x+y)\,dx}_{\text{Step 1}=(\frac{1}{2}+y)}\,dy \\ &= \int_0^1 \left(\frac{1}{2} + y\right)\,dy\end{aligned} \tag{4.34}$$

Step 2: Now, integrate the resulting with respect to y. Break the integral into separate terms and integrate each term separately:

$$\begin{aligned}\int_0^1 \left(\frac{1}{2}+y\right) dy &= \int_0^1 \frac{1}{2}\,dy + \int_0^1 y\,dy \\ &= \left[\frac{1}{2}y\right]_{y=0}^{y=1} + \left[\frac{y^2}{2}\right]_{y=0}^{y=1} \\ &= \left(\frac{1}{2}(1) - \frac{1}{2}(0)\right) + \left(\frac{1^2}{2} - \frac{0^2}{2}\right) \\ &= \left(\frac{1}{2} - 0\right) + \left(\frac{1}{2} - 0\right) \\ &= \frac{1}{2} + \frac{1}{2} \\ &= 1\end{aligned} \tag{4.35}$$

Hence,

$$\iint_R (x+y)\,dA = 1 \tag{4.36}$$

Double integration can be performed efficiently and conveniently using MATLAB and Python, as illustrated with numerical examples later in this chapter.

4.4.3 TRIPLE INTEGRALS

Finally, consider a three-dimensional space, such as the interior of a room, where a quantity—such as temperature or density—varies throughout the volume. A triple integral accounts for such quantities by summing their contributions over every infinitesimal volume element within the region. This process enables the accumulation of a scalar field across a three-dimensional domain. Mathematically, a triple integral is expressed as:

$$\iiint_D f(x,y,z)\,dV = F(x,y,z) + (\text{arbitrary functions of one or two variables}) \tag{4.37}$$

where D is the three-dimensional region of the plane and dV represents an infinitesimal volume element, typically written as $dV = dx\,dy\,dz$, or $dV = dz\,dy\,dx$.

Example: Evaluate the definite triple integral of the polynomial function $f(x,y,z) = x^2 + y^2 + z^2$ over the region D, where $x \in (0,1)$, $y \in (0,1)$, and $z \in (0,1)$:

$$\iiint_D (x^2+y^2+z^2)\,dV = \int_0^1\int_0^1\int_0^1 (x^2+y^2+z^2)\,dx\,dy\,dz \tag{4.38}$$

Solution:

Three successive integrations are required. We may begin with respect to any of the variables x, y, or z. Here, we start by integrating with respect to x, followed by y, and finally z:

$$\iiint_D (x^2 + y^2 + z^2)\, dV = \underbrace{\int_0^1 \underbrace{\int_0^1 \underbrace{\int_0^1 (x^2 + y^2 + z^2)\, dx}_{\text{Step 1}}\, dy}_{\text{Step 2}}\, dz}_{\text{Step 3}} \tag{4.39}$$

Step 1: Integrate with respect to x:

$$\begin{aligned}\int_0^1 (x^2 + y^2 + z^2)\, dx &= \int_0^1 x^2\, dx + \int_0^1 y^2\, dx + \int_0^1 z^2\, dx \\ &= \left[\frac{x^3}{3}\right]_0^1 + (y^2)x|_0^1 + (z^2)x|_0^1 \\ &= \left(\frac{1^3}{3} - \frac{0^3}{3}\right) + y^2(1-0) + z^2(1-0) \\ &= \frac{1}{3} + y^2 + z^2\end{aligned} \tag{4.40}$$

Step 2: Integration with respect to y results in:

$$\begin{aligned}\int_0^1 \left(\frac{1}{3} + y^2 + z^2\right) dy &= \int_0^1 \frac{1}{3}\, dy + \int_0^1 y^2\, dy + \int_0^1 z^2\, dy \\ &= \frac{1}{3}y|_0^1 + \left[\frac{y^3}{3}\right]_0^1 + (z^2)y|_0^1 \\ &= \frac{1}{3}(1-0) + \left(\frac{1^3}{3} - \frac{0^3}{3}\right) + (z^2)(1-0) \\ &= \frac{1}{3} + \frac{1}{3} + z^2 \\ &= \frac{2}{3} + z^2\end{aligned} \tag{4.41}$$

Step 3: Finally, integrate with respect to z:

$$\begin{aligned}\int_0^1 \left(\frac{2}{3} + z^2\right) dz &= \int_0^1 \frac{2}{3}\, dz + \int_0^1 z^2\, dz \\ &= \frac{2}{3}z|_0^1 + \left[\frac{z^3}{3}\right]_0^1 \\ &= \frac{2}{3}(1-0) + \left(\frac{1^3}{3} - \frac{0^3}{3}\right) \\ &= \frac{2}{3} + \frac{1}{3} = 1\end{aligned} \tag{4.42}$$

Hence,

$$\iiint_D (x^2 + y^2 + z^2)\, dV = 1 \tag{4.43}$$

Example: Evaluate the following integral involving a combination of polynomial and trigonometric functions:

$$\int (3x^2 + \sin x)\, dx \tag{4.44}$$

Solution:

By breaking it into separate terms and integrating each term, we obtain

$$\begin{aligned} \int (3x^2 + \sin x)\, dx &= \int 3x^2\, dx + \int \sin x\, dx \\ &= 3\frac{x^3}{3} + (-\cos x) + C \\ &= x^3 - \cos x + C \end{aligned} \tag{4.45}$$

where C is a constant.

If the integral were definite—for example, over the interval $(0, \pi)$—we could compute its exact value as follows:

$$\begin{aligned} \int_0^\pi (3x^2 + \sin x)\, dx &= \int_0^\pi 3x^2\, dx + \int_0^\pi \sin x\, dx \\ &= \left[3\frac{x^3}{3}\right]_0^\pi + [-\cos x]_0^\pi \\ &= x^3|_0^\pi - \cos x|_0^\pi \\ &= (\pi^3 - 0) - (\cos\pi - \cos 0) \\ &= \pi^3 - (-1 - 1) \\ &= \pi^3 - (-2) \\ &= \pi^3 + 2 \end{aligned} \tag{4.46}$$

Example: Evaluate the following double integral:

$$\int\int (2x + \cos y)\, dx\, dy \tag{4.47}$$

Solution:

Beginning with the inner integral (step 1 in previous examples), the integration with respect to x yields:

$$\begin{aligned} \int (2x + \cos y)\, dx &= \int 2x\, dx + \int \cos y\, dx \\ &= 2\frac{x^2}{2} + x\cos y + C \\ &= x^2 + x\cos y + C \end{aligned} \tag{4.48}$$

where C_1 is a constant. Move to the outer integral and integrate with respect to y:

$$\int (x^2 + x\cos y + C_1)\, dy \tag{4.49}$$

Treating x^2 and x as constants with respect to y, the integral becomes:

$$\begin{aligned}\int (x^2 + x\cos y + C_1)\, dy &= \int x^2\, dy + \int x\cos y\, dy + \int C_1 dy \\ &= x^2 y + x\sin y + C_1 y + C_2\end{aligned} \tag{4.50}$$

where C_2 is a constant. Combine the results:

$$\int\int (2x + \cos y)\, dx\, dy = x^2 y + x\sin y + C_1 y + C_2 \tag{4.51}$$

As demonstrated through numerical examples later in this chapter, MATLAB and Python provide efficient and convenient tools for performing triple integration.

4.4.4 USING MATLAB AND PYTHON FOR SINGLE, DOUBLE AND TRIPLE INTEGRATION

In MATLAB, the command `int(integrand_single, x, 0, 2)` is used to compute a **single integral** of a function with respect to x over the defined interval [0, 2]. For **double integration**, the process is done iteratively: first using `int(integrand_double, y, 0, 2)` to integrate with respect to y, and then `int(inner_double, x, 0, 1)` to integrate the result with respect to x. Similarly, **triple integration** involves nested commands: `int(integrand_triple, x, 0, 1)` integrates with respect to x, followed by `int(inner_triple, y, 0, 2)` and finally `int(middle_triple, z, 0, 3)` to evaluate the integral over a three-dimensional domain.

In Python, using the `sympy` library, symbolic integration is performed using the `integrate` command. For a single integral, the command `integrate(integrand_single, (x, 0, 2))` evaluates the definite integral of the function with respect to x over the interval $[0, 2]$.

Double integration is carried out in two stages: first, `integrate(integrand_double, (y, 0, 2))` integrates the expression with respect to y, and the resulting expression is then integrated with respect to x using `integrate(inner_double, (x, 0, 1))`.

Triple integration follows a similar approach, with the first integration with respect to x, then with respect to y, and finally with respect to z, using the nested commands:

`integrate(integrand_triple, (x, 0, 1))`
`integrate(inner_triple, (y, 0, 2))`, and
`integrate(middle_triple, (z, 0, 3))`.

Symbolic results are displayed using the `print` function, while numerical approximations are obtained by applying `evalf()` to each symbolic result and converting it to a floating-point value using `float()`.

Example: Apply MATLAB and Python to evaluate the following single, double and triple integrals:

$$\text{Definite single integral:} \quad \int_0^2 (x^2+3)\,dx \tag{4.52}$$

$$\text{Double integral:} \quad \int_0^1 \int_0^2 (xy+\mathrm{e}^x)\,dy\,dx \tag{4.53}$$

$$\text{Triple integral:} \quad \int_0^3 \int_0^2 \int_0^1 \mathrm{e}^{x+y+z}\,dx\,dy\,dz \tag{4.54}$$

Solution:

Example 4.5 contains the MATLAB script `CH04Mcode3.m`, which successively evaluates a definite single integral ($\int_0^2(x^2+3)\,dx$), a double integral ($\int_{x=0}^1 \int_{y=0}^2 (xy+\mathrm{e}^x\,dy\,dx)$) and a triple integral ($\int_{z=0}^3 \int_{y=0}^2 \int_{x=0}^1 \mathrm{e}^{(x+y+z)}\,dx\,dy\,dz$).

Example 4.6 supplies the equivalent Python script, `CH04Pcode3.py`.

Example 4.5

MATLAB code. This MATLAB script, named `CH04Mcode3.m`, evaluates single integral $\int_0^2(x^2+3)\,dx$, double integral $\int_{x=0}^1 \int_{y=0}^2 (xy+\mathrm{e}^x\,dy\,dx)$, and triple integral $\int_{z=0}^3 \int_{y=0}^2 \int_{x=0}^1 \mathrm{e}^{(x+y+z)}\,dx\,dy\,dz$.

MATLAB Input:

```
syms x y z
%
%% 1) Single integral
integrand_single = x^2 + 3;
result_single = int(integrand_single, x, 0, 2);
%
%% 2) Double integral
integrand_double = x*y + exp(x);
% Integrate w.r.t y
inner_double = int(integrand_double, y, 0, 2);
% Integrate w.r.t x
result_double = int(inner_double, x, 0, 1);
%
%% 3) Triple integral
integrand_triple = exp(x + y + z);
% Integrate w.r.t x
inner_triple = int(integrand_triple, x, 0, 1);
% Integrate w.r.t y
middle_triple = int(inner_triple, y, 0, 2);
```

```
% Integrate w.r.t z
result_triple = int(middle_triple, z, 0, 3);
%
%% Display symbolic results
disp('Symbolic results:')
disp(['1) Single integral = ', char(result_single)]);
disp(['2) Double integral = ', char(result_double)]);
disp(['3) Triple integral = ', char(result_triple)]);
%
%% Display numerical approximations
disp(' ')
disp('Numerical approximations:')
disp(['1) Single integral ~= ', ...
num2str(double(result_single))]);
disp(['2) Double integral ~= ', ...
num2str(double(result_double))]);
disp(['3) Triple integral ~= ', ...
num2str(double(result_triple))]);
```

MATLAB Output:

```
Symbolic results:
1) Single integral = 26/3
2) Double integral = 2*exp(1) - 1
3) Triple integral = (exp(1) - 1)^3*(2*exp(1) + 2*exp(2) + exp
    (3) + 1)

Numerical approximations:
1) Single integral ~= 8.6667
2) Double integral ~= 4.4366
3) Triple integral ~= 209.5248
```

Example 4.6

Python code. This Python code, named `CH04Pcode3.py`, evaluates single integral $\int_0^2 (x^2 + 3)\, dx$, double integral $\int_{x=0}^{1} \int_{y=0}^{2} (xy + e^x\, dy\, dx)$, and triple integral $\int_{z=0}^{3} \int_{y=0}^{2} \int_{x=0}^{1} e^{(x+y+z)}\, dx\, dy\, dz$.

Python Input:

```
from sympy import symbols, exp, integrate
```

```
#
# Define the symbols
x, y, z = symbols('x y z', real=True)
#
# 1) Single integral
integrand_single = x**2 + 3
result_single = integrate(integrand_single, (x, 0, 2))
#
# 2) Double integral
integrand_double = x*y + exp(x)
# Integrate w.r.t y
inner_double = integrate(integrand_double, (y, 0, 2))
# Integrate w.r.t x
result_double = integrate(inner_double, (x, 0, 1))
#
# 3) Triple integral
integrand_triple = exp(x + y + z)
# Integrate w.r.t x
inner_triple = integrate(integrand_triple, (x, 0, 1))
# Integrate w.r.t y
middle_triple = integrate(inner_triple, (y, 0, 2))
# Integrate w.r.t z
result_triple = integrate(middle_triple, (z, 0, 3))
#
# Print symbolic results
print("Symbolic results:")
print(f"1) Single integral = {result_single}")
print(f"2) Double integral = {result_double}")
print(f"3) Triple integral = {result_triple}")
#
# Print numerical approximations
print("\nNumerical approximations:")
print(f"1) Single integral ~= {float(result_single.evalf())}")
print(f"2) Double integral ~= {float(result_double.evalf())}")
print(f"3) Triple integral ~= {float(result_triple.evalf())}")
```

Python Output:

```
Symbolic results:
1) Single integral = 26/3
2) Double integral = -1 + 2*E
3) Triple integral = -exp(3) - 1 + E + exp(2) + (-exp(2) - E +
    1 + exp(3))*exp(3)
```

```
Numerical approximations:
1) Single integral ~= 8.666666666666666
2) Double integral ~= 4.43656365691809
3) Triple integral ~= 209.52482228440397
```

Example: Evaluate the following triple integral using MATLAB and Python:

$$\int_0^3 \int_0^2 \int_0^1 \left(x^2 + y \sin z\right)\, dx\, dy\, dz \tag{4.55}$$

Solution:

Example 4.7 presents the MATLAB routine `CH04Mcode4.m` that computes the triple integral $\int_0^3\int_0^2\int_0^1 (x^2 + y\sin z)\, dx\, dy\, dz$. Example 4.8 shows the Python counterpart, `CH04Pcode4.py`.

Example 4.7

MATLAB code. This MATLAB script, named `CH04Mcode4.m`, evaluate triple integral $\int_0^3 \int_0^2 \int_0^1 (x^2 + y \sin z)\, dx\, dy\, dz$.

MATLAB Input:

```
% MATLAB script to evaluate the triple integral
syms x y z;
% Define the integrand
integrand = x^2 + y*sin(z);
%
% Perform the integrations step by step
% Integrate w.r.t x from 0 to 1
inner_integral = int(integrand, x, 0, 1);
% Integrate w.r.t y from 0 to 2
middle_integral = int(inner_integral, y, 0, 2);
% Integrate w.r.t z from 0 to 3
outer_integral = int(middle_integral, z, 0, 3);
%
% Display the result
disp('The value of the triple integral is:');
disp(double(outer_integral));
```

MATLAB Output:

```
The value of the triple integral is:
    5.9800
```

Example 4.8

Python code. This Python code, named CH04Pcode4.py, evaluate triple integral $\int_0^3 \int_0^2 \int_0^1 (x^2 + y \sin z)\, dx\, dy\, dz$.

Python Input:

```
from sympy import symbols, sin, integrate
#
# Define symbols
x, y, z = symbols('x y z')
#
# Define the integrand
integrand = x**2 + y*sin(z)
#
# Perform the integrations step by step
# Integrate w.r.t x from 0 to 1
inner_integral = integrate(integrand, (x, 0, 1))
# Integrate w.r.t y from 0 to 2
middle_integral = integrate(inner_integral, (y, 0, 2))
# Integrate w.r.t z from 0 to 3
outer_integral = integrate(middle_integral, (z, 0, 3))
#
# Print the result
print(f"The value of the triple integral is: \
{float(outer_integral)}")
```

Python Output:

```
The value of the triple integral is: 5.979984993200891
```

4.5 CHANGE OF VARIABLES AND JACOBIANS

When solving mathematical problems—particularly those involving integrals, differential equations, or coordinate transformations—it is often necessary to simplify the problem to make it more manageable. One common technique for achieving this is the **Change of Variables**, which involves transforming one set of variables into another to facilitate the solution process.

Let $x_1, x_2, x_3, \ldots, x_n$ denote the variables in the original coordinate system. Introducing new variables $u_1, u_2, u_3, \ldots, u_n$, defined as functions of the original variables, involves expressing the old variables in terms of the new ones or vice versa. Moreover, all mathematical expressions, including integrals, must be reformulated to reflect the transformation.

For example, in evaluating a double integral, one can take advantage of symmetry by changing from Cartesian coordinates (x, y) to polar coordinates (r, θ). Consider the following double integral:

$$\int\int_R f(x,y)\, dx\, dy \tag{4.56}$$

By switching the original variables (x, y) to new variables (u, v), the integral adjusts to:

$$\int\int_R f\left(x(u,v), y(u,v)\right)\, dR \tag{4.57}$$

where the area element dR transforms as follows:

$$dR = |\det(J)|\, du\, dv \tag{4.58}$$

where J is Jacobian (defined later). Therefore, the transformed integral becomes:

$$\iint_R f(x,y)\, dx\, dy = \iint_S f\left(x(u,v), y(u,v)\right)\, |\det(J)|\, du\, dv \tag{4.59}$$

Here, R is the region in the original xy-plane, and S is the corresponding region in the uv-plane under the given transformation. The transformation maps points (u, v) to points (x, y) via the functions $x = x(u, v)$ and $y = y(u, v)$, and the Jacobian determinant $\det(J)$ accounts for the local change in area under this mapping. $\det(J)$ is the determinant of the Jacobian matrix J and accounts for the distortion of space due to the transformation.

The Jacobian matrix J describes the relationship between the old and new variables and is formed as follows:

$$J = \begin{bmatrix} \frac{\partial x}{\partial u} & \frac{\partial x}{\partial v} \\ \frac{\partial y}{\partial u} & \frac{\partial y}{\partial v} \end{bmatrix} \tag{4.60}$$

For a transformation $(u_1, u_2, \ldots, u_n)$ as functions of $(x_1, x_2, \ldots, x_n)$, the Jacobian matrix J is defined as:

$$J = \begin{bmatrix} \frac{\partial u_1}{\partial x_1} & \frac{\partial u_1}{\partial x_2} & \cdots & \frac{\partial u_1}{\partial x_n} \\ \frac{\partial u_2}{\partial x_1} & \frac{\partial u_2}{\partial x_2} & \cdots & \frac{\partial u_2}{\partial x_n} \\ \vdots & \vdots & \ddots & \vdots \\ \frac{\partial u_n}{\partial x_1} & \frac{\partial u_n}{\partial x_2} & \cdots & \frac{\partial u_n}{\partial x_n} \end{bmatrix} \tag{4.61}$$

when $\det(J) = 0$ the transformation locally collapses, which can indicate singularities.

Example: Evaluate the following double integral:

$$\int\int_R (x+y)\, dx\, dy \tag{4.62}$$

where R is a non-rectangular region in the xy-plane bounded by the parallelogram with vertices at $(0,0)$, $(2,1)$, $(4,3)$, and $(2,2)$.

Solution:

Due to the non-rectangular shape of R, direct evaluation of the integral can be cumbersome. To simplify the region, we apply the following coordinate transformation:

$$u = x - y, \quad v = x + y \tag{4.63}$$

This transformation straightens out the parallelogram into a rectangle in the uv-plane. From this equation, we can express x and y in terms of u and v:

$$\begin{aligned}
&u = x - y \Rightarrow y = x - u \\
&v = x + \underbrace{y}_{x-u} \Rightarrow v = x + (x - u) \Rightarrow v = 2x - u \Rightarrow x = \frac{u+v}{2} \\
&y = \underbrace{x}_{\frac{u+v}{2}} - u = \frac{u+v}{2} - u = \frac{v-u}{2}
\end{aligned} \tag{4.64}$$

Therefore, we have:

$$x = \frac{u+v}{2} \tag{4.65}$$

$$y = \frac{v-u}{2} \tag{4.66}$$

Compute the partial derivatives to form the Jacobian matrix:

$$J = \begin{bmatrix} \frac{\partial x}{\partial u} & \frac{\partial x}{\partial v} \\ \frac{\partial y}{\partial u} & \frac{\partial y}{\partial v} \end{bmatrix} = \begin{bmatrix} \frac{1}{2} & \frac{1}{2} \\ -\frac{1}{2} & \frac{1}{2} \end{bmatrix} \tag{4.67}$$

The determinant of the Jacobian matrix J is given by:

$$\det(J) = \left(\frac{1}{2} \cdot \frac{1}{2}\right) - \left(\frac{1}{2} \cdot -\frac{1}{2}\right) = \frac{1}{4} + \frac{1}{4} = \frac{1}{2} \tag{4.68}$$

Under the change of variables, we have:

The vertex $(x = 0, y = 0)$ becomes $(u = 0, v = 0)$.
The vertex $(x = 2, y = 1)$ becomes $(u = 1, v = 3)$.
The vertex $(x = 4, y = 3)$ becomes $(u = 1, v = 7)$.
The vertex $(x = 2, y = 2)$ becomes $(u = 0, v = 4)$.

The region R in the uv-plane is now a parallelogram with:

$$0 \leq u \leq 1, \quad 3u \leq v \leq 3u + 4. \tag{4.69}$$

Using equation (4.64), $x + y$ can be replaced with v:

$$x + y = \frac{u+v}{2} + \frac{v-u}{2} = v \tag{4.70}$$

Rewrite the integral:

$$\begin{aligned}\iint_R (x+y)\,|\det(J)|\,dv\,du &= \int_0^1 \int_{3u}^{3u+4} v\,\frac{1}{2}\,dv\,du \\ &= \frac{1}{2}\int_0^1 \underbrace{\int_{3u}^{3u+4} v\,dv}_{\left[\frac{v^2}{2}\right]_{3u}^{3u+4}=\frac{(3u+4)^2-(3u)^2}{2}}\,du \\ &= \frac{1}{4}\int_0^1 (24u+16)\,du \\ &= \frac{1}{4}(12u^2+16u)|_0^1 \\ &= \frac{1}{4}(28-0) \\ &= 7\end{aligned} \tag{4.71}$$

Transforming coordinates (e.g., from Cartesian to polar or another suitable system) and using the Jacobian determinant can greatly simplify integrals and differential equations when analyzing systems with complex geometries, which is common in advanced architectural and industrial applications . Figure 4.8 illustrates the practical relevance of coordinate transformations in structural analysis. Figure 4.8a shows a real-world example—a metal and glass roof with a complex curved geometry—highlighting the kind of non-Cartesian surfaces encountered in advanced architectural design. Figure 4.8b provides a schematic view of how a change of variables, along with the Jacobian determinant, can be employed to simplify the calculation of surface area and stress distribution on such curved geometries.

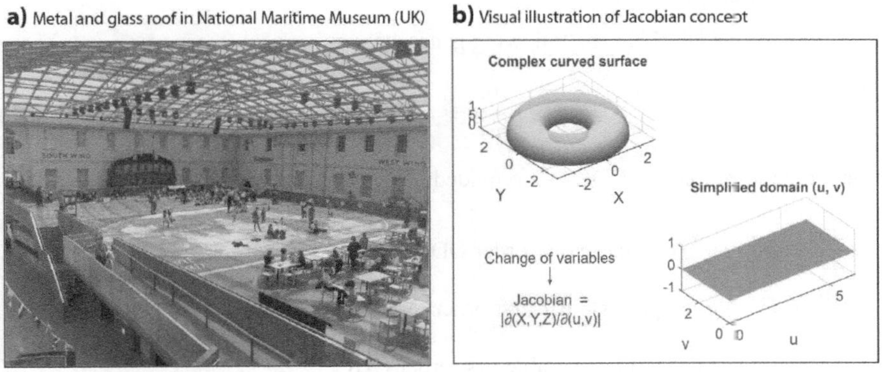

FIGURE 4.8 Geometric transformations in structural design. **a)** A photo of a metal and glass roof with a complex curved geometry in National Maritime Museum in Greenwich, London, UK. **b)** A schematic illustrating how change of variables and the Jacobian are used to simplify surface area and stress calculations on curved designs, where a curved surface is transferred to a simple rectangular surface.

4.6 FIRST-ORDER DIFFERENTIAL EQUATIONS

A **differential equation** is an equation involving an unknown function and its derivatives. An **ordinary differential equation (ODE)** involves a function of a single variable and its derivatives with respect to that variable. The **order** of a differential equation is the order of the highest derivative present. Examples of the order of differential equations are:

- First-order ODE:
$$\frac{dy}{dx} + 2y = x$$
where the highest derivative is the first-order derivative, $\frac{dy}{dx}$.
- Second-order ODE:
$$\frac{d^2y}{dx^2} + 3\frac{dy}{dx} + y = \sin(x)$$
where the highest derivative is the second-order derivative, $\frac{d^2y}{dx^2}$.
- Third-order ODE:
$$\frac{d^3y}{dx^3} + x\frac{d^2y}{dx^2} + 4y = e^x$$
where the highest derivative is the third-order derivative, $\frac{d^3y}{dx^3}$.

A differential equation is **linear** if it can be expressed in the form of:

$$a_n(x)\frac{d^ny}{dx^n} + a_{n-1}(x)\frac{d^{n-1}y}{dx^{n-1}} + \cdots + a_1(x)\frac{dy}{dx} + a_0(x)y = g(x) \tag{4.72}$$

where the functions $a_i(x)$ and $g(x)$ are known, and the unknown function $y(x)$ is not multiplied by any of its derivatives and there are no nonlinear functions of y or its derivatives such as $\sin y$ or $(\frac{dy}{dx})^2$.

A **first-order ODE** has the following general form:

$$\frac{dy}{dx} = f(x, y). \tag{4.73}$$

Some other examples of ODEs are provided below:

- $\dfrac{dy}{dx} = x^2 + y^2$ (nonlinear first-order ODE)
- $\dfrac{dy}{dx} + P(x)\,y = Q(x)$ (linear first-order ODE)
- $\dfrac{dy}{dx} = (x+1)(y-1)$ (separable first-order ODE)

Engineering application: Many natural phenomena—such as chemical reactions, population dynamics, and simple mechanical systems—are governed by first-order differential equations. These equations describe the rate of change of a quantity with respect to time or another independent variable and form the foundation of numerous simplified models in engineering and the physical sciences.

Figure 4.9 presents a clear example of how mathematical models can describe real-world motion on water. In Figure 4.9a, we see a boat moving along the River Thames in London—a familiar and tangible scenario. Figure 4.9b translates this physical situation into a simplified mathematical model using a first-order ODE, such as $\frac{dx}{dt} = v(x)$. This equation captures how the boat's position changes over time, where the velocity $v(x)$ reflects the combined effects of river flow and engine thrust. In this way, the figure bridges the gap between observation and analysis, illustrating how differential equations are used to model and understand dynamic systems in nature.

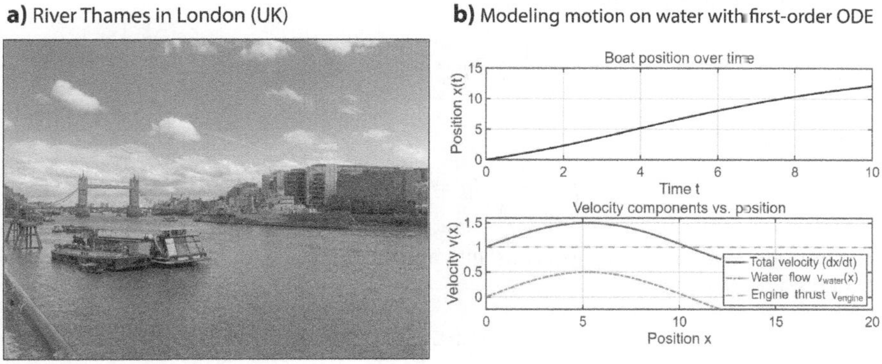

FIGURE 4.9 Modeling motion on water. **a)** A photo of a river with a boat in motion in River Thames, London (UK). **b)** A basic diagram showing how a first-order ODE, e.g., $\frac{dx}{dt} = v(x)$, can describe the boat's position and velocity, influenced by water flow and engine thrust.

4.6.1 SEPARABLE FIRST-ORDER ODE

A first-order ODE is called separable if it can be written in the following form:

$$\frac{dy}{dx} = g(x)\,h(y) \tag{4.74}$$

where g is a function of x only and h is a function of y only. The solution method involves the following steps:

1. Rewrite the equation so that all y-dependent terms are on one side and all x-dependent terms are on the other:

$$\frac{1}{h(y)}\,dy = g(x)\,dx$$

2. Integrate both sides:

$$\int \frac{1}{h(y)}\,dy = \int g(x)\,dx$$

3. Solve for y in terms of x if possible.

Example: Solve the following ODE by applying the concept of separable first-order ODEs:

$$\frac{dy}{dx} = (x+1)(y-1).$$

Solution:

$$\frac{dy}{dx} = (x+1)(y-1).$$

We rewrite this in separable form:

$$\frac{1}{y-1}\,dy = (x+1)\,dx.$$

Integrate both sides:

$$\int \frac{1}{y-1}\,dy = \int (x+1)\,dx.$$

The left side is: $\ln|y-1|$ (up to a constant), and the right side is: $\frac{x^2}{2}+x$ (plus a constant). Therefore, we have:

$$\ln|y-1| = \frac{x^2}{2}+x+C,$$

where C is an integration constant. Exponentiate both sides to obtain the following:

$$|y-1| = \mathrm{e}^{\frac{x^2}{2}+x+C} = \mathrm{e}^{C}\cdot \mathrm{e}^{\frac{x^2}{2}+x}$$

Let $K = \pm\mathrm{e}^{C}$ be a new constant. It can be any non-zero constant of any form. Thus:

$$y-1 = K\,\mathrm{e}^{\frac{x^2}{2}+x}$$

Therefore, the general solution of the problem becomes:

$$y = 1 + K\,\mathrm{e}^{\frac{x^2}{2}+x}$$

4.6.2 THE METHOD OF INTEGRATING FACTOR

A linear first-order ODE can be written in the following standard form:

$$\frac{dy}{dx} + P(x)\,y = Q(x) \tag{4.75}$$

The method of integrating factor is used to solve such equations by following the four steps listed below:

1. Identify $P(x)$. The integrating factor (IF) is given by:

$$\mu(x) = \mathrm{e}^{\int P(x)\,dx}.$$

2. Multiply both sides of the ODE by $\mu(x)$. The left-hand side becomes the derivative of the product $\mu(x)\, y$:

$$\mu(x)\frac{dy}{dx} + \mu(x)P(x)\, y = \frac{d}{dx}\left(\mu(x)\, y\right)$$

Hence, the equation becomes:

$$\frac{d}{dx}\left(\mu(x)\, y\right) = \mu(x)\, Q(x)$$

3. Integrate both sides with respect to x:

$$\mu(x)\, y = \int \mu(x)\, Q(x)\, dx + C$$

4. Solve for y:

$$y(x) = \frac{1}{\mu(x)}\left[\int \mu(x)\, Q(x)\, dx + C\right]$$

Example: Solve the following ODE using the method of integrating factor:

$$\frac{dy}{dx} + x\, y = x^3$$

Solution:

$$\frac{dy}{dx} + x\, y = x^3$$

Here $P(x) = x$ and $Q(x) = x^3$. First, compute the integrating factor:

$$\mu(x) = \mathrm{e}^{\int x\, dx} = \mathrm{e}^{\frac{x^2}{2}}$$

Multiply the entire ODE by $\mathrm{e}^{\frac{x^2}{2}}$:

$$\mathrm{e}^{\frac{x^2}{2}}\frac{dy}{dx} + x\,\mathrm{e}^{\frac{x^2}{2}}\, y = x^3\,\mathrm{e}^{\frac{x^2}{2}}$$

The left-hand side is the derivative of $\mathrm{e}^{\frac{x^2}{2}}\, y$:

$$\frac{d}{dx}\left(\mathrm{e}^{\frac{x^2}{2}}\, y\right) = x^3\,\mathrm{e}^{\frac{x^2}{2}}$$

Integrate both sides with respect to x:

$$\mathrm{e}^{\frac{x^2}{2}}\, y = \int x^3\,\mathrm{e}^{\frac{x^2}{2}}\, dx + C_1$$

To integrate $\int x^3\,\mathrm{e}^{\frac{x^2}{2}}\, dx$, we can use substitution: let $u = \frac{x^2}{2}$, so $du = x\, dx$. Notice that $x^3 = x^2 \cdot x$, and $x^2 = 2u$. Hence,

$$\int x^3\mathrm{e}^{\frac{x^2}{2}}\, dx = \int x^2 \cdot x\,\mathrm{e}^{u}\, dx = \int (2u)\,\mathrm{e}^{u}\, du = 2\int u\,\mathrm{e}^{u}\, du$$

Now use integration by parts on $\int u\,\mathrm{e}^u\,du$:

$$\int u\,\mathrm{e}^u\,du = u\mathrm{e}^u - \int \mathrm{e}^u\,du = u\mathrm{e}^u - \mathrm{e}^u + C_2 = \mathrm{e}^u(u-1) + C_2$$

So,

$$\int x^3 \mathrm{e}^{\frac{x^2}{2}}\,dx = 2\,\mathrm{e}^u(u-1) + C_3 = 2\,\mathrm{e}^{\frac{x^2}{2}}\left(\frac{x^2}{2} - 1\right) + C_3$$

Thus,

$$\mathrm{e}^{\frac{x^2}{2}}\,y = 2\,\mathrm{e}^{\frac{x^2}{2}}\left(\frac{x^2}{2} - 1\right) + C_4$$

Divide by $\mathrm{e}^{\frac{x^2}{2}}$:

$$y = 2\left(\frac{x^2}{2} - 1\right) + C_4\,\mathrm{e}^{-\frac{x^2}{2}}$$

By simplifying the previous equation, the general solution of the differential equation is obtained as follows:

$$y = x^2 - 2 + C_4\,\mathrm{e}^{-\frac{x^2}{2}}$$

4.6.3 EXACT DIFFERENTIAL EQUATION

A first-order ODE of the following form:

$$M(x,y)\,dx + N(x,y)\,dy = 0 \tag{4.76}$$

is called an **exact differential equation** if there exists a scalar function $\Psi(x,y)$ such that:

$$\frac{\partial \Psi}{\partial x} = M(x,y), \quad \frac{\partial \Psi}{\partial y} = N(x,y). \tag{4.77}$$

In this case, the differential form $M\,dx + N\,dy$ is said to be **exact**, and the general solution can be found by determining the potential function $\Psi(x,y)$.

To find $\Psi(x, y)$, follow the steps below:

1. Integrate $M(x, y)$ with respect to x (treating y as a constant):

$$\Psi(x,y) = \int M(x,y)\,dx + h(y), \tag{4.78}$$

where $h(y)$ is an arbitrary function of y only.
2. Differentiate this expression with respect to y and compare the result with $N(x, y)$:

$$\frac{\partial \Psi}{\partial y} = \frac{\partial}{\partial y}\left(\int M(x,y)\,dx\right) + h'(y) = N(x,y). \tag{4.79}$$

Solve this equation to find $h'(y)$, and integrate to determine $h(y)$.

Once $\Psi(x, y)$ is found, the general implicit solution to the differential equation is:

$$\Psi(x,y) = C, \tag{4.80}$$

where C is an arbitrary constant. For the equation to be exact, it is necessary that:

$$\frac{\partial M}{\partial y} = \frac{\partial N}{\partial x} \tag{4.81}$$

in the region of interest. If this condition does not hold, the equation is not exact. However, it may sometimes be possible to make the equation exact by multiplying through by a suitable integrating factor.

Example: Solve the following ODE by applying the concept of exact differential equations:

$$(y+x)\,dx + (y-x)\,dy = 0$$

Solution:

$$M(x,y) = y + x, \quad N(x,y) = y - x$$

Check exactness:

$$\frac{\partial M}{\partial y} = 1, \quad \frac{\partial N}{\partial x} = -1.$$

Since $\frac{\partial M}{\partial y} \neq \frac{\partial N}{\partial x}$, the equation is not exact. However, we might attempt to see if an integrating factor $\mu(x, y)$ can make it exact. But let us proceed with a simpler approach first. We rearrange the equation as follows:

$$(y+x)\,dx = -(y-x)\,dy$$

or

$$\frac{dx}{dy} = -\frac{y-x}{y+x}$$

You might handle this equation through alternative methods. For illustration here, let us look for an integrating factor of the form $\mu(x)$ or $\mu(y)$.

We now examine the condition under which an integrating factor depends on x alone:

$$\frac{1}{N}\left(\frac{\partial M}{\partial y} - \frac{\partial N}{\partial x}\right) = f(x)$$

Let us see if that leads to a function of x only:

$$\frac{1}{y-x}(1-(-1)) = \frac{2}{y-x}$$

This depends on $y - x$, which is not purely a function of x. Similarly, for $\mu(y)$ we check:

$$\frac{1}{M}\left(\frac{\partial N}{\partial x} - \frac{\partial M}{\partial y}\right) = g(y)$$

$$\frac{1}{y+x}((-1)-1) = \frac{-2}{y+x}$$

This is not purely a function of y either (it depends on x as well). In the next section, we introduce a new method to solve this ODE.

4.6.4 HOMOGENEOUS DIFFERENTIAL EQUATIONS

A more straightforward approach to solve certain differential equations is to observe whether the equation is **homogeneous** and attempt a suitable substitution. A function $f(x, y)$ is said to be *homogeneous of degree* n if, for any real number $t > 0$,

$$f(tx, ty) = t^n f(x, y),$$

where n is the degree of homogeneity. For example:

- $f(x, y) = x + y$ is homogeneous of degree 1, because $f(tx, ty) = tx + ty = t(x + y) = tf(x, y)$.
- $f(x, y) = x^2 + y^2$ is homogeneous of degree 2, since $f(tx, ty) = t^2(x^2 + y^2) = t^2 f(x, y)$.

A first-order differential equation of the form

$$M(x, y)\, dx + N(x, y)\, dy = 0$$

is called a *homogeneous differential equation* if the functions $M(x, y)$ and $N(x, y)$ are both homogeneous of the same degree. In such cases, a common method of solution is to apply the substitution

$$v = \frac{y}{x}, \quad \text{so that} \quad y = vx \quad \text{and} \quad dy = v\, dx + x\, dv,$$

which transforms the original equation into a new equation in terms of v and x. This often reduces the equation to a separable form, which can then be integrated directly.

Example: Consider the differential equation presented in the previous example. Solve it using the method for homogeneous differential equations.

$$(x + y)\,dx + (y - x)\,dy = 0.$$

Solution:

Here, $M(x, y) = x + y$ and $N(x, y) = y - x$. Both are homogeneous functions of degree 1, because:

$$M(tx, ty) = t(x + y), \quad N(tx, ty) = t(y - x).$$

Therefore, the equation is homogeneous and can be solved using the substitution $v = \frac{y}{x}$, provided $x \neq 0$.

Let $y = vx$, so $dy = v\,dx + x\,dv$. Substituting into the original equation:

$$\begin{aligned}(x + y)\,dx + (y - x)\,dy &= 0\\ \Rightarrow (x + vx)\,dx + (vx - x)(v\,dx + x\,dv) &= 0\\ \Rightarrow x(1 + v)\,dx + x(v - 1)(v\,dx + x\,dv) &= 0\end{aligned}$$

Now expand the second term:

$$x(1 + v)\,dx + x(v - 1)\,(v\,dx + x\,dv) =$$
$$x(1 + v)\,dx + x(v - 1)v\,dx + x^2(v - 1)\,dv = 0$$

Factor terms:

$$\begin{aligned}x\left[(1 + v) + v(v - 1)\right]\,dx + x^2(v - 1)\,dv &= 0\\ x\left(1 + v + v^2 - v\right)\,dx + x^2(v - 1)\,dv &= 0\\ x(1 + v^2)\,dx + x^2(v - 1)\,dv &= 0\end{aligned}$$

Divide both sides by x^2 (assuming $x \neq 0$) to simplify:

$$\frac{1 + v^2}{x}\,dx + (v - 1)\,dv = 0$$

Now separate variables:

$$(v - 1)\,dv = -\frac{1 + v^2}{x}\,dx$$

Integrate both sides:

$$\int (v - 1)\,dv = -\int \frac{1 + v^2}{x}\,dx$$

The left side integrates easily:

$$\frac{v^2}{2} - v = -(1 + v^2)\ln|x| + C$$

Finally, substitute back $v = \frac{y}{x}$ to express the solution in terms of x and y:

$$\frac{1}{2}\left(\frac{y}{x}\right)^2 - \frac{y}{x} = -\left(1 + \left(\frac{y}{x}\right)^2\right)\ln|x| + C$$

This is the implicit general solution.

4.6.5 USING MATLAB AND PYTHON TO SOLVE ODES

MATLAB offers both numerical and symbolic methods for solving ODEs. The function `ode45` provides a numerical solution using a Runge-Kutta method, suitable for non-stiff problems, while the function `dsolve` returns an exact symbolic solution when one exists. These tools enable users to handle a wide range of ODEs efficiently within a single platform.

In Python, ODEs can be solved both symbolically and numerically. For symbolic solutions, the `sympy.dsolve` function is used, which provides an exact, analytical solution to the ODE given the equation and initial conditions. For numerical solutions, the `scipy.integrate.solve_ivp` function is commonly employed, offering flexibility with different integration methods such as Runge-Kutta, and can handle stiff equations, systems of ODEs, and event-driven problems. Another option is `scipy.integrate.odeint`, a widely used function for solving ODEs numerically, though `solve_ivp` is generally preferred in newer codes due to its enhanced features and flexibility.

Example: Apply various ODE solvers in MATLAB and Python to solve the following ODE:

$$\frac{dy}{dx} = x^2, \quad \text{with } y(0) = 0$$

Solution:

In the following examples, this ODE is solved using both MATLAB and Python. Examples 4.9 and 4.10 illustrate the solutions in MATLAB, while Examples 4.11 and 4.12 correspond to the Python implementations.

In MATLAB, Example 4.9 uses the numerical solver `ode45` to solve this ODE (`CH04Mcode5.m`), and Example 4.10 employs the symbolic solver `dsolve` for the same equation (`CH04Mcode6.m`).

The Python counterparts are provided in Example 4.11 (`CH04Pcode5.py`, numerical solution) and Example 4.12 (`CH04Pcode6.py`, symbolic solution). In the Python implementations, the numerical solver `scipy.integrate.solve_ivp` is used in Example 4.11, while the symbolic solver `sympy.dsolve` is applied in Example 4.12.

The numerical solution obtained in Example 4.9 closely approximates the exact solution of the ODE, $y = \int x^2\,dx = \frac{x^3}{3} + C = \frac{x^3}{3}$, with the condition $y(0) = 0$.

Example 4.9

MATLAB code. This MATLAB script, named `CH04Mcode5.m`, solves the ODE $\frac{dy}{dx} = x^2$ subject to $y(0) = 0$ using ode45 solver.

MATLAB Input:

```
function solve_ode1_numerical
```

```
% ODE: y'(x) = x^2, y(0) = 0
% Using ode45 for a numerical solution.
% Define the ODE as a function handle
% dydx = x^2
odeFunc = @(x,y) x.^2;
%
% Define the interval of integration and the initial condition
xspan = [0 2]; % Solve from x=0 to x=2 (example interval)
y0 = 0; % y(0) = 0
%
% Call ode45
[xSol,ySol] = ode45(odeFunc, xspan, y0);
%
% Display the results
disp('Numerical solution (x, y):');
result = table(xSol, ySol, 'VariableNames', {'x','y'});
disp(result);
%
% Optional: plot the solution
plot(xSol, ySol, 'o-');
title('Numerical Solution of y''(x) = x^2');
xlabel('x');
ylabel('y');
end
```

MATLAB Output:

```
Numerical solution (x, y):
    x y
   ____ __________

     0 0
   0.05 4.1667e-05
    0.1 0.00033333
   0.15 0.001125
    0.2 0.0026667
   0.25 0.0052083
    0.3 0.009
   0.35 0.014292
    0.4 0.021333
   0.45 0.030375
    0.5 0.041667
   0.55 0.055458
    0.6 0.072
```

```
0.65 0.091542
 0.7 0.11433
0.75 0.14062
 0.8 0.17067
0.85 0.20471
 0.9 0.243
0.95 0.28579
   1 0.33333
1.05 0.38588
 1.1 0.44367
1.15 0.50696
 1.2 0.576
1.25 0.65104
 1.3 0.73233
1.35 0.82012
 1.4 0.91467
1.45 1.0162
 1.5 1.125
1.55 1.2413
 1.6 1.3653
1.65 1.4974
 1.7 1.6377
1.75 1.7865
 1.8 1.944
1.85 2.1105
 1.9 2.2863
1.95 2.4716
   2 2.6667
```

Example 4.10

MATLAB code. This MATLAB script, named CH04Mcode6.m, solves the ODE $\frac{dy}{dx} = x^2$ subject to $y(0) = 0$ using dsolve solver.

MATLAB Input:

```
function solve_ode1_symbolic
% ODE: y'(x) = x^2, y(0) = 0
% Using dsolve for a symbolic solution.
%
syms x y(x)
```

```
%
% Define the ODE symbolically
odeEqn = diff(y, x) == x^2;
%
% Solve with the initial condition y(0) = 0
sol = dsolve(odeEqn, y(0) == 0);
%
% Display the symbolic solution
disp('Symbolic solution for y(x):');
disp(sol);
end
```

MATLAB Output:

```
Symbolic solution for y(x):
x^3/3
```

Example 4.11

Python code. This Python code, named `CH04Pcode5.py`, solves the ODE $\frac{dy}{dx} = x^2$ subject to $y(0) = 0$ using solve_ivp solver.

Python Input:

```
import numpy as np
from scipy.integrate import solve_ivp

def solve_ode1_numerical():
    """
    ODE: dy/dx = x^2, y(0) = 0
    Using solve_ivp for a numerical solution.
    """
    # Define the ODE function in the form f(t, y)
    def ode_func(t, y):
        # dy/dt = t^2
        return [t**2]

    # Integration interval
    x_span = (0, 2)
    # Initial condition (as a list/array of length 1)
    y0 = [0]
```

```
    # Points at which to store the solution
    t_eval = np.linspace(0, 2, 11)

    # Solve the ODE
    sol = solve_ivp(ode_func, x_span, y0, t_eval=t_eval)

    # Print results
    print("x values:", sol.t)
    print("y values:", sol.y[0])

    # Return the solution arrays
    return sol.t, sol.y[0]

if __name__ == "__main__":
    x_vals, y_vals = solve_ode1_numerical()
```

Python Output:

```
x values: [0. 0.2 0.4 0.6 0.8 1. 1.2 1.4 1.6 1.8 2. ]
y values: [0.00000000e+00 2.66666667e-03 2.13333333e-02
   7.20000000e-02
 1.70666667e-01 3.33333333e-01 5.76000000e-01 9.14666667e-01
 1.36533333e+00 1.94400000e+00 2.66666667e+00]
```

Example 4.12

Python code. This Python code, named CH04Pcode6.py, solves the ODE $\frac{dy}{dx} = x^2$ subject to $y(0) = 0$ using sympy.dsolve solver.

Python Input:

```
import sympy as sp
#
def solve_ode1_symbolic():
    """
    ODE: dy/dx = x^2, y(0) = 0
    Using sympy.dsolve for a symbolic solution.
    """
    x = sp.Symbol('x', real=True)
    y = sp.Function('y')(x)

    # Define the ODE
```

```
        ode_equation = sp.Eq(sp.diff(y, x), x**2)

        # Solve with the initial condition y(0)=0
        solution = sp.dsolve(ode_equation, ics={y.subs(x, 0): 0})

        print("Symbolic solution for y(x):")
        print(solution)
    #
    if __name__ == "__main__":
        solve_ode1_symbolic()
```

Python Output:

```
Symbolic solution for y(x):
Eq(y(x), x**3/3)
```

Example: Apply various ODE solvers in MATLAB and Python to solve the following ODE:

$$\frac{dy}{dx} + y(x) = x^2, \quad \text{subject to } y(0) = 1$$

Solution:

Example 4.13 solves this linear ODE numerically in MATLAB through the script `CH04Mcode7.m` and Example 4.14 provides the symbolic MATLAB solution using script `CH04Mcode8.m`. Output values in Example 4.13 approximate the exact solution of this differential equation. For this example, the analytic solution can be found via integrating factor:

$$\frac{dy}{dx} + y = x^2 \implies y(x)\mathrm{e}^x = \int x^2\mathrm{e}^x\, dx + \text{const} \tag{4.82}$$

By solving the differential equation, the general solution takes the following form:

$$y = (x^2 - 2x + 2) - \mathrm{e}^{-x} \tag{4.83}$$

which can also be expressed in various equivalent forms, such as $y = (x^2 - 2x + 2) + C\mathrm{e}^{-x}$, with the constant determined by $y(0) = 1$.

The corresponding Python implementations are presented in Examples 4.15 and 4.16. Example 4.15 (`CH04Pcode7.py`) employs the numerical solver `scipy.integrate.solve_ivp`, while Example 4.16 (`CH04Pcode8.py`) uses the symbolic solver `sympy.dsolve`.

Example 4.13

MATLAB code. This MATLAB script, named CH04Mcode7.m, solves the ODE $\frac{dy}{dx} + y(x) = x^2$ subject to $y(0) = 1$ using ode45 solver.

MATLAB Input:

```
function solve_ode2_numerical
% ODE: y'(x) + y(x) = x^2, y(0) = 1
% Numerical solution with ode45.
%
% Define the ODE: y'(x) = x^2 - y(x)
odeFunc = @(x,y) x.^2 - y;
xspan = [0 2]; % example interval
y0 = 1; % initial condition
%
[xSol, ySol] = ode45(odeFunc, xspan, y0);
% Display the results
disp('Numerical solution (x, y):');
result = table(xSol, ySol, 'VariableNames', {'x','y'});
disp(result);
% Optional plotting
plot(xSol, ySol, 'o-');
title('Numerical Solution of y''(x) + y(x) = x^2');
xlabel('x');
ylabel('y');
end
```

MATLAB Output:

```
Numerical solution (x, y):
   x y
   ____ _______

     0 1
  0.05 0.95127
   0.1 0.90516
  0.15 0.86179
   0.2 0.82127
  0.25 0.7837
   0.3 0.74918
  0.35 0.71781
   0.4 0.68968
  0.45 0.66487
   0.5 0.64347
```

```
0.55 0.62555
 0.6 0.61119
0.65 0.60045
 0.7 0.59341
0.75 0.59013
 0.8 0.59067
0.85 0.59508
 0.9 0.60343
0.95 0.61576
   1 0.63212
1.05 0.65256
 1.1 0.67713
1.15 0.70586
 1.2 0.73881
1.25 0.776
 1.3 0.81747
1.35 0.86326
 1.4 0.9134
1.45 0.96793
 1.5 1.0269
1.55 1.0903
 1.6 1.1581
1.65 1.2305
 1.7 1.3073
1.75 1.3887
 1.8 1.4747
1.85 1.5653
 1.9 1.6604
1.95 1.7602
   2 1.8647
```

Example 4.14

MATLAB code. This MATLAB script, named CH04Mcode8.m, solves the ODE $\frac{dy}{dx} + y(x) = x^2$ subject to $y(0) = 1$ using dsolve solver.

MATLAB Input:

```
function solve_ode2_symbolic
% ODE: y'(x) + y(x) = x^2, y(0) = 1
% Symbolic solution with dsolve.
```

```
syms x y(x)
odeEqn = diff(y, x) + y == x^2;
sol = dsolve(odeEqn, y(0) == 1);
disp('Symbolic solution for y(x):');
disp(sol);
end
```

MATLAB Output:

```
Symbolic solution for y(x):
x^2 - exp(-x) - 2*x + 2
```

Example 4.15

Python code. This Python code, named CH04Pcode7.py, solves the ODE $\frac{dy}{dx} + y(x) = x^2$ subject to $y(0) = 1$ using solve_ivp solver.

Python Input:

```
import numpy as np
from scipy.integrate import solve_ivp

def solve_ode2_numerical():
    """
    ODE: dy/dx + y = x^2, y(0) = 1
    Rewritten as dy/dx = x^2 - y.
    """
    # Define the right-hand side f(t, y)
    def ode_func(t, y):
        return t**2 - y

    # Integration interval and initial condition
    t_span = (0, 2)
    y0 = [1]

    # Points at which to store the solution
    t_eval = np.linspace(0, 2, 11)

    # Solve!
    sol = solve_ivp(ode_func, t_span, y0, t_eval=t_eval)

    # Report results
```

```
    print("t values:", sol.t)
    print("y values:", sol.y[0])
    return sol.t, sol.y[0]

if __name__ == "__main__":
    solve_ode2_numerical()
```

Python Output:

```
t values: [0. 0.2 0.4 0.6 0.8 1. 1.2 1.4 1.6 1.8 2. ]
y values: [1. 0.82121191 0.68939118 0.61088702 0.59067553
   0.63227985
 0.73859482 0.91302706 1.15806662 1.47505452 1.86496009]
```

Example 4.16

Python code. This Python code, named CH06Pcode8.py, solves the ODE $\frac{dy}{dx} + y(x) = x^2$ subject to $y(0) = 1$ using sympy.dsolve solver.

Python Input:

```
import sympy as sp
#
def solve_ode2_symbolic():
    """
    ODE: dy/dx + y = x^2, y(0) = 1
    Symbolic solution using sympy.
    """
    x = sp.symbols('x', real=True)
    y = sp.Function('y')
    ode = sp.Eq(sp.diff(y(x), x) + y(x), x**2)

    # Solve with initial condition y(0) = 1
    solution = sp.dsolve(ode, ics={y(0): 1})

    print("Symbolic solution for y(x):")
    print(solution)
#
if __name__ == "__main__":
    solve_ode2_symbolic()
```

Python Output:

```
Symbolic solution for y(x):
Eq(y(x), x**2 - 2*x + 2 - exp(-x))
```

4.7 SECOND-ORDER LINEAR DIFFERENTIAL EQUATIONS

A second-order differential equation is an equation involving the second derivative of an unknown function. In these notes, we focus on **linear** second-order ODEs. A general second-order linear ODE can be written as:

$$a_2(x)\,\frac{d^2y}{dx^2} + a_1(x)\,\frac{dy}{dx} + a_0(x)\,y = g(x), \tag{4.84}$$

where $a_2(x)$, $a_1(x)$, $a_0(x)$, and $g(x)$ are given functions of the independent variable x, and $a_2(x)$ is not zero.

Second-order differential equations arise in systems where acceleration and force are key, such as mass-spring-damper systems, structural vibrations, and mechanical resonances. Engineers use these models to predict how buildings, vehicles, and machines behave under dynamic loads (see Figure 4.10). Figure 4.10a shows a burned structure with weakened elements after the January 2024 Noto earthquake in Japan. Figure 4.10b presents a schematic of beam deflection under oscillatory loads, modeled by the following second-order linear differential equation:

$$m\frac{d^2x}{dt^2} + c\frac{dx}{dt} + kx = F(t),$$

where m is the mass, c is the damping coefficient, k is the stiffness of the beam, $x(t)$ is the displacement as a function of time, and $F(t)$ is the external force applied to the system.

4.7.1 HOMOGENEOUS SECOND-ORDER EQUATIONS

When $g(x) = 0$, the Equation 4.84 is called homogeneous:

$$a_2(x)\,\frac{d^2y}{dx^2} + a_1(x)\,\frac{dy}{dx} + a_0(x)\,y = 0 \tag{4.85}$$

One of the most common and important classes of second-order linear ODEs is:

$$a\,\frac{d^2y}{dx^2} + b\,\frac{dy}{dx} + c\,y = 0, \tag{4.86}$$

where a, b, c are constants, and $a \neq 0$. To solve this, we look for solutions of the form $y = \mathrm{e}^{rx}$. Substituting $y = \mathrm{e}^{rx}$ into the homogeneous equation gives:

$$a\,r^2\,\mathrm{e}^{rx} + b\,r\,\mathrm{e}^{rx} + c\,\mathrm{e}^{rx} = 0 \tag{4.87}$$

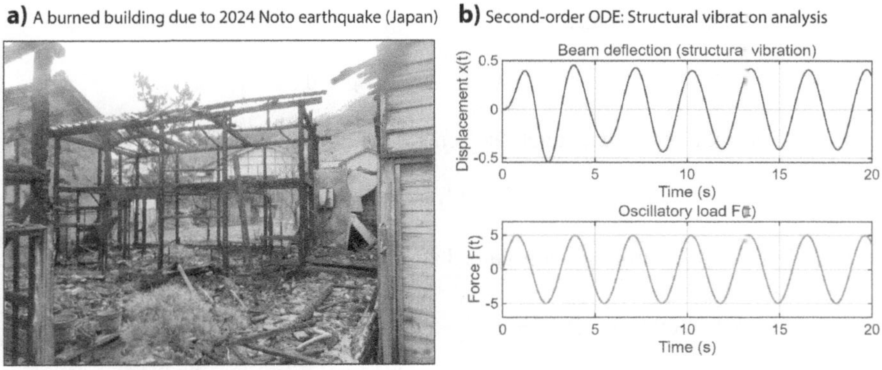

FIGURE 4.10 Analyzing structural vibration in a damaged building. **a)** A photo of a burned building with compromised structural elements following the January 2024 Noto earthquake in Japan. **b)** A schematic model of the beam deflection under oscillatory loads, governed by second-order linear DE: $m\frac{d^2x}{dt^2} + c\frac{dx}{dt} + kx = F(t)$.

Since $e^{rx} \neq 0$ for all real x, we may divide both sides of the equation by e^{rx} to obtain the characteristic equation:

$$a\,r^2 + b\,r + c = 0 \tag{4.88}$$

The nature of the solutions depends on the discriminant $\Delta = b^2 - 4ac$:

- **Case 1:** $\Delta > 0$ (two distinct real roots, r_1 and r_2). The general solution becomes:

$$y(x) = C_1 e^{r_1 x} + C_2 e^{r_2 x} \tag{4.89}$$

- **Case 2:** $\Delta = 0$ (one real repeated root, $r_1 = r_2 = r$). The general solution is:

$$y(x) = C_1 e^{rx} + C_2 x\, e^{rx} \tag{4.90}$$

- **Case 3:** $\Delta < 0$ (two complex conjugate roots, $r = \alpha \pm i\beta$). The general solution (in real form) takes the following form:

$$y(x) = e^{\alpha x}\big(C_1 \cos(\beta x) + C_2 \sin(\beta x)\big) \tag{4.91}$$

Example: Solve the following second-order homogeneous ordinary differential equation:

$$y'' - 3y' + 2y = 0$$

Solution:

To solve this second-order linear homogeneous differential equation with constant coefficients, we begin by forming the characteristic equation:

$$r^2 - 3r + 2 = 0$$

Factoring the quadratic yields:

$$(r-1)(r-2)=0 \quad \Rightarrow \quad r_1=1, \quad r_2=2$$

Since the roots are real and distinct, the general solution is given by:

$$y(x)=C_1\mathrm{e}^{r_1 x}+C_2\mathrm{e}^{r_2 x}=C_1\mathrm{e}^{x}+C_2\mathrm{e}^{2x}$$

where C_1 and C_2 are arbitrary constants determined by initial or boundary conditions.

4.7.2 REDUCTION OF ORDER

If one solution $y_1(x)$ to a homogeneous second-order equation is known, a second linearly independent solution can be found by setting $y_2(x) = v(x)\, y_1(x)$ and solving for $v(x)$. This method is often useful when we already know one particular solution from inspection or prior analysis.

For non-homogeneous second-order linear ODEs of the form:

$$a_2(x)\, y'' + a_1(x)\, y' + a_0(x)\, y = g(x), \tag{4.92}$$

the **general solution** is given by:

$$y_{\text{general}}(x) = y_{\text{hom}}(x) + y_{\text{part}}(x), \tag{4.93}$$

where $y_{\text{hom}}(x)$ is the general solution of the corresponding homogeneous equation ($g(x) = 0$), and $y_{\text{part}}(x)$ is any **particular solution** to the non-homogeneous equation.

4.7.3 SOLVING SECOND-ORDER ODES USING MATLAB AND PYTHON

In this section, we apply MATLAB and Python to solve second-order ODEs.

Example: Solve the following second-order ordinary differential equation using MATLAB and Python, subject to the initial conditions $y(0) = 2$ and $\frac{dy}{dx}(0) = 1$:

$$\frac{d^2y}{dx^2} = 6x$$

Solution:

In MATLAB, two approaches are employed. The first approach, implemented in script `CH04Mcode9.m` (Example 4.17) uses the numerical solver `ode45`. This example reformulates the second-order ODE into a system of first-order equations and solves it numerically. The second approach, presented in script `CH04Mcode10.m` (Example 4.18), uses the symbolic solver `dsolve` to obtain the exact analytical solution.

Equivalent Python implementations are provided in Examples 4.19 and 4.20. Example 4.19 (`CH04Pcode9.py`) employs a numerical approach using the `solve_ivp` function from the `scipy.integrate` module. Example 4.20 (`CH04Pcode10.py`) provides the exact analytical solution using the symbolic solver `dsolve` from the `sympy` library.

The numerical solution provided in Example 4.17 agrees with the exact solution of the second-order ODE. To obtain the exact solution analytically, we proceed by integrating the differential equation step by step.

First, integrate the second-order equation:

$$\frac{dy}{dx} = \int 6x\,dx = 3x^2 + C_1$$

Apply the initial condition $\frac{dy}{dx}(0) = 1$:

$$\frac{dy}{dx}(0) = 3(0)^2 + C_1 = 1 \quad \Longrightarrow \quad C_1 = 1$$

Thus, the first derivative becomes:

$$\frac{dy}{dx} = 3x^2 + 1$$

Integrate again to obtain $y(x)$:

$$y(x) = \int (3x^2 + 1)\,dx = x^3 + x + C_2$$

Apply the second initial condition $y(0) = 2$:

$$y(0) = (0)^3 + 0 + C_2 = 2 \quad \Longrightarrow \quad C_2 = 2$$

Therefore, the exact solution is:

$$y(x) = x^3 + x + 2$$

Example 4.17

MATLAB code. This MATLAB script, named `CH04Mcode9.m`, solves the ODE $\frac{d^2y}{dx^2} = 6x$ subject to the initial conditions $y(0) = 2$ and $\frac{dy}{dx}(0) = 1$. using ode45 solver.

MATLAB Input:

```
% For a second order ODE, we typically reduce it to a system of
    two first-order ODEs:
% Let v(x) = y'(x). Then dv/dx=6x. So we have:
% y'(x) = v(x) & v'(x) = 6x with y(0)=2 & v(0)=1.
%
function solve_ode1_numerical
% ODE: y''(x) = 6x, with y(0)=2, y'(0)=1.
% Convert to system:
% y'(x) = v(x)
% v'(x) = 6x
% System function
% State vector: Y = [ y; v ]
```

```
% Y'(x) = [ v; 6x ]
odeFunc = @(x,Y) [Y(2); 6*x];
% Initial conditions
y0 = 2; % y(0)
v0 = 1; % y'(0)
Y0 = [y0; v0];
% Interval of integration
xspan = [0 2];
% Solve numerically
[xSol, YSol] = ode45(odeFunc, xspan, Y0);
% YSol(:,1) = y values, YSol(:,2) = v (= y') values
ySol = YSol(:,1);
vSol = YSol(:,2);
% Display the results
disp('Numerical solution (x, y, y'') in terms of (x, y, v):');
result = table(xSol, ySol, vSol, 'VariableNames', ...
{'x','y','v'});
disp(result);
% Optional: plot the solution for y(x)
figure;
plot(xSol, ySol, 'o-');
title('Numerical Solution of y''''(x) = 6x');
xlabel('x');
ylabel('y');
end
```

MATLAB Output:

```
Numerical solution (x, y, y') in terms of (x, y, v):
    x y v
   ____ ______ ______

     0 2 1
   0.05 2.0501 1.0075
    0.1 2.101 1.03
   0.15 2.1534 1.0675
    0.2 2.208 1.12
   0.25 2.2656 1.1875
    0.3 2.327 1.27
   0.35 2.3929 1.3675
    0.4 2.464 1.48
   0.45 2.5411 1.6075
    0.5 2.625 1.75
   0.55 2.7164 1.9075
```

```
 0.6 2.816 2.08
0.65 2.9246 2.2675
 0.7 3.043 2.47
0.75 3.1719 2.6875
 0.8 3.312 2.92
0.85 3.4641 3.1675
 0.9 3.629 3.43
0.95 3.8074 3.7075
   1 4 4
1.05 4.2076 4.3075
 1.1 4.431 4.63
1.15 4.6709 4.9675
 1.2 4.928 5.32
1.25 5.2031 5.6875
 1.3 5.497 6.07
1.35 5.8104 6.4675
 1.4 6.144 6.88
1.45 6.4986 7.3075
 1.5 6.875 7.75
1.55 7.2739 8.2075
 1.6 7.696 8.68
1.65 8.1421 9.1675
 1.7 8.613 9.67
1.75 9.1094 10.187
 1.8 9.632 10.72
1.85 10.182 11.268
 1.9 10.759 11.83
1.95 11.365 12.407
   2 12 13
```

Example 4.18

MATLAB code. This MATLAB script, named CH04Mcode10.m, solves the ODE $\frac{d^2y}{dx^2} = 6x$ subject to the initial conditions $y(0) = 2$ and $\frac{dy}{dx}(0) = 1$, using dsolve solver.

MATLAB Input:

```
function solve_ode1_symbolic
% ODE: y''(x) = 6x, with y(0)=2, y'(0)=1.
syms x y(x)
```

```
%
% Define the ODE:
odeEqn = diff(y, 2) == 6*x;
%
% Define the derivative:
Dy = diff(y, x);
%
% Initial conditions:
conds = [y(0) == 2, Dy(0) == 1];
%
% Solve the ODE with the initial conditions:
sol = dsolve(odeEqn, conds);
disp('Symbolic solution for y(x):');
disp(sol);
end
```

MATLAB Output:

```
Symbolic solution for y(x):
x^3 + x + 2
```

Example 4.19

Python code. This Python code, named `CH04Pcode9.py`, solves the ODE $\frac{d^2y}{dx^2} = 6x$ subject to the initial conditions $y(0) = 2$ and $\frac{dy}{dx}(0) = 1$, using solve_ivp solver.

Python Input:

```
import numpy as np
from scipy.integrate import solve_ivp
#
def solve_ode1_numerical():
    """
    ODE: y''(x) = 6x, with y(0)=2, y'(0)=1.
    Convert to system of first order:
      y'(x) = v
      v'(x) = 6x
    """

    def ode_system(x, Y):
        y, v = Y
        return [v, 6*x]
```

```
#
    # Initial conditions
    Y0 = [2, 1] # [y(0), v(0)=y'(0)]
    # Domain and output points
    x_span = (0, 2)
    t_eval = np.linspace(0, 2, 11)
#
    # Solve the system
    sol = solve_ivp(ode_system, x_span, Y0, t_eval=t_eval)

    # Unpack
    x_vals = sol.t
    y_vals = sol.y[0]
    v_vals = sol.y[1]
#
    # Show
    print("x values:", x_vals)
    print("y values:", y_vals)
    print("v = y' values:", v_vals)

    return x_vals, y_vals, v_vals
#
if __name__ == "__main__":
    solve_ode1_numerical()
```

Python Output:

```
x values: [0. 0.2 0.4 0.6 0.8 1. 1.2 1.4 1.6 1.8 2. ]
y values: [ 2. 2.208 2.464 2.816 3.312 4. 4.928 6.144 7.696
   9.632
 12. ]
v = y' values: [ 1. 1.12 1.48 2.08 2.92 4. 5.32 6.88 8.68 10.72
   13. ]
```

Example 4.20

Python code. This Python code, named `CH04Pcode10.py`, solves the CDE $\frac{d^2y}{dx^2} = 6x$ subject to the initial conditions $y(0) = 2$ and $\frac{dy}{dx}(0) = 1$, using sympy.dsolve solver.

Python Input:

```python
import sympy as sp
#
def solve_ode1_symbolic():
    """
    ODE: y''(x) = 6x, y(0)=2, y'(0)=1
    Use sympy.dsolve to get a symbolic solution.
    """
    x = sp.symbols('x', real=True)
    y = sp.Function('y')(x)
#
    # Define the ODE y'' = 6*x
    ode_equation = sp.Eq(y.diff(x, 2), 6*x)
#
    # Solve with initial conditions y(0)=2, y'(0)=1
    sol = sp.dsolve(ode_equation,
                    ics={y.subs(x, 0): 2,
                         y.diff(x).subs(x, 0): 1})
#
    print("Symbolic solution for y(x):")
    print(sol) # prints: Eq(y(x), x**3 + x + 2)
#
if __name__ == "__main__":
    solve_ode1_symbolic()
```

Python Output:

```
Symbolic solution for y(x):
Eq(y(x), x**3 + x + 2)
```

Example: Solve the following second-order linear ordinary differential equation with constant coefficients and a polynomial on the right-hand side (non-homogeneous ODE), subject to the initial conditions $y(0) = 0$ and $y'(0) = 0$, using MATLAB and Python:

$$\frac{d^2y}{dx^2} + 2\frac{dy}{dx} + y = x^2$$

Solution:

This non-homogeneous ODE is examined in both MATLAB and Python using numerical and symbolic approaches in the following four scripts:

- Example 4.21 (script `CH04Mcode11.m`) solves the equation numerically in MATLAB using the function `ode45`.
- Example 4.22 (script `CH04Mcode12.m`) derives the exact analytical solution in MATLAB using symbolic computation via function `dsolve`.

- Example 4.23 (script `CH04Pcode11.py`) provides the equivalent numerical solution in Python using `solve_ivp` function from the `scipy.integrate` module.
- Example 4.24 (script `CH04Pcode12.py`) presents the symbolic solution using Python's `sympy` library and `sp.dsolve` function.

Let us now determine the analytical solution of the given non-homogeneous differential equation. We begin by solving the corresponding homogeneous equation:

$$y'' + 2y' + y = 0$$

The characteristic equation associated with this homogeneous equation is:

$$r^2 + 2r + 1 = 0 \quad \Longrightarrow \quad (r+1)^2 = 0$$

This yields a repeated root $r = -1$, so the general solution to the homogeneous equation is:

$$y_{\text{hom}}(x) = C_1 \mathrm{e}^{-x} + C_2 x \mathrm{e}^{-x}$$

To find a particular solution, we assume a polynomial form:

$$y_{\text{part}}(x) = Ax^2 + Bx + C$$

Substituting y_{part} into the differential equation:

$$\begin{aligned} y'' + 2y' + y &= (2A) + 2(2Ax + B) + (Ax^2 + Bx + C) \\ &= Ax^2 + (4A + B)x + (2A + 2B + C) \end{aligned}$$

Set this equal to the right-hand side x^2:

$$Ax^2 + (4A + B)x + (2A + 2B + C) = x^2$$

By matching coefficients, we have:

$$\begin{aligned} A &= 1 \\ 4A + B &= 0 \quad \Rightarrow \quad B = -4 \\ 2A + 2B + C &= 0 \quad \Rightarrow \quad C = 6 \end{aligned}$$

Therefore, the particular solution is:

$$y_{\text{part}}(x) = x^2 - 4x + 6$$

Hence, the general solution to the original non-homogeneous differential equation is:

$$y(x) = y_{\text{hom}}(x) + y_{\text{part}}(x) = C_1 \mathrm{e}^{-x} + C_2 x \mathrm{e}^{-x} + x^2 - 4x + 6$$

Example 4.21

MATLAB code. This MATLAB script, named `CH04Mcode11.m`, solves the ODE $y''(x) + 2y'(x) + y(x) = x^2$ subjected to $y(0) = 0$ and $y'(0) = 0$, using `ode45` solver.

MATLAB Input:

```
function solve_ode2_numerical
% ODE: y'' + 2y' + y = x^2, y(0)=0, y'(0)=0.
% Convert to system of first order:
% y1' = y2
% y2' = x^2 - 2y2 - y1
odeFunc = @(x, Y)[ Y(2); x^2 - 2*Y(2) - Y(1) ];
Y0 = [0; 0]; % [y(0), y'(0)]
xspan = [0 2];
[xSol, YSol] = ode45(odeFunc, xspan, Y0);
ySol = YSol(:,1);
vSol = YSol(:,2); % y' = v
% Display result
disp('Numerical solution (x, y, y''):');
result = table(xSol, ySol, vSol, 'VariableNames', ...
{'x','y','yprime'});
disp(result);
% Optional: plot
figure;
plot(xSol, ySol, 'o-');
xlabel('x');
ylabel('y');
title('Solution of y'''' + 2y'' + y = x^2');
end
```

MATLAB Output:

```
Numerical solution (x, y, y'):
    x y yprime
   ____ __________ __________

     0 0 0
   0.05 4.1818e-07 4.0779e-05
    0.1 7.9898e-06 0.00031718
   0.15 3.9819e-05 0.0010442
    0.2 0.00012318 0.0024153
   0.25 0.00029497 0.0046035
    0.3 0.00059995 0.0077635
```

```
0.35 0.00109 0.012034
 0.4 0.0018237 0.017536
0.45 0.0028659 0.024378
 0.5 0.0042857 0.032653
0.55 0.0061565 0.042444
 0.6 0.0085562 0.053821
0.65 0.011566 0.066842
 0.7 0.015269 0.08156
0.75 0.019751 0.098016
 0.8 0.0251 0.11624
0.85 0.031405 0.13626
 0.9 0.038757 0.1581
0.95 0.047246 0.18177
   1 0.056964 0.20728
1.05 0.068004 0.23462
 1.1 0.080458 0.2638
1.15 0.094415 0.29481
 1.2 0.10997 0.32764
1.25 0.12721 0.36228
 1.3 0.14623 0.39871
1.35 0.16711 0.43691
 1.4 0.18995 0.47686
1.45 0.21482 0.51853
 1.5 0.24183 0.56191
1.55 0.27104 0.60696
 1.6 0.30255 0.65366
1.65 0.33644 0.70196
 1.7 0.37278 0.75186
1.75 0.41165 0.8033
 1.8 0.45313 0.85627
1.85 0.4973 0.91073
 1.9 0.54423 0.96663
1.95 0.59399 1.024
   2 0.64665 1.0827
```

Example 4.22

MATLAB code. This MATLAB script, named `CH04Mcode12.m`, solves the ODE $y''(x)+2y'(x)+y(x) = x^2$ subjected to $y(0) = 0$ and $y'(0) = 0$, using `dsolve` solver.

MATLAB Input:

```
function solve_ode2_symbolic
% ODE: y'' + 2y' + y = x^2, y(0)=0, y'(0)=0
syms x y(x)
Dy = diff(y, x);
D2y = diff(y, x, 2);
odeEqn = D2y + 2*Dy + y == x^2;
% Initial conditions
conds = [y(0) == 0, Dy(0) == 0];
% Solve the ODE
sol = dsolve(odeEqn, conds);
disp('Symbolic solution for y(x):');
disp(sol);
end
```

MATLAB Output:

```
Symbolic solution for y(x):
x^2 - 6*exp(-x) - 2*x*exp(-x) - 4*x + 6
```

Example 4.23

Python code. This Python code, named `CH04Pcode11.py`, solves the ODE $y''(x)+2y'(x)+y(x)=x^2$ subjected to $y(0)=0$ and $y'(0)=0$, using `solve_ivp` solver.

Python Input:

```
import numpy as np
from scipy.integrate import solve_ivp
#
def solve_ode2_numerical():
    """
    ODE: y'' + 2y' + y = x^2, y(0)=0, y'(0)=0.
    Converted to first-order system:
      Y = [y, v],
      Y' = [v, x^2 - 2v - y].
    """
    def ode_system(t, Y):
        y, v = Y
        dydt = v
```

```
        dvdt = t**2 - 2*v - y
        return [dydt, dvdt]
#
    Y0 = [0, 0] # initial conditions: y(0)=0, y'(0)=0
    t_span = (0, 2)
    t_eval = np.linspace(0, 2, 11) # 11 points between 0 and 2
#
    sol = solve_ivp(ode_system, t_span, Y0, t_eval=t_eval)
    if not sol.success:
        raise RuntimeError("ODE solver failed: " + sol.message)
#
    t_vals = sol.t
    y_vals = sol.y[0]
    v_vals = sol.y[1]
#
    print("t:", t_vals)
    print("y:", y_vals)
    print("y':", v_vals)
    return t_vals, y_vals, v_vals
#
if __name__ == "__main__":
    solve_ode2_numerical()
```

Python Output:

```
t: [0. 0.2 0.4 0.6 0.8 1. 1.2 1.4 1.6 1.8 2. ]
y: [0.00000000e+00 1.23258832e-04 1.82366990e-03 8.56684091e-03
 2.50958995e-02 5.69835416e-02 1.09958500e-01 1.90001073e-01
 3.02569966e-01 4.53095382e-01 6.46618194e-01]
y': [0. 0.00241527 0.01753626 0.0538063 0.11624715 0.20725099
 0.32765555 0.47678689 0.65362785 0.85631647 1.08271969]
```

Example 4.24

Python code. This Python code, named `CH04Pcode12.py`, solves the ODE $y''(x) + 2y'(x) + y(x) = x^2$ subjected to $y(0) = 0$ and $y'(0) = 0$, using `sympy.dsolve` solver.

Python Input:

```
import sympy as sp
#
def solve_ode2_symbolic():
```

```
    """
    ODE: y'' + 2y' + y = x^2, y(0)=0, y'(0)=0
    Use sympy to solve symbolically.
    """
    # Declare symbols and function
    x = sp.Symbol('x', real=True)
    y = sp.Function('y')(x)
#
    # Define the ODE
    ode_equation = sp.Eq(y.diff(x, 2) + 2*y.diff(x) + y, x**2)
#
    # Solve with initial conditions
    sol = sp.dsolve(
       ode_equation,
       ics={
          y.subs(x, 0): 0,
          y.diff(x).subs(x, 0): 0
       }
    )
#
    print("Symbolic solution for y(x):")
    print(sol)
#
if __name__ == "__main__":
    solve_ode2_symbolic()
```

Python Output:

```
Symbolic solution for y(x):
Eq(y(x), x**2 - 4*x + (-2*x - 6)*exp(-x) + 6)
```

4.7.4 METHOD OF UNDETERMINED COEFFICIENTS

When the coefficients a_2, a_1, and a_0 are constant and $g(x)$ is of a special (usually polynomial, exponential, or trigonometric) form, we can guess a form for y_{part} and solve for its coefficients. This procedure is known as **method of undetermined coefficients**.

Example: Solve the following non-homogeneous ODE using the method of undetermined coefficients:

$$y'' + y = \cos x$$

Solution:

The associated homogeneous equation becomes:

$$y'' + y = 0$$

Its characteristic equation takes the following form:

$$r^2 + 1 = 0 \quad \Rightarrow \quad r = \pm i$$

Therefore, the homogeneous solution is expressed as:

$$y_{\text{hom}}(x) = C_1 \cos x + C_2 \sin x$$

To find a particular solution (y_{part}) for this differential equation, note that the right-hand side is $\cos x$. Usually, one would guess:

$$y_{\text{part}}(x) = A \cos x + B \sin x$$

However, since $\cos x$ and $\sin x$ are already solutions of the corresponding homogeneous equation, the initial guess for the particular solution fails due to duplication. To resolve this, we multiply the guess by x, resulting in a modified trial function:

$$y_{\text{part}}(x) = x(A \cos x + B \sin x)$$

Substitute this guess into the ODE to solve for A and B. First compute $y'_{\text{part}}(x)$ and $y''_{\text{part}}(x)$:

$$y'_{\text{part}}(x) = A \cos x + B \sin x + x(-A \sin x + B \cos x)$$
$$y''_{\text{part}}(x) = -2A \sin x + 2B \cos x + x(-A \cos x - B \sin x)$$

Now substitute y_{part}, y''_{part} into the left-hand side of $y'' + y = \cos x$:

$$y''_{\text{part}} + y_{\text{part}} = \left[-2A \sin x + 2B \cos x + x(-A \cos x - B \sin x)\right] + \left[x(A \cos x + B \sin x)\right]$$

Notice that the x terms cancel:

$$x(-A \cos x - B \sin x) + x(A \cos x + B \sin x) = 0$$

Thus, we are left with:

$$y''_{\text{part}} + y_{\text{part}} = -2A \sin x + 2B \cos x$$

We want this expression to equal $\cos x$. To achieve this, we equate the coefficients of like terms in the following equation:

$$-2A \sin x + 2B \cos x = \cos x$$

This gives the following system of equations:

$$-2A = 0 \quad \Rightarrow \quad A = 0$$

$$2B = 1 \quad \Rightarrow \quad B = \tfrac{1}{2}$$

So $y_{\text{part}}(x) = x\left(0 \cdot \cos x + \frac{1}{2} \sin x\right) = \frac{x}{2} \sin x$.

Therefore, the general solution is

$$y(x) = y_{\text{hom}}(x) + y_{\text{part}}(x) = C_1 \cos x + C_2 \sin x + \frac{x}{2} \sin x$$

4.7.5 METHOD OF VARIATION OF PARAMETERS

When $g(x)$ is more complicated or the coefficient functions $a_2(x)$, $a_1(x)$, and $a_0(x)$ are not all constants, the method of undetermined coefficients may not be applicable. In such cases, **variation of parameters** is a more general method. It proceeds as follows for the general second-order linear ODE:

$$y'' + p(x)\,y' + q(x)\,y = g(x). \tag{4.94}$$

Assume we have two linearly independent solutions $y_1(x)$ and $y_2(x)$ of the associated homogeneous equation. Then we seek a particular solution of the following form:

$$y_{\text{part}}(x) = u_1(x)\,y_1(x) + u_2(x)\,y_2(x), \tag{4.95}$$

where $u_1(x)$ and $u_2(x)$ are new unknown functions. Imposing conditions to simplify the expression for y'_{part} leads to a system of equations for $u'_1(x)$ and $u'_2(x)$. Solving for $u'_1(x)$ and $u'_2(x)$, and then integrating, gives $u_1(x)$ and $u_2(x)$, hence $y_{\text{part}}(x)$.

4.8 PROBLEMS

Problem 4.1: A local business finds its daily profit $P(x)$ (in dollars) for selling x custom T-shirts can be modeled by:

$$P(x) = 800\ +\ 15x\ -\ 0.2\,x^2.$$

Find the instantaneous rate of change of profit with respect to the number of T-shirts sold.

Answer: $P'(x)\ =\ 15\ -\ 0.4\,x$

Problem 4.2: The temperature T (in degrees Celsius) at a point (x, y) on a metal plate is given by:

$$T(x, y) = 100\ +\ 2\,x^2\ +\ y^3.$$

Determine the partial derivatives $\dfrac{\partial T}{\partial x}$ and $\dfrac{\partial T}{\partial y}$, which indicate how temperature changes with small variations in x and y.

Answer: $\frac{\partial T}{\partial x}\ =\ 4x,\quad \frac{\partial T}{\partial y}\ =\ 3y^2$

Problem 4.3: A car moving along a straight highway has velocity $v(t) = 15 + 2t$ (in m/s), where t is in seconds. If the car starts from rest at $t = 0$, use an integral to find the distance it travels from $t = 0$ to $t = 10$ s.

Answer: Distance $= 250$ m

Problem 4.4: A rectangular field extends $0 \leq x \leq 50$ m and $0 \leq y \leq 20$ m. The rainfall intensity $R(x, y)$ (in cm) at each point (x, y) is modeled by

$$R(x, y) = 0.01\,x\ +\ 0.02\,y.$$

Find the total amount of rainfall (in appropriate volume units) over the entire field using a double integral.

Answer: 4.5 m^3

Problem 4.5: A rectangular solid occupies the following region:

$$0 \leq x \leq 2, \quad 0 \leq y \leq 1, \quad 0 \leq z \leq 3.$$

Its density at any point (x, y, z) is:

$$\rho(x, y, z) = 2\,x\ +\ 4\,y\ +\ z \quad \mathrm{kg/m^3}.$$

Compute the total mass of the solid.

Answer: The total mass is 33 kg.

Problem 4.6: Find the volume of the solid ball (sphere) of radius 2 by using a triple integral in spherical coordinates. In spherical coordinates (ρ, θ, ϕ), the volume element is $\rho^2 \sin(\phi)\, \mathrm{d}\rho\, \mathrm{d}\phi\, \mathrm{d}\theta$ and the region is:

$$0 \leq \rho \leq 2, \quad 0 \leq \theta \leq 2\pi, \quad 0 \leq \phi \leq \pi.$$

Answer: The volume is $\dfrac{32\pi}{3}$ (cubic units).

Problem 4.7: A cup of coffee is initially at 90°C in a room held at 20°C. By applying the Newton's Law of Cooling given below:

$$\frac{\mathrm{d}T}{\mathrm{d}t} = -k\,(T - 20),$$

where $k = 0.05\,\mathrm{min}^{-1}$, calculate $T(t)$ given the initial condition $T(0) = 90$.

Answer: $T(t) = 20 + 70\,\mathrm{e}^{-0.05t}$

Problem 4.8: A bacterial population $P(t)$, measured in thousands of organisms, grows at a rate proportional to its size:

$$\frac{\mathrm{d}P}{\mathrm{d}t} = 0.3\,P,$$

with $P(0) = 200$. Solve the differential equation to find $P(t)$.

$$\textit{Answer}: P(t) \;=\; 200\,\mathrm{e}^{0.3t} \text{ (population in thousands)}$$

Problem 4.9: A mass-spring system with mass $m = 1\,\mathrm{kg}$ and spring constant $k = 4\,\mathrm{N/m}$ (no damping) obeys the following equation:

$$\frac{\mathrm{d}^2x}{\mathrm{d}t^2} + 4\,x \;=\; 0,$$

where $x(t)$ is the displacement from equilibrium. If $x(0) = 1$ and $x'(0) = 0$, solve for $x(t)$.

Answer: $x(t) = \cos(2t)$

Problem 4.10: Consider a damped oscillator with $m = 2\,\mathrm{kg}$, damping coefficient $c = 4\,\mathrm{kg/s}$, and spring constant $k = 8\,\mathrm{N/m}$. The motion $x(t)$ satisfies the following equation:

$$2\,\frac{\mathrm{d}^2x}{\mathrm{d}t^2} + 4\,\frac{\mathrm{d}x}{\mathrm{d}t} + 8\,x \;=\; 0.$$

Solve for the general form of $x(t)$.

$$\textit{Answer}: x(t) \;=\; \mathrm{e}^{-t}\left[C_1\,\cos(\sqrt{3}\,t) \;+\; C_2\,\sin(\sqrt{3}\,t)\right]$$

REFERENCES

[1] Mohammad Heidarzadeh and Siamak Feizi. "A cascading risk model for the failure of the concrete spillway of the Toddbrook dam, England during the August 2019 flooding." In: *International Journal of Disaster Risk Reduction* 80 (2022), p. 103214.

[2] Mohammad Heidarzadeh, Ali A Mirghasemi, and Habib Niroomand. "Construction of relief wells under artesian flow conditions at dam toes: engineering experiences from Karkheh earth dam, Iran." In: *International Journal of Civil Engineering* 13(1) (2015). https://doi.org/10.22068/IJCE.13.1.73, p. 73–80.

[3] Payam Khazaeinejad and Asif S. Usmani. "On thermo-mechanical nonlinear behaviour of shallow shells." In: *International Journal of Non-Linear Mechanics* 82 (2016), p. 114–123.

[4] Payam Khazaeinejad and Asif S. Usmani. "Temperature-dependent nonlinear analysis of shallow shells: A theoretical approach." In: *Composite Structures* 141 (2016), p. 1–13.

[5] Payam Khazaeinejad, Asif S. Usmani, and O. Laghrouche. "Temperature-dependent nonlinear behaviour of thin rectangular plates exposed to through-depth thermal gradients." In: *Composite Structures* 132 (2015), p. 652–664.

5 Root Finding, Interpolation, and Curve Fitting

Root finding, interpolation, and curve fitting are fundamental topics in engineering, offering powerful tools for analyzing and solving complex problems. Engineers often encounter scenarios where precise solutions are essential—such as determining stress points in structures, optimizing system designs, or predicting fluid behavior. Root finding methods help locate zeros of equations critical in system stability analysis and optimization tasks.

Interpolation provides a means to estimate unknown values within a dataset, enabling engineers to model material properties or fill gaps in experimental data. For example, in the design of dams and reservoirs (Figure 5.1a), a limited number of permeability tests are conducted in various parts of the reservoir and dam foundation, which are then interpolated to develop models for predicting potential seepage through the dam foundation and reservoir [3, 4].

Curve fitting, on the other hand, facilitates the creation of mathematical models that describe trends in data, aiding in predictive analysis and decision-making. For example, curve fitting can be employed to develop empirical models for wave runups on coastal areas during storms and tsunamis (Figure 5.1b), allowing engineers to design resilient coastal defenses and mitigate disaster risks [5]. Together, these methods form a cornerstone for innovation, ensuring accurate simulations, improved designs, and better understanding across diverse engineering disciplines.

FIGURE 5.1 Examples of infrastructure systems where root finding, interpolation and curve fitting are used for their design. **a)** The Llandegfedd dam and reservoir, UK. **b)** Sea defenses in Nagashima Island, Japan.

DOI: 10.1201/9781003511854-5

5.1 ROOT FINDING

Overview. In engineering, we often encounter mathematical equations that cannot be solved explicitly. Instead, we rely on numerical methods to approximate the solutions. These solutions, called roots, are the values of a variable that satisfy the given equation. For example, solving $f(x) = 0$ involves finding the value(s) of x at which the function $f(x)$ intersects the horizontal axis.

Root-finding methods are essential tools in engineering, as they are used to analyze systems, design components, and optimize processes. Examples include determining the load at which a structural beam fails in civil engineering, the critical speed at which a rotating shaft experiences resonance in mechanical engineering, and the temperature at which a chemical reaction reaches equilibrium in chemical engineering. This section will help students to understand the concept of roots and their significance in engineering problems, learn basic root-finding methods such as the **Bisection** method and the **Newton-Raphson** method, and develop a systematic approach to applying numerical methods to practical engineering problems.

Mathematical definition of a root. The root of a function $f(x)$ is defined as a value $x = r$ such that:

$$f(r) = 0 \tag{5.1}$$

Roots can be categorized as follows:

- **Real or complex:** Roots can be real numbers or complex numbers, depending on the type of equation. For example, the roots of $x^2 - 4 = 0$ are real, while the roots of $x^2 + 4 = 0$ are complex.

- **Single or multiple:** A *single root* occurs when a function equals zero at a point and the graph crosses the x-axis at that point. A *multiple root* occurs when the function equals zero at a point, but the graph only touches the x-axis and turns back without crossing it. For example, $f(x) = (x - 2)^2$ has a multiple root at $x = 2$.

Example: Determine the roots of the following quadratic equation and represent them graphically (Figure 5.2):

$$f(x) = x^2 - 4 = 0 \tag{5.2}$$

Solution:

The roots of this equation are:

$$x = 2 \quad \text{and} \quad x = -2$$

as these are the values where $f(x) = 0$. Figure 5.2 shows the two roots of this quadratic function.

In practical engineering problems, equations are often complex, and analytical solutions may not always exist. For example, determining the load at which a beam bends

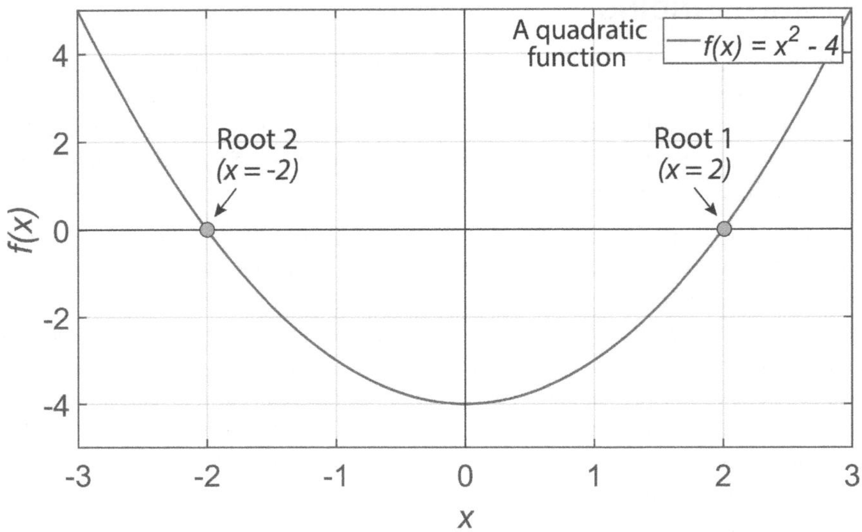

FIGURE 5.2 The roots of a quadratic function: $f(x) = x^2 - 4$.

under stress often requires solving equations that cannot be explicitly resolved. Numerical methods provide an efficient way to find approximate solutions to such problems. In the following sections, a selection of numerical methods for root-finding problems is discussed, with implementations demonstrated using MATLAB and Python.

5.1.1 THE BISECTION METHOD

The Bisection method is one of the simplest and most reliable numerical techniques for finding roots of a continuous function. It is based on the Intermediate Value Theorem, which states that if a function $f(x)$ is continuous on an interval $[a, b]$ and $f(a) \cdot f(b) < 0$, then there exists at least one root c in the interval $[a, b]$ such that $f(c) = 0$.

STEPS OF THE BISECTION METHOD

The Bisection method is an iterative technique, meaning that certain steps must be repeated until the desired outcome is achieved. The following four steps outline the process for finding the roots of a function using the Bisection method.

i. **Define the interval:** Start with an initial interval $[a, b]$ such that $f(a) \cdot f(b) < 0$, meaning the function has opposite signs at a and b.

ii. **Compute the midpoint:** Calculate the midpoint of the interval,

$$c = \frac{a+b}{2} \tag{5.3}$$

iii. **Evaluate the function at the midpoint:**
- If $f(c) = 0$, then c is the root.
- If $f(c) \cdot f(a) < 0$, the root lies in the interval $[a, c]$. Update $b = c$.
- If $f(c) \cdot f(b) < 0$, the root lies in the interval $[c, b]$. Update $a = c$.

iv. **Repeat:** Continue the process until the interval $[a, b]$ is sufficiently small, or the value of $f(c)$ is close enough to zero.

Example: Find the roots of the following equation using the Bisection method:

$$f(x) = x^2 - 2x - 3 = 0$$

Solution:

It is known that this function has two roots at $x = -1$ and $x = 3$. In this example, we focus on finding the root within the interval $[0, 4]$, where the function changes sign.

- **Initial interval:** $[a, b] = [0, 4]$

$$f(0) = -3, \quad f(4) = 1.$$

  Since $f(0) \cdot f(4) < 0$, a root exists in $[0, 4]$.
- **Iteration 1:**

$$c = \frac{0+4}{2} = 2, \quad f(2) = -1.$$

  Since $f(2) \cdot f(4) < 0$, the root lies in $[2, 4]$.
- **Iteration 2:**

$$c = \frac{2+4}{2} = 3, \quad f(3) = 0.$$

Since $f(3) = 0$, the root is found at $x = 3$.

To find the **second root** using the Bisection method, you will need to repeat the process in the interval $[-4, 0]$ where the function changes sign, as follows:

- **Initial interval:** $[a, b] = [-4, 0]$

$$f(-4) = 23, \quad f(0) = -3.$$

  Since $f(-4) \cdot f(0) < 0$, a root exists in $[-4, 0]$.
- **Iteration 1:**

$$c = \frac{-4+0}{2} = -2, \quad f(-2) = 1.$$

  Since $f(-2) \cdot f(0) < 0$, the root lies in $[-2, 0]$.

- **Iteration 2:**

$$c = \frac{-2+0}{2} = -1, \quad f(-1) = 0.$$

Since $f(-1) = 0$, the root is found at $x = -1$.

The iterative process used to find the roots of $f(x) = x^2 - 2x - 3$ using the Bisection method can be implemented in both MATLAB and Python. Neither MATLAB nor Python has a built-in function specifically for the Bisection method. However, it is straightforward to implement the method using basic programming constructs such as "`if statement`" and "`while loop`." Example 5.1 (script `CH05Mcode1.m`) and Example 5.2 (script `CH05Pcode1.py`) provide MATLAB and Python scripts, respectively, to find the roots of $f(x) = x^2 - 2x - 3$. All MATLAB scripts presented in this book are available at: `https://github.com/prog871/CRC-book2025`

Example 5.1

MATLAB code. This MATLAB script, named `CH05Mcode1.m`, finds the roots of the function $f(x) = x^2 - 2x - 3$. This MATLAB script, together with all other scripts presented in this book, is available at:

`https://github.com/prog871/CRC-book2025`

MATLAB Input:

```
% This code finds the roots
% of: f(x) = x^2 - 2x -3
% using 'Bisection Method'
clc; clear; close all;
f = @(x) x^2 - 2*x - 3;
% First root in interval [0,4]
a1 = 0; b1 = 4;
tol=1e-6; %Tolerance for stopping
X2= 500; % Maximum iterations
% Bisection Method for first root
iter = 0;
while (b1 - a1) / 2 > tol
   c1 = (a1 + b1) / 2;
   if f(c1) == 0; break;
   elseif f(a1) * f(c1) < 0; b1=c1;
   else; a1 = c1;
   end
   iter = iter + 1;
   if iter > X2
    disp('Max iterations');break;
   end
```

```
end
root1 = (a1 + b1) / 2;
disp(['First root: ',num2str(root1)]);
% Second root in the interval [-4,0]
a2 = -4; b2 = 0; iter = 0;
while (b2 - a2) / 2 > tol
   c2 = (a2 + b2) / 2;
   if f(c2) == 0; break;
   elseif f(a2)*f(c2)<0; b2=c2;
   else; a2 = c2;
   end
   iter = iter + 1;
   if iter > X2
    disp('Max iterations');break;
   end
end
root2 = (a2 + b2) / 2;
disp(['Second root: ',num2str(root2)]);
% end of M - file
```

MATLAB Output:

```
First root: 3
Second root: -1
```

Example 5.2

Python code. This Python script, named `CH05Pcode1.py`, finds the roots of the function $f(x) = x^2 - 2x - 3$.

Python Input:

```
# This code finds the roots
# of: f(x) = x^2 - 2x - 3
# using 'Bisection Method'
def f(x):
   return x**2 - 2*x - 3
# First root in domain [0,4]
a1, b1 = 0, 4
tol= 1e-6 #Stoping tolerance
X2 = 500 #Maximum iterations
#Bisection Method for 1st root
```

```
iter_count = 0
while (b1 - a1) / 2 > tol:
    c1 = (a1 + b1) / 2
    if f(c1) == 0: break
    elif f(a1)*f(c1)<0: b1=c1
    else: a1 = c1
    iter_count += 1
    if iter_count > X2:
     print('Max iterations');break
root1 = (a1 + b1) / 2
print(f'First root: {root1}')
# Second root in domain [-4,0]
a2, b2 = -4, 0; iter_count =0
# for second root
while (b2 - a2) / 2 > tol:
    c2 = (a2 + b2) / 2
    if f(c2) == 0: break
    elif f(a2)*f(c2)<0: b2=c2
    else: a2 = c2
    iter_count += 1
    if iter_count > X2:
     print('Max iterations'); break
root2 = (a2 + b2) / 2
print(f'Second root: {root2}')
# End of Python script
```

Python Output:

```
First root: 3.0
Second root: -1.0
```

Engineering application: In open channel hydraulics, Manning's equation is a fundamental formula used to calculate the flow rate or velocity in channels where the flow is gravity-driven (see Figure 5.3). The equation is expressed as:

$$Q = \frac{1}{n} A R^{2/3} S_0^{1/2} \tag{5.4}$$

where:

- Q is the flow rate (m^3/s),
- n is Manning's roughness coefficient (dimensionless),
- A is the cross-sectional area of flow (m^2),
- R is the hydraulic radius (m), given by $\frac{A}{P}$ where P is the wetted perimeter,
- S_0 is the bed slope (dimensionless).

For a rectangular channel, as illustrated in Figure 5.3b, the following relationship applies:

- $A = by$, where b is the width of the channel and y is the flow depth,
- $P = b + 2y$, where P is the wetted perimeter.

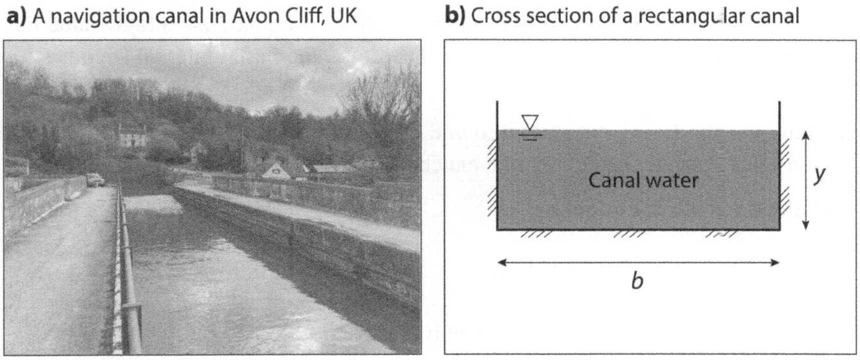

FIGURE 5.3 **a)** A navigation canal located in Avoncliff, UK. **b)** Cross section of a rectangular canal with a flow depth of y and a canal width of b.

The problem here is determining the flow depth (y) for a given flow rate (Q) in an open channel. In practical scenarios, especially under non-uniform flow conditions, determining the flow depth for a known flow rate can be complex and often requires numerical methods. Here, we will use the Bisection method to find the depth corresponding to a given flow rate.

Consider a rectangular channel with:

- Channel Width, b = 3 m
- Channel Slope, S_0 = 0.001 (dimensionless)
- Manning's Roughness Coefficient, n = 0.013 (for concrete lining of the canal)
- Flow Rate, Q = 10 m³/s

The objective is to find the flow depth, y, using the Bisection method. Equation 5.4 is rearranged to the following form to develop a function for y:

$$f(y) = \frac{1}{n} A R^{2/3} S_0^{1/2} - Q \tag{5.5}$$

Finding the flow depth, y, means finding the roots of Equation 5.5, as follows:

$$f(y) = 0 \tag{5.6}$$

which leads to the following equation after substituting the relationships for A and R:

$$f(y) = \frac{1}{n} by \left(\frac{by}{b + 2y} \right)^{2/3} S_0^{1/2} - Q = 0 \tag{5.7}$$

By substituting the values of n, b, Q and S_0 in the above equation, we have:

$$f(y) = \frac{1}{0.013} \cdot 3 \cdot y \left(\frac{3y}{3+2y}\right)^{2/3} \cdot (0.001)^{1/2} - 10 = 0 \tag{5.8}$$

Initial interval selection: We choose an initial interval for y. Since y must be positive, start with a low value like 0.1 m where flow would be minimal, and a high value like 4 m where flow would exceed the given rate. Therefore, the initial interval becomes $[a = 0.1, b = 4]$.

Iteration: Use the Bisection method to find y where $f(y) = 0$ or is very close to zero, indicating that the calculated flow rate matches the given flow rate.

Check the signs at the endpoints:

$$f(0.1) = -9.85, \quad \text{and} \quad f(4) = 20.93$$

Since $f(0.1) \cdot f(4) < 0$, there is a root in [0.1, 4].

Set tolerance: Let us set a tolerance of 0.04. This implies that the solution for y is acceptable if $f(y)$ lies within ± 0.04.

Bisection method iterations:

- **Initial domain:**
  - $f(0.1) = -9.85$
  - $f(4) = 20.93$
  - Since $f(0.1) \cdot f(4) < 0$, there is a root in [0.1, 4]
- **Iteration 1:**
  - Midpoint $c = \frac{0.1+4}{2} = 2.05$
  - $f(2.05) = 3.59$ (root in [0.1, 2.05])
- **Iteration 2:**
  - $c = \frac{0.1+2.05}{2} = 1.075$
  - $f(1.075) = -4.26$ (root in [1.075, 2.05])
- **Iteration 3:**
  - $c = \frac{1.075+2.05}{2} = 1.5625$
  - $f(1.5625) = -0.46$ (root in [1.5625, 2.05])
- **Iteration 4:**
  - $c = \frac{1.5625+2.05}{2} = 1.8062$
  - $f(1.8062) = 1.54$ (root in [1.5625, 1.8062])
- **Iteration 5:**
  - $c = \frac{1.5625+1.8062}{2} = 1.6844$
  - $f(1.6844) = 0.53$ (root in [1.5625, 1.6844])
- **Iteration 6:**
  - $c = \frac{1.5625+1.6844}{2} = 1.6235$
  - $f(1.6235) = 0.036$ (root in [1.5625, 1.6235])

As $f(1.6235) = 0.036 <$ tolerance of 0.04, the final solution is $y = 1.62$ m. This indicates that the flow depth in the canal is $y = 1.62$ m under the specified flow conditions.

The MATLAB and Python scripts presented in Example 5.3 (script `CH05Mcode2.m`) and Example 5.4 (script `CH05Pcode2.py`), respectively, can be used to solve the aforementioned engineering example.

Example 5.3

MATLAB code. This MATLAB script, named `CH05Mcode2.m`, solves flow depth in a rectangular channel based on the Manning's equation using the Bisection method.

MATLAB Input:

```
% This code finds the flow
% depth (y) in a channel
% based on Manning's equation
% using 'Bisection Method'
clc; clear; close all;
% Define the function
f = @(y) (1/0.013) * 3 * y *...
((3 * y) / (3 + 2 * y))^(2/3) *...
sqrt(0.001) - 10;
% Initial interval
a = 0.1; b = 4;
tol = 0.001; % Tolerance for stopping
X2 = 100; % Maximum iterations
% Check if initial interval is valid
if f(a) * f(b) > 0
   error('No root in [a, b]');
end
iter = 0;
while (b - a) / 2 > tol
  c = (a + b) / 2; fc = f(c);
  % Check if midpoint is the root
  if abs(fc) < tol; break;
  elseif f(a)*fc<0; b=c; %Root in [a,c]
  else a = c; % Root in [c,b]
  end
   iter = iter + 1;
  if iter>X2; error('Max iterations');
  end
end
root=(a+b)/2; %Final root approximation
```

```
% Display results
fprintf('Root found: y = %.4f \n',root);
fprintf('f(y) at root is %.4f\n',f(root));
% end of M - file
```

MATLAB Output:

```
Root found: y = 1.6187
f(y) at root is -0.0033
```

Example 5.4

Python code. This Python script, named `CH05Pcode2.py`, solves flow depth in a rectangular channel based on the Manning's equation using the Bisection method.

Python Input:

```
# This code finds the flow depth (y)
# in a channel # based on Manning's
# equation using 'Bisection Method'
import math
def f(y):
  return (1/0.013)*3*y*((3*y)\
   /(3+2*y))**(2/3)*math.sqrt(0.001)-10
# Initial interval
a = 0.1; b = 4
tol = 0.001 # Tolerance for stopping
X2= 100 # Maximum iterations
# Check if initial interval is valid
if f(a) * f(b) > 0:
 raise ValueError("No root in [a, b]")
# Bisection Method
iter_count = 0
while (b - a) / 2 > tol:
   c = (a + b) / 2 # Midpoint
   fc = f(c)
   # Check if midpoint is the root
   if abs(fc) < tol:
      break
   elif f(a)*fc<0: b=c #Root in [a,c]
   else: a = c # Root in [c,b]
   iter_count += 1
```

```
    if iter_count > X2:
        raise ValueError("Maximum iterations reached")
root = (a + b) / 2 # Final root
# Display results
print(f"Root found: y = {root:.4f}")
print(f"f(y) at root is {f(root):.4f}")
# End of Python script
```

Python Output:

```
Root found: y = 1.6187
f(y) at root is -0.0033
```

5.1.2 NEWTON-RAPHSON METHOD

The Newton-Raphson method, often referred to as Newton's method, is a numerical technique for finding approximate solutions to equations of the form $f(x) = 0$. It is a powerful tool for locating the roots (or zeros) of a function. The method is widely used due to its fast convergence when applied to problems with well-behaved functions and good initial guesses. Compared to the bisection method, which ensures convergence but is slower, Newton-Raphson achieves faster convergence but requires the calculation of the derivative and may fail if the derivative is zero or near zero.

The concept. The Newton-Raphson method begins with an initial guess, x_0, which is the user's estimate of the root. Using this guess, the method calculates a better approximation of the root through repeated iterations based on the formula:

$$x_{n+1} = x_n - \frac{f(x_n)}{f'(x_n)} \tag{5.9}$$

Here:

- x_n is the current approximation of the root.
- $f'(x_n)$ is the derivative of the function $f(x)$ evaluated at x_n.
- x_{n+1} is the next, improved approximation.

The process is repeated until the change between successive approximations is smaller than a predefined tolerance or the value of $f(x_n)$ is close to zero.

Step-by-step procedure:

i. **Function and its derivative:** Begin by selecting the function $f(x)$ for which you aim to find the root. Compute the derivative of this function, denoted as $f'(x)$.
ii. **Initial approximation (initial guess):** Choose an initial approximation, x_0, anticipated to be in proximity to the true root of $f(x)$.

iii. **Iterative refinement:** Employ the Newton-Raphson formula iteratively to refine your approximation:

$$x_{n+1} = x_n - \frac{f(x_n)}{f'(x_n)}$$

This equation adjusts the current estimate x_n towards the root by following the tangent line at x_n to its intersection with the x-axis.

iv. **Convergence criterion:** Continue the iterations until the difference between successive approximations, $\epsilon = |x_{n+1} - x_n|$, is less than a predetermined tolerance level, or until $|f(x_n)|$ approaches zero, indicating that the approximation has sufficiently converged to the root.

Example: Determine the roots of the following function using the Newton-Raphson method considering a convergence criterion of $\epsilon = 0.05$:

$$f(x) = x^2 - 2x - 3$$

Solution:

In the previous section, the Bisection method was employed to determine the roots of this function, which were found to be $x = -1$ and $x = 3$. Now, we apply the Newton-Raphson method to solve the same equation and compare its performance with the Bisection method. The solution follows the four steps outlined earlier, as detailed below. A convergence criterion of $\epsilon = 0.05$ is considered for this example.

i. **Define the function and obtain its derivative:** $f(x) = x^2 - 2x - 3$, so its derivative takes the following form:

$$f'(x) = 2x - 2 \tag{5.10}$$

ii. **Initial guess:** Let $x_0 = 4$.

iii. **Iterations:**

$$x_{n+1} = x_n - \frac{f(x_n)}{f'(x_n)} = x_n - \frac{x_n^2 - 2x_n - 3}{2x_n - 2} \tag{5.11}$$

Perform iterations as follows:

- **Iteration 1:** for the first iteration ($n = 0$), Equation 5.11, becomes:

$$x_1 = x_0 - \frac{x_0^2 - 2x_0 - 3}{2x_0 - 2} \tag{5.12}$$

- Here, $x_0 = 4$ and x_1 is the next, improved approximation which is going to be obtained using the Newton-Raphson formula:

$$x_1 = 4 - \frac{4^2 - 2 \cdot 4 - 3}{2 \cdot 4 - 2} = 4 - \frac{5}{6} = 3.1667 \tag{5.13}$$

- Check the convergence criterion: $\epsilon = |x_1 - x_0| = 0.8333$ which is bigger than the allowed convergence criteria of $\epsilon = 0.05$ and therefore iterations are continued.

- **Iteration 2:** for the second iteration ($n = 1$), Equation 5.11, becomes:

$$x_2 = x_1 - \frac{x_1^2 - 2x_1 - 3}{2x_1 - 2} \tag{5.14}$$

- Here, $x_1 = 3.1667$ and x_2 is the next, improved approximation:

$$x_2 = 3.1667 - \frac{3.1667^2 - 2 \cdot 3.1667 - 3}{2 \cdot 3.1667 - 2} = 3.0064 \tag{5.15}$$

- Check the convergence criterion: $\epsilon = |x_2 - x_1| = 0.16 > 0.05$; therefore, iterations are continued.
- **Iteration 3:**

$$x_3 = 3.0064 - \frac{3.0064^2 - 2 \cdot 3.0064 - 3}{2 \cdot 3.0064 - 2} = 3.0000 \tag{5.16}$$

- Check the convergence criterion: $\epsilon = |x_3 - x_2| = 0.0064 < 0.05$; therefore, iterations are terminated and the final answer is $x = 3.0000$.

iv. **Stop when close to a root:** It can be seen that after three iterations, $\epsilon = 0.0064 < 0.05$, and therefore $x = 3.0000$ is one of the roots of $f(x) = x^2 - 2x - 3$.

Note 1: The Newton-Raphson method converges to one root at a time. The other root of the equation, $x = -1$, can be found by starting with an initial guess closer to it, such as $x_0 = -2$.

Note 2: It is also noted that this method works best when:

- The function is continuous and differentiable.
- The initial guess is close to the actual root.
- However, The Newton-Raphson method may fail or converge slowly if the derivative is zero or very small near the root.

Engineering application: In transportation engineering, understanding traffic flow through models like Greenshields's model is crucial for urban traffic management as it helps in predicting congestion points, optimizing road capacity, and planning effective traffic control strategies. By identifying the density at which traffic flow is maximized, urban planners can design roads and traffic signals to manage peak-hour traffic more efficiently, reducing travel times and fuel consumption, thereby enhancing the overall livability and sustainability of urban areas (Figure 5.4).

The Greenshields model relates traffic flow (q), density (k), free-flow speed (v_f), and jam density (k_j) through the following mathematical equation [1, 7]:

$$q = k \cdot v_f \left(1 - \frac{k}{k_j}\right) \tag{5.17}$$

The example presented here aims to determine the density k at which traffic flow is maximized for a given traffic flow (q). The known and unknown parameters for this problem are listed below:

a) An urban road in London, UK **b)** Motorway M4, west of the UK

FIGURE 5.4 **a)** An urban road in London, UK. **b)** A motorway in the west of the UK, named as M4.

- Free-flow speed, $v_f = 100$ km/hr
- Jam density, $k_j = 100$ veh/km
- Traffic flow, $q = 2500$ veh/hr
- Density, $k =$ the parameter to be calculated

By substituting these values into Equation 5.17, it becomes:

$$2500 = 100k\left(1 - \frac{k}{100}\right) \tag{5.18}$$

To solve this problem using the Newton-Raphson method, we take the following steps:

i. **Define the function:** By moving $q = 2500$ to the right side of the Equation 5.18, the function takes the following form:

$$f(k) = 100k\left(1 - \frac{k}{100}\right) - 2500 \tag{5.19}$$

ii. **Simplify the function:** Expand $f(k)$:

$$\begin{aligned} f(k) &= 100k\left(1 - \frac{k}{100}\right) - 2500 \\ &= 100k - k^2 - 2500 \end{aligned}$$

iii. **Find the derivative:** Compute $f'(k)$:

$$\begin{aligned} f'(k) &= \frac{d}{dk}\left(100k - k^2 - 2500\right) \\ &= 100 - 2k \end{aligned}$$

iv. **Write the Newton-Raphson formula:** The iterative formula for Newton-Raphson is:

$$k_{n+1} = k_n - \frac{f(k_n)}{f'(k_n)} \tag{5.20}$$

By substituting the values of $f(k)$ and $f'(k)$ into Equation 5.20, we obtain:

$$k_{n+1} = k_n - \frac{100k_n - k_n^2 - 2500}{100 - 2k_n} \tag{5.21}$$

v. **Choose an initial guess:** Select an initial value $k_0 = 60$.
vi. **Perform iterations:** Substitute k_n into the formula to calculate k_{n+1} until the convergence criterion $\epsilon = |k_{n+1} - k_n| < 2$ is satisfied.

- **Iteration 1:** Using $k_0 = 60$:

$$\begin{aligned} f(k_0) &= 100(60) - 60^2 - 2500 \\ &= 6000 - 3600 - 2500 = -100 \\ f'(k_0) &= 100 - 2(60) = -20 \\ k_1 &= 60 - \frac{-100}{-20} = 60 - 5 = 55 \end{aligned}$$

Check convergence: $|k_1 - k_0| = |55 - 60| = 5$, which is greater than $\epsilon = 2$. Continue.

- **Iteration 2:** Using $k_1 = 55$:

$$\begin{aligned} f(k_1) &= 100(55) - 55^2 - 2500 \\ &= 5500 - 3025 - 2500 = -25 \\ f'(k_1) &= 100 - 2(55) = -10 \\ k_2 &= 55 - \frac{-25}{-10} = 55 - 2.5 = 52.5 \end{aligned}$$

Check convergence: $|k_2 - k_1| = |52.5 - 55| = 2.5$, which is greater than $\epsilon = 2$. Continue.

- **Iteration 3:** Using $k_2 = 52.5$:

$$\begin{aligned} f(k_2) &= 100(52.5) - 52.5^2 - 2500 \\ &= 5250 - 2756.25 - 2500 = -6.25 \\ f'(k_2) &= 100 - 2(52.5) = -5 \\ k_3 &= 52.5 - \frac{-6.25}{-5} = 52.5 - 1.25 = 51.25 \end{aligned}$$

Check convergence: $|k_3 - k_2| = |51.25 - 52.5| = 1.25$, which is less than $\epsilon = 2$. Convergence achieved.

Final solution: The density k at which the traffic flow is maximized is $k = 51$, based on the chosen convergence criterion of $\epsilon = 2$. It is noted that the final solution could be further refined if a smaller convergence criterion (ϵ) were selected.

We now use MATLAB and Python to perform the above calculations, employing the scripts presented in Example 5.5 (`CH05Mcode3.m`) and Example 5.6 (`CH05Pcode3.py`), respectively. In these scripts, the convergence criterion is set to $\epsilon = 0.1$. As a result, the final solution obtained is $k = 50$, which differs slightly from the manually-obtained solution above, where $\epsilon = 2$ was used.

Example 5.5

MATLAB code. This MATLAB script, named `CH05Mcode3.m`, finds density (k) at which a traffic flow (q) is maximized.

MATLAB Input:

```
% This code finds density k at
% which traffic flow is maximized
% f(k) = 100k - k^2 - 2500
% using 'ewton-Raphson'
clc; clear; close all;
f=@(k)100*k-k^2-2500; %f(k)
f_prime=@(k)100-2*k; %f'(k)
k0 = 60; % Initial guess
epsilon=0.1;%Convergence criterion
X2 = 500;% Maximum iterations
% Newton-Raphson iterations
for iter = 1:X2
  f_k = f(k0);
  f_prime_k=f_prime(k0);
  k_next=k0-f_k/f_prime_k;%next value
  diff=abs(k_next-k0);%convergence check
  if diff<epsilon %Check convergence
    final=round(k_next);%Round
    fprintf('Iterations = %d\n', iter);
    fprintf('Solution, k = %d\n',final);
    break;
  end
  k0 = k_next; % Update k0
end
if iter==X2
    fprintf('\nMaximum iterations\n');
end
```

```
% end of M - file
```

MATLAB Output:

```
Iterations = 7
Solution, k = 50
```

Example 5.6

Python code. This Python script, named `CH05Pcode3.py`, finds density (k) at which a traffic flow (q) is maximized.

Python Input:

```
# This code finds density k
# at which traffic flow is
# maximized for traffic flow q
# f(k) = 100k - k^2 - 2500
# using Newton-Raphson Method
def newton_raphson(f, f_prime, k0,\
        epsilon, max_it):
  for iter in range(1, max_it + 1):
    f_k = f(k0)
    f_prime_k = f_prime(k0)
    k_next=k0-f_k/f_prime_k #Next value
    diff=abs(k_next-k0) #Convergence
    if diff<epsilon:#Check convergence
      final=round(k_next)#Round answer
      print(f"Iterations = {iter}")
      print(f"Solution, k = {final}")
      return final
    k0=k_next #Update k0
  print("Maximum iterations")
  return None
#Define functions f(k),f'(k)
f=lambda k: 100*k-k**2-2500
f_prime= lambda k: 100-2*k
# Initial guess, parameters
k0 = 60
epsilon = 0.1
max_it = 500
# Perform Newton-Raphson method
```

```
solution=newton_raphson(f,f_prime,\
    k0, epsilon, max_it)
# End of Python script
```

Python Output:

```
Iterations = 7
Solution, k = 50
```

5.1.3 BUILT-IN ROOT-FINDING FUNCTIONS IN MATLAB AND PYTHON: FZERO AND FSOLVE

We have previously explored fundamental methods for finding the roots of equations, such as the Bisection and Newton-Raphson methods. We now explore powerful built-in tools in MATLAB and Python that streamline this process, as outlined in the text box below:

- In MATLAB, the **fzero** function is used to find the root of a function, starting from an initial guess provided by the user.
- In Python, the **fsolve** function serves a similar purpose, allowing for the solution of a single equation or even a system of nonlinear equations.

These functions significantly simplify the root-finding process by utilizing advanced algorithms. Rather than manually performing the iterative steps, you simply input the equation, and the software determines the root. This enables engineers to focus on solving practical, real-world problems, rather than getting bogged down in the computational details.

Engineering application: Consider the suspension bridge shown in Figure 5.5. In suspension bridge engineering, determining cable tension (T) is fundamental to ensuring structural integrity and achieving design accuracy. Suspension bridge cables often approximate a parabolic profile under uniform load, described by the classic equation [8]:

$$d = \frac{wL^2}{8T} \tag{5.22}$$

where d is the sag (see Figure 5.5), w is the uniform load, L is the span length, and T is the cable tension. However, in this example, we consider a special case where the relationship between cable tension and sag incorporates an additional non-linear factor, which may arise due to complex material behavior or external influences. The sag (d) of the cable is described by the following equation:

$$d = \frac{wL^2}{8T} + \frac{TL}{EI} \tag{5.23}$$

where E is the modulus of elasticity and I is the second moment of area.

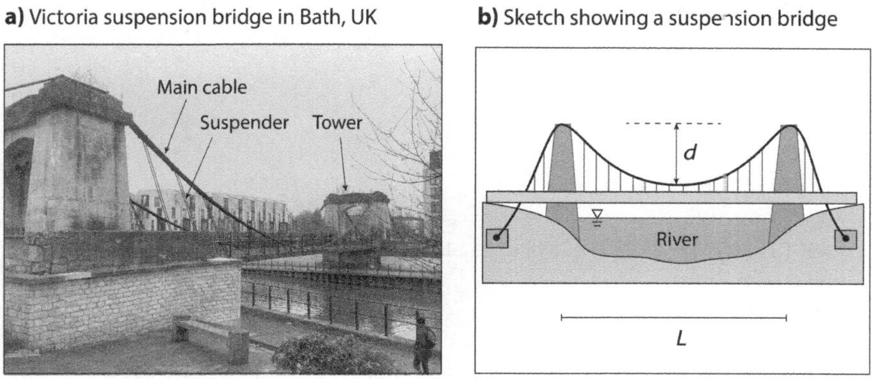

FIGURE 5.5 a) The Victoria suspension bridge in Bath, UK. b) Sketch of a suspension bridge showing some of the parameters used for its design.

As a numerical example, we consider the following parameter values for the suspension bridge shown in Figure 5.5:

- Span (L) = 45 m
- Load per unit length (w) = 15 kN/m
- Sag (d) = 6 m
- Modulus of elasticity (E) = 200 GPa ($= 200 \times 10^6\ kN/m^2$)
- Second moment of area (I) = 3.976×10^{-5} m^4

By substituting the above values in Equation 5.23, it becomes:

$$6 = \frac{15 \times (45)^2}{8T} + \frac{T \times 45}{200 \times 10^6 \times 3.976 \times 10^{-5}} \tag{5.24}$$

By rearranging the previous equation, the function $f(T)$ is formed as follows:

$$f(T) = \frac{30375}{8T} + \frac{T}{176711.11} - 6 \tag{5.25}$$

The objective is to compute the tension T at the lowest point of the cable by solving for the roots of Equation 5.25 using MATLAB's `fzero` and Python's `fsolve` functions. The MATLAB and Python scripts provided in Examples 5.7 (script `CH05Mcode4.m`) and 5.8 (script `CH05Pcode4.py`) can be used to solve this problem. By examining the outputs of the MATLAB and Python codes, we find that the solution is $T = 633.2$ kN.

Example 5.7

MATLAB code. This MATLAB script, named `CH05Mcode4.m`, computes the tension T at the lowest point of the cable of a suspension bridge.

MATLAB Input:

```
%This code computes tension (T)
%at the lowest point of cable
%of a suspension bridge.
% using 'fzero' function
clc; clear; close all;
% Define the function f(T)
f=@(T)(30375/(8*T))+(T/176711.11)-6;
%Provide an initial guess for T
ini_T=400; %Adjust if necessary
% Solve for T using 'fzero'
[T_sol,fval,exitflag]=fzero(f,ini_T);
% Display the results
if exitflag == 1
 fprintf('T = %.4f kN\n', T_sol);
 fprintf('f(T) = %.4f\n',fval);
else
 fprintf('fzero did not converge.\n');
end
% end of M - file
```

MATLAB Output:

```
T = 633.1906 kN
f(T) = 0.0000
```

Example 5.8

Python code. This Python script, named CH05Pcode4.py, computes the tension T at the lowest point of the cable of a suspension bridge.

Python Input:

```
#This code computes tension (T)
#at the lowest point of cable
#using 'fsolve' function
import numpy as np
from scipy.optimize import fsolve
# Define the function f(T)
def f(T):
 return (30375/(8*T))+(T/176711.11)-6
```

```
# Provide initial guess for T
ini_T=400 #Adjust if needed
# Solve for T using 'fsolve'
T_sol = fsolve(f, ini_T)
# Display the results
print(f'T = {T_sol[0]:.4f} kN')
print(f'f(T) = {f(T_sol[0]):.4f}')
# End of Python script
```

Python Output:

```
T = 633.1906 kN
f(T) = 0.0000
```

5.2 LEAST SQUARES AND CURVE FITTING

Curve fitting is a method of finding a mathematical function that best represents a set of observation (or measured) data points. In engineering and science, data points are often obtained from experiments or field observations, and curve fitting helps in understanding underlying patterns or relationships.

The **least squares method** is a widely used technique for curve fitting. It minimizes the error between the observed data points (y_i) and the predicted model (y_{model}) obtained by curve fitting. Specifically, the method aims to minimize the **sum of squared errors (SSE)**, where the error for each data point is the vertical difference between the observed and predicted values:

$$\text{Error} = y_i - y_{\text{model}} \tag{5.26}$$

By squaring the errors and summing them, we ensure that both positive and negative deviations contribute equally to the total error, and we emphasize larger deviations. The least squares method identifies the model parameters that result in the smallest possible SSE, thus providing the best fit.

5.2.1 LINEAR CURVE FITTING CASE: $Y = A + MX$

In its simplest form, the least squares method fits a straight line to the observation (or measured) data points. The line is defined as:

$$y = a + mx \tag{5.27}$$

where:

- y: dependent variable (e.g., measured response),
- x: independent variable (e.g., time, distance, or some other parameter),
- a: intercept of the line,
- m: slope of the line.

To find a and m, the following formulas are derived by minimizing the error represented in Equation 5.26 (see references such as [6]):

$$m = \frac{n\sum(x_i y_i) - \sum x_i \sum y_i}{n\sum(x_i^2) - (\sum x_i)^2} \tag{5.28}$$

$$a = \frac{\sum y_i - m\sum x_i}{n} \tag{5.29}$$

Here:

- n: number of data points,
- $\sum x_i$: sum of all x-values,
- $\sum y_i$: sum of all y-values,
- $\sum(x_i y_i)$: sum of the products of corresponding x- and y-values,
- $\sum(x_i^2)$: sum of the squares of x-values.

Engineering application: A civil engineer is measuring the settlement of a bridge pier over time. They have recorded the settlement (y, in mm) at different times (x, in years). The data is provided in Table 5.1. Here, the **objective** is to fit a straight line in the form of $y = a + mx$ to this observation data using the least squares method.

Step-by-step solution:

TABLE 5.1

Data of settlement of a bridge foundation over time

Time (years) x_i	Settlement (mm) y_i
1	2.1
2	4.5
3	6.3
4	8.0
5	10.1
6	12.2
7	14.5

i) Compute necessary sums: According to Equations 5.28 and 5.29, the following sums are required to be calculated:

$$\text{Compute: } \sum x_i, \sum y_i, \sum(x_i^2), \sum(x_i y_i) \tag{5.30}$$

Based on calculations presented in Table 5.2, the results are:

$$\sum x_i = 28, \quad \sum y_i = 57.7, \quad \sum(x_i^2) = 140, \quad \sum(x_i y_i) = 287.2$$

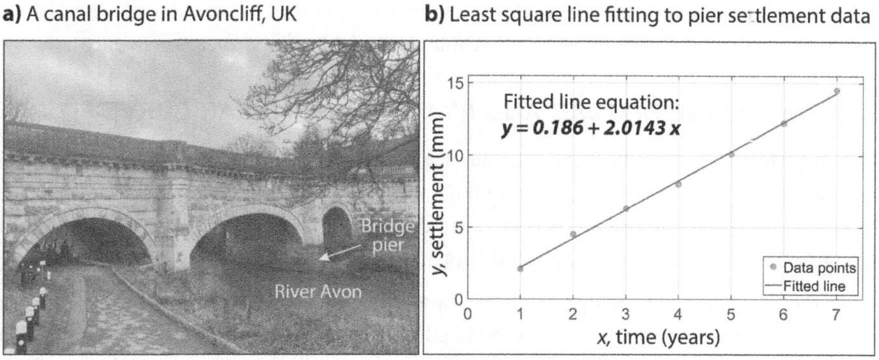

FIGURE 5.6 **a)** A canal bridge in Avoncliff (UK) over the River Avon. **b)** A least-square line fitting to data of bridge pier settlement.

TABLE 5.2
Computed values for least squares calculations

x_i	y_i	x_i^2	x_iy_i
1	2.1	1	2.1
2	4.5	4	9.0
3	6.3	9	18.9
4	8.0	16	32.0
5	10.1	25	50.5
6	12.2	36	73.2
7	14.5	49	101.5

---

ii) Calculate m and a using Equations 5.28 and 5.29:

$$m = \frac{n\sum(x_iy_i) - \sum x_i \sum y_i}{n\sum(x_i^2) - (\sum x_i)^2} \tag{5.31}$$

Substitute $n = 7$, $\sum x_i = 28$, $\sum y_i = 57.7$, $\sum(x_i^2) = 140$, $\sum(x_iy_i) = 287.2$:

$$m = \frac{7(287.2) - (28)(57.7)}{7(140) - (28)^2} = \frac{2010.4 - 1615.6}{980 - 784} = 2.0143$$

$$a = \frac{\sum y_i - m\sum x_i}{n} \tag{5.32}$$

$$a = \frac{57.7 - 2.015(28)}{7} = 0.186$$

iii) Equation of the line: The fitted line is:

$$y = 0.186 + 2.0143x \tag{5.33}$$

iv) Plot and interpretation: The measured data points (x_i, y_i) and the fitted line $y = 0.186 + 2.0143x$ are plotted on the graph in Figure 5.6b. The line represents the trend of pier settlement over time. From the slope of the fitted line ($m = 2.0143$), we observe that the pier settlement increases approximately 2.0143 mm per year on average.

5.2.2 APPLYING MATLAB AND PYTHON FOR LINEAR CURVE FITTING

The example above can be solved using MATLAB and Python. In MATLAB, a linear least squares fit can be performed using the backslash operator (\) or the built-in function `polyfit`. Here, we utilize the `polyfit` function. Example 5.9 provides the MATLAB script `CH05Mcode5.m` to accomplish this task.

In Python, a linear least squares fit can be performed using the `numpy.polyfit` function, which functions similarly to MATLAB's `polyfit`. Additionally, Python provides the `scipy.stats.linregress` function for a more comprehensive statistical analysis, offering the slope, intercept, and additional metrics such as the correlation coefficient and p-value. Here, we use the `numpy.polyfit` function, as demonstrated in Example 5.10 with the Python script `CH05Pcode5.py`.

Example 5.9

MATLAB code. This MATLAB script, named `CH05Mcode5.m`, fits a linear least square line to fit data points.

MATLAB Input:

```
%This code fits a least-square
%fitted line to fit data points.
%using 'polyfit' function
clc; clear; close all;
set(0,'defaultaxesfontsize',28);
% Data points
x=[1, 2, 3, 4, 5, 6, 7]';
y=[2.1,4.5,6.3,8.0,10.1,12.2,14.5]';
% Fit a linear model
f=fit(x,y,'poly1'); % Linear fit
% Extract coefficients
m = f.p1; % Slope
a = f.p2; % Intercept
y2=m.*x+a; % y2 is from fitted line
figure;
plot(x,y,'o','MarkerSize',10);
hold on; % Retain current plot
plot(x,y2,'LineWidth',2);%Plot line
```

```
xlabel('x'); ylabel('y');
xlim([0 7.5]); ylim([0 15]);
grid on;set(gca,'linewidth',1.5);
fprintf('Slope (m): %.3f\n',m);
fprintf('Intercept(a): %.3f\n',a);
% end of M - file
```

MATLAB Output:

```
Slope (m): 2.014
Intercept(a): 0.186
- Also, this script plots the measured data points alcng with
    the fitted line.
```

Example 5.10

Python code. This Python script, named `CH05Pcode5.py`, fits a linear least square line to fit data points.

Python Input:

```
#This code fits a linear least
#square line to fit data points.
#using 'numpy.polyfit' function
import numpy as np
import matplotlib.pyplot as plt
# Data points
x=np.array([1,2,3,4,5,6,7])
y=np.array([2.1,4.5,6.3,8.0,10.1,12.2,14.5])
# Fit a linear model
coeff=np.polyfit(x,y,1) #Linear fit
m = coeff[0] # Slope
a = coeff[1] # Intercept
#Calculate fitted line values
y2=m*x+a #y2 is from fitted line
# Plotting
plt.figure(figsize=(8, 6))
plt.plot(x,y,'o',markersize=10)
plt.plot(x,y2,linewidth=2)
plt.xlabel('x'); plt.ylabel('y')
plt.xlim([0,7.5]);plt.ylim([0,15])
plt.gca().tick_params(axis='both')
```

```
# Print results
print(f"Slope (m): {m:.3f}")
print(f"Intercept (a): {a:.3f}")
plt.show()
# End of Python script
```

Python Output:

```
Slope (m): 2.014
Intercept (a): 0.186
- Also, this script plots the measured data points along with
   the fitted line.
```

5.2.3 NON-LINEAR LEAST-SQUARE CURVE FITTING

In many real-world engineering problems, data may not follow a simple linear relationship. Instead, the relationship between the dependent and independent variables could be more complex and nonlinear. Nonlinear curve fitting is a powerful technique used to model such data, where the model involves nonlinear functions of the parameters.

For example, in civil engineering, it is common to encounter data that follows polynomial, exponential, logarithmic, or other complex functional forms. Both MATLAB and Python provide effective tools for performing nonlinear curve fitting, as outlined in the information box below:

> In MATLAB, the **`lsqcurvefit`** function provides an easy and effective way to perform nonlinear curve fitting, while Python's **`scipy.optimize.curve_fit`** function serves a similar purpose.

Nonlinear curve fitting involves finding the best parameters of a nonlinear model that minimizes the difference between the predicted values from the model and the actual observed data. The goal is to determine the parameters that yield the most accurate predictions based on the data.

In many cases, you will need to define the nonlinear model yourself. Both `lsqcurvefit` in MATLAB and `curve_fit` in Python allow you to optimize these parameters to find the best-fit curve for the data.

5.2.4 NONLINEAR CURVE FITTING WITH LSQCURVEFIT IN MATLAB

The `lsqcurvefit` function in MATLAB is designed to solve nonlinear curve fitting problems by minimizing the sum of squared errors between the observed data and the model's predictions. It allows you to define a custom nonlinear model, provide initial guesses for the parameters, and let MATLAB automatically optimize these parameters to fit the data. In this section, we will demonstrate how to use `lsqcurvefit` to fit a polynomial model to a set of data points. This approach can easily be adapted to other types of nonlinear models.

The general syntax of the `lsqcurvefit` function is:

```
[params, resnorm] = lsqcurvefit(modelFunc, initialGuess,
xData, yData);
```

where:

- `modelFunc`: An anonymous function or function handle that defines the model. This function must take the parameters to be fitted and the independent variable as inputs.
- `initialGuess`: A vector containing the initial guesses for the model parameters.
- `xData` and `yData`: The observed data points for the independent and dependent variables, respectively.
- `params`: The output, which contains the fitted parameters of the model.
- `resnorm`: A measure of the fitting error, representing the residual sum of squares.

Engineering application: In hydraulic engineering, understanding and predicting the flow through a dam's spillway is crucial for managing flood risks and ensuring dam safety. Spillways are designed to release water from a dam when the reservoir level reaches a certain height, preventing overtopping (Figure 5.7). One common method to model this flow is by using a power law relationship, which captures the nonlinear behavior of water discharge over the spillway crest. This approach simplifies the complex physics of fluid dynamics into a manageable form for practical engineering applications.

In practice, the exact discharge can be influenced by many factors, including the shape

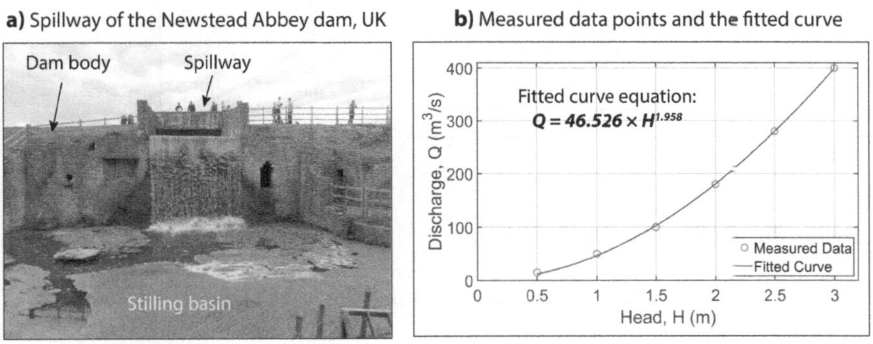

FIGURE 5.7 **a)** Photo showing the spillway of the Newstead Abbey dam in the UK. **b)** Plot of the measured data points of water head (H) and discharge (Q) along with the fitted curve using the least-square method.

of the spillway, the condition of the spillway's surface, the approach velocity of the water, and the geometry of the weir or crest (Figure 5.7). The equation often used is [2]:

$$Q = C_d L H^{3/2} \tag{5.34}$$

where Q is discharge capacity of the spillway, C_d is the discharge coefficient, L is the length of the weir crest, and H is the head over the crest. However, for our example, we simplify Equation 5.34 to:

$$Q = CH^n \tag{5.35}$$

to focus on the nonlinear fitting aspect. This model helps engineers predict how much water will be discharged for any given head, aiding in design and operational decisions.

Problem description and objective:
The discharge through a spillway of a dam can often be modeled with a power law relationship as described in Equation 5.35:

$$Q = CH^n$$

where Q is the discharge (flow rate) in m³/s, H is the head over the spillway crest in meters, and C and n are coefficients to be determined. The data in Table 5.3 represent measurements of water head (H) and discharge (Q) for a specific spillway. The **objective** is to fit a nonlinear equation, as described in Equation 5.35, to these measurement data.

Fitting process: The following four steps are used to fit a nonlinear curve to the measured data in Table 5.3.

TABLE 5.3
Input data for head over spillway crest and corresponding discharge

Head, H_i (m)	**Discharge, Q_i $(m^3$/s)**
0.5	15
1.0	50
1.5	100
2.0	180
2.5	280
3.0	400

i. **Define the nonlinear model function:**

```
modelFunc = @(params, H) params(1) * H.^params(2)
```

The above line is a command line in MATLAB (see the full script in Example 5.11). Here, `params(1)` represents C, and `params(2)` represents n.

ii. **Initial guess:** Provide initial guesses for the parameters C and n. These guesses help MATLAB start the optimization process.

```
initialGuess = [1, 1]; % Initial guesses for C and n
```

iii. **Fit the nonlinear model:** Use the `lsqcurvefit` function to find the best-fit parameters by minimizing the sum of squared errors.

```
params = lsqcurvefit(modelFunc, initialGuess, H, Q);
```

The MATLAB script CH05Mcode6.m, presented in Example 5.11, can be used to perform this analysis whose results are shown in Figure 5.7b. Based on the outputs of Example 5.11 and Figure 5.7b, the fitted nonlinear curve is given by:

$$Q = 46.5257\, H^{1.9575} \tag{5.36}$$

Example 5.11

MATLAB code. This MATLAB script, named CH05Mcode6.m, fits a nonlinear least-square curve to the measured data points.

MATLAB Input:

```
%This code fits a nonlinear least
%square curve to fit data points.
%using 'lsqcurvefit' function
clc; clear; close all;
set(0,'defaultaxesfontsize',29);
% Input data, head (H), discharge (Q)
H=[0.5,1.0,1.5,2.0,2.5,3.0]; %in meter
Q=[15,50,100,180,280,400]; %in m^3/s
%Define function % Q=C*H^n
modelF= @(par,H) par(1)*H.^par(2);
ini_Guess=[1,1];% Initial guess for C&n
%Fit nonlinear model using lsqcurvefit
par=lsqcurvefit(modelF,ini_Guess,H,Q);
% Display the results
fprintf(['Parameters: C=%.4f, ' ...
    'n=%.4f \n'],par(1),par(2));
%predictions using fitted model
H_pred=linspace(min(H),max(H),100);
Q_pred=modelF(par, H_pred);%Calculate Q
% Plot original data and fitted curve
figure; subplot('position',[.2 .2 .6 .7]);
plot(H,Q,'ro','MarkerSize',12,'linewidth',2);
hold on; set(gca,'linewidth',2);
plot(H_pred,Q_pred,'b-','LineWidth',2);%Fitted
grid on; xlim([0 3.2]);ylim([0 410]);
xlabel('Head, H (m)');
ylabel('Discharge, Q (m^3/s)');
legend('Measured Data','Fitted Curve', ...
    'Location','southeast');
% End of M - file
```

MATLAB Output:

```
Parameters: C=46.5257, n=1.9575
- Also, this script plots the measured data points along with
  the fitted curve.
```

5.2.5 NONLINEAR CURVE FITTING WITH SCIPY.OPTIMIZE.CURVE_FIT IN PYTHON

The Python equivalent of MATLAB's `lsqcurvefit` is the `scipy.optimize.curve_-fit` function from the `SciPy` library. This function fits a curve to a set of data points by minimizing the sum of squared residuals between the data and the model. The script `CH05Pcode6.py`, presented in Example 5.12, performs the same task of fitting a nonlinear curve to the measured data in Table 5.3. The outcome of Example 5.12 shows that the fitted line is identical to the one obtained using MATLAB, as given in Equation 5.36.

Example 5.12

Python code. This Python script, named `CH05Pcode6.py`, fits a nonlinear least-square curve to the measured data points.

Python Input:

```
#This code fits a nonlinear least
#square curve to fit data points
#using 'scipy.optimize.curve_fit'
import numpy as np
from scipy.optimize import curve_fit
import matplotlib.pyplot as plt
#Input data: head (H), discharge (Q)
H=np.array([.5,1,1.5,2,2.5,3]) #meters
Q=np.array([15,50,100,180,280,400]) #m^3/s
# Define the model function: Q=C*H^n
def modelF(H, C, n):
   return C * H**n
# Initial guess for C and n
ini_guess = [1, 1]
# Fit nonlinear model using curve_fit
par, _ =curve_fit(modelF,H,Q,p0=ini_guess)
# Extract fitted parameters
C, n = par
print(f"Results: C= {C:.4f}, n= {n:.4f}")
# Predictions using fitted model
H_pred=np.linspace(min(H),max(H),100)
```

```
Q_pred = modelF(H_pred, C, n)
# Plot original data & fitted curve
plt.figure(figsize=(8, 6))
plt.plot(H,Q,'ro', markersize=8,\
  label='Measured Data')
plt.plot(H_pred, Q_pred, 'b-', \
  linewidth=2,label='Fitted Curve')
plt.grid(True, linewidth=0.5)
plt.xlim([0,3.2]); plt.ylim([0,410])
plt.xlabel('Head, H (m)')
plt.ylabel('Discharge, Q (m3/s)')
plt.legend(loc='lower right')
plt.tight_layout(); plt.show()
# End of Python script
```

Python Output:

```
Results: C= 46.5257, n= 1.9575
- Also, this script plots the measured data points along with
    the fitted curve.
```

5.3 NUMERICAL DIFFERENTIATION AND INTEGRATION TECHNIQUES

5.3.1 NUMERICAL DIFFERENTIATION

In many engineering applications, the precise mathematical expressions for the derivatives of functions are not available, making it impossible to compute derivatives using analytical methods. In such situations, numerical differentiation is employed to approximate the derivatives. This approach is fundamental in various scientific and engineering disciplines where analytical derivatives are inaccessible.

The derivative of a function at a point measures how one variable changes in response to another. Analytical differentiation allows us to calculate exact derivatives using mathematical formulas. However, in many practical situations, numerical differentiation becomes essential. These situations include:

i) When the function is defined only at discrete data points, such as data obtained from experiments or observations.
ii) When the function is too complex to differentiate using standard analytical methods.

5.3.1.1 Basic Forms of Numerical Differentiation

Numerical differentiation commonly employs finite difference methods to approximate derivatives. These methods are particularly useful when dealing with discrete data or complex functions. The three basic finite difference approximations are as follows:

1. **Forward difference approximation:**

$$f'(x) \approx \frac{f(x + \Delta x) - f(x)}{\Delta x} \tag{5.37}$$

where Δx is a small increment in x, and the expression $f'(x)$ represents the derivative of the function $f(x)$. This method estimates the derivative using the function value at $x + \Delta x$.

2. **Central difference approximation:**

$$f'(x) \approx \frac{f(x + \Delta x) - f(x - \Delta x)}{2\Delta x} \tag{5.38}$$

which generally provides a more accurate approximation than the forward difference method for the same step size Δx. This method averages the change in function values around x.

3. **Backward difference approximation:**

$$f'(x) \approx \frac{f(x) - f(x - \Delta x)}{\Delta x} \tag{5.39}$$

This method uses the function value at $x - \Delta x$ to estimate the derivative at x.

Engineering application: Flow velocity and acceleration are important engineering variables used in the design of various structures in hydraulic engineering. This example focuses on the application of numerical differentiation to determine canal flow acceleration.

Acceleration is the derivative of velocity with respect to distance or time. In many real-world scenarios, such as in river and canal flow analysis (Figure 5.8), the velocity data is available only at discrete points, and there is no exact equation for the velocity. In such cases, numerical differentiation can be applied to approximate the derivative of velocity which is acceleration.

The velocity $v(x)$ of water flowing through a river is measured at discrete distances along the river's length. The data is provided in Table 5.4. We aim to estimate the rate of change of velocity (= acceleration) at $x = 40$ m using the central difference approximation.

Solution:

Acceleration is given by the derivative of velocity with respect to distance. Using the central difference method, the formula for the first derivative is:

$$v'(x) \approx \frac{v(x + \Delta x) - v(x - \Delta x)}{2\Delta x} \tag{5.40}$$

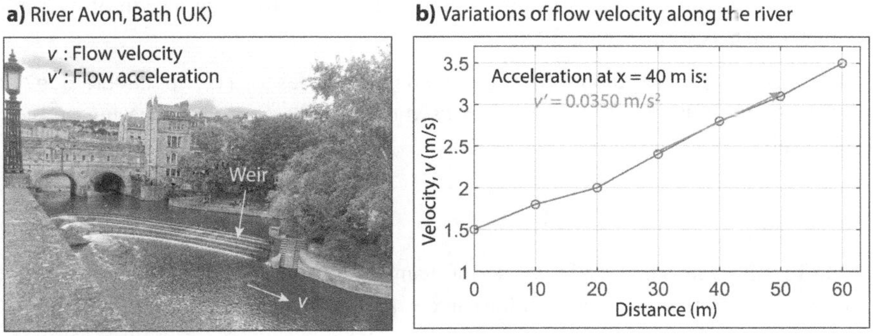

FIGURE 5.8 **a)** River Avon in Bath, UK. **b)** Variations of flow velocity along the river and flow acceleration at $x = 40\ m$.

TABLE 5.4

Velocity measurements along the river

Distance, x (m)	Velocity, $v(x)$ (m/s)
0	1.5
10	1.8
20	2.0
30	2.4
40	2.8
50	3.1
60	3.5

For this problem, we have the following:

- $\Delta x = 10$ m (the spacing between data points. See Table 5.4),
- $x = 40$ m (the desired location for calculating acceleration),
- $v(x + \Delta x) = v(50) = 3.1$ m/s (from Table 5.4),
- $v(x - \Delta x) = v(30) = 2.4$ m/s (from Table 5.4).

By substituting these values into Equation (5.40), we obtain:

$$v'(40) \approx \frac{3.1 - 2.4}{2 \times 10} \tag{5.41}$$

Simplify:

$$v'(40) \approx \frac{0.7}{20} = 0.035\ \text{m/s}^2 \tag{5.42}$$

Final Answer:

The acceleration of the water at $x = 40$ m is approximately $0.035\ \text{m/s}^2$.

Interpretation:

This result shows that the velocity of water is increasing at a rate of $0.035\ \text{m/s}^2$ for every meter along the river at $x = 40$ m. Numerical differentiation allows us to estimate the rate of change in velocity, which is useful for analyzing and optimizing design of hydraulic structures such as weir (Figure 5.8a).

The MATLAB and Python scripts presented in Examples 5.13 (script `CH05Mcode7.m`) and 5.14 (script `CH05Pcode7.py`) can be used to solve the above engineering problem, respectively.

Example 5.13

MATLAB code. This MATLAB script, named `CH05Mcode7.m`, estimates the rate of change of velocity (=acceleration) at x = 40 m using central difference approximation.

MATLAB Input:

```
%This code estimates the rate of
%change of velocity (=acceleration)
% at x=40m using central difference
clc; clear; close all;
set(0,'defaultaxesfontsize',28);
x=[0,10,20,30,40,50,60];%Location(m)
v=[1.5,1.8,2.0,2.4,2.8,3.1,3.5];%Velocity
%Acceleration using central difference
delta_x=10; %Spacing (m)
v_for=3.1; %v(x+delta_x) at x=50m
v_back=2.4; %v(x-delta_x) at x=30m
acc=(v_for - v_back)/(2*delta_x);
% Plot velocity vs. location
subplot('position',[0.15 0.2 0.5 0.6]);
plot(x,v,'o-','LineWidth',2,'MarkerSize',12);
xlim([0 63]);ylim([1 3.7]);
set(gca,'linewidth',1.5); grid on;
% Print calculated acceleration
fprintf(['Calculated acceleration at ' ...
    'x = 40 m: %.4f m/s^2\n'], acc);
% End of M - file
```

MATLAB Output:

```
Calculated acceleration at x = 40 m: 0.0350 m/s^2.
This script also produces the plot shown in Figure 5.8b.
```

Example 5.14

Python code. This Python script, named `CH05Pcode7.py`, estimates the rate of change of velocity (=acceleration) at x = 40 m using central difference approximation.

Python Input:

```
#This code estimates the rate of
#change of velocity (=acceleration)
#at x=40m using central difference
import numpy as np
import matplotlib.pyplot as plt
x=np.array([0,10,20,30,40,50,60]) #(m)
v=np.array([1.5,1.8,2.0,2.4,2.8,3.1,3.5]) #(m/s)
#Acceleration using central difference
delta_x = 10 # Spacing (m)
v_for = 3.1 # v(x+delta_x) at x=50m
v_back = 2.4 # v(x-delta_x) at x=30m
acc = (v_for - v_back) / (2 * delta_x)
#Plot velocity vs. location
plt.figure(figsize=(8, 6))
plt.plot(x,v,'o-',linewidth=2,markersize=12)
plt.xlim(0, 63)
plt.ylim(1, 3.7)
plt.grid(True,linestyle='--',linewidth=1)
plt.xlabel('Location (m)',fontsize=14)
plt.ylabel('Velocity (m/s)', fontsize=14)
print(f'Calculated acceleration at \
x = 40 m: {acc:.4f} m/s^2')
plt.tight_layout(); plt.show()
#End of Python script
```

Python Output:

```
Calculated acceleration at x = 40 m: 0.0350 m/s^2.
This script also produces the plot shown in Figure 5.8b.
```

5.3.2 NUMERICAL INTEGRATION

Numerical integration is a technique used to estimate the area under a curve when analytical methods are impractical. It involves dividing a complex shape into smaller segments, such as rectangles or trapezoids (Figure 5.9), to approximate the total area. In engineering, many problems involve irregular data or functions without simple mathematical expressions. Numerical integration provides an efficient way to handle such cases, enabling accurate estimations where exact solutions are unattainable.

Numerical integration is an essential tool in engineering, enabling the resolution of complex real-world problems with precision and reliability. Its applications include calculating beam deflections under variable loads, estimating excavation or fill volumes for road and railway projects, and determining total forces resulting from non-uniform loads on structures.

5.3.2.1 Performing Numerical Integration

Numerical integration involves breaking a complex area under a curve into smaller, simpler shapes, like rectangles or trapezoids (Figure 5.9), and summing their areas to approximate the total.

For a curve described by $y = f(x)$, numerical integration can be performed by following the two steps below using rectangle or trapezoidal approximations:

- **Step 1**: Divide the interval $[a, b]$ into n equal parts of width $\Delta x = \frac{b-a}{n}$.
- **Step 2**: Approximate the area under the curve. This can be done using one of the following three methods, depending on the chosen approach:

1. **Left-hand rule:** The height of each rectangle is taken at the left endpoint of each subinterval. The approximate area is:

$$\text{Area}_\text{l} \approx \sum_{i=0}^{n-1} f(x_i)\Delta x \tag{5.43}$$

   This method may **underestimate** the area for increasing functions or **overestimate** for decreasing functions (see Figure 5.9a). Note that in Equation 5.43, the index of summation starts from $i = 0$ and goes up to $n - 1$.

2. **Right-hand rule:** The height of each rectangle is taken at the right endpoint of each subinterval. The approximate area is:

$$\text{Area}_\text{r} \approx \sum_{i=1}^{n} f(x_i)\Delta x \tag{5.44}$$

   This method behaves oppositely to the left-hand rule: it **overestimates** for increasing functions and **underestimates** for decreasing functions (see Figure 5.9b). In Equation 5.44, the summation index starts from $i = 1$ and goes up to n.

3. **Trapezoidal rule:** The trapezoidal rule averages the left- and right-hand rules by approximating the area under the curve with trapezoids instead of rectangles. The formula is:

$$\text{Area}_\text{trap} \approx \frac{\Delta x}{2}\left[f(x_0)+2f(x_1)+2f(x_2)+\cdots+2f(x_{n-1})+f(x_n)\right] \tag{5.45}$$

   This method is often more accurate as it accounts for the slope of the curve between subintervals.

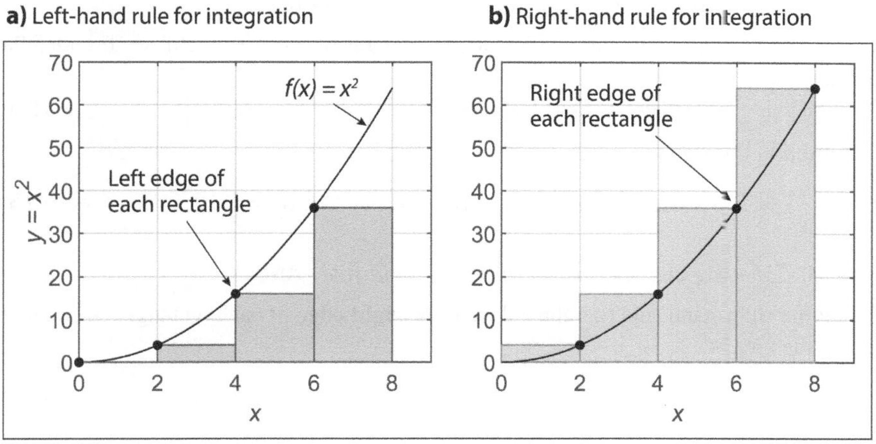

FIGURE 5.9 **a)** Numerical integration of $y = x^2$ using left-hand rule and considering $\Delta x = 2$. **b)** Numerical integration of $y = x^2$ using right-hand rule and considering $\Delta x = 2$.

By increasing the number of subintervals (n), all these methods yield progressively better approximations, converging toward the true area under the curve.

Example: Numerically integrate the following function in the interval $[0, 8]$ by estimating the area under the curve using four rectangles (Figure 5.9):

$$y = x^2 \tag{5.46}$$

Solution:

Step 1: Divide the interval, and decide Δx

The interval $[0, 8]$ is divided into 4 equal parts (Figure 5.9), so:

$$\Delta x = \frac{8 - 0}{4} = 2 \tag{5.47}$$

The points are: $x_0 = 0$, $x_1 = 2$, $x_2 = 4$, $x_3 = 6$, $x_4 = 8$.

Step 2: Evaluate the function

The values of the function $y = x^2$ at these points are:

$$f(0) = 0^2 = 0, \tag{5.48}$$

$$f(2) = 2^2 = 4, \tag{5.49}$$

$$f(4) = 4^2 = 16, \tag{5.50}$$

$$f(6) = 6^2 = 36, \tag{5.51}$$

$$f(8) = 8^2 = 64. \tag{5.52}$$

Step 3: Calculate the area using the left-hand rule ($Area_l$)

Using the left-hand rule (heights taken at the **left edge** of each rectangle- see Figure 5.9a), the area is:

$$\text{Area}_\text{l} \approx \Delta x\big[f(x_0) + f(x_1) + f(x_2) + f(x_3)\big] \tag{5.53}$$

Substituting the values:

$$\text{Area}_\text{l} \approx 2\big[0 + 4 + 16 + 36\big] = 2 \times 56 = 112. \tag{5.54}$$

Step 4: Calculate the area using the right-hand rule ($Area_r$)

Using the right-hand rule (heights taken at the **right edge** of each rectangle—see Figure 5.9b), the area is:

$$\text{Area}_\text{r} \approx \Delta x\big[f(x_1) + f(x_2) + f(x_3) + f(x_4)\big] \tag{5.55}$$

Substituting the values:

$$\text{Area}_\text{r} \approx 2\big[4 + 16 + 36 + 64\big] = 2 \times 120 = 240. \tag{5.56}$$

Step 5: Calculate the area using the trapezoidal rule ($Area_{trap}$)

The trapezoidal rule averages the left-hand and right-hand heights for each subinterval, effectively forming trapezoids. The formula is:

$$\text{Area}_\text{trap} \approx \frac{\Delta x}{2}\big[f(x_0) + 2f(x_1) + 2f(x_2) + 2f(x_3) + f(x_4)\big] \tag{5.57}$$

Substituting the values:

$$\text{Area}_\text{trap} \approx \frac{2}{2}\big[0+2(4)+2(16)+2(36)+64\big] = 1\big[0+8+32+72+64\big] = 176. \tag{5.58}$$

Conclusion

The estimated areas are as follows:

- Left-hand rule: $\text{Area}_\text{l} = 112$
- Right-hand rule: $\text{Area}_\text{r} = 240$
- Trapezoidal rule: $\text{Area}_\text{trap} = 176$

This illustrates how the choice of method influences the accuracy of the estimate, with the trapezoidal rule generally yielding a more precise result. The accuracy improves further as the number of rectangles increases, reducing the width of each interval Δx.

5.3.2.2 Applying MATLAB and Python for Numerical Integration

We can use MATLAB and Python to perform numerical integration using the functions presented in the information box below:

> In MATLAB and Python, numerical integration can be performed using the commands **`trapz`** and **`np.trapz`** (in Python with NumPy), as well as **`scipy.integrate.trapz`** (in Python with SciPy).

Both functions implement the trapezoidal rule, which approximates definite integrals by summing the areas of trapezoids formed under the curve. These tools are particularly efficient for numerical integration of discrete data points or when an analytical solution to the integral is not feasible. They are widely used in engineering, physics, and applied mathematics for approximating definite integrals with high accuracy, especially when the function is sampled at non-uniform intervals.

Example: Apply MATLAB and Python to numerically integrate the following function over the interval $[0, 8]$ using a step size of $\Delta x = 0.05$:

$$y = x^2 \tag{5.59}$$

Solution:

The calculations are implemented via scripts `CH05Mcode8.m` (in MATLAB, see Example 5.15) and `CH05Pcode8.py` (in Python, see Example 5.16). Both scripts calculate the integral by discretizing the domain, applying the trapezoidal rule, and summing up the contributions of all subintervals. The outputs of the MATLAB and Python codes reveal that the value of the integral is 170.67 (see the outputs of Examples 5.15 and 5.16). The reader is encouraged to rerun these scripts by applying varied values of the step size (Δx) to examine its impact on the results.

Example 5.15

MATLAB code. This MATLAB script, titled `CH05Mcode8.m`, computes the integral of the function $y = x^2$ over the interval $[0, 8]$, using a step size of $\Delta x = 0.05$.

MATLAB Input:

```
%This code estimates the integral
%of f(x)=x^2 at the interval [0,8]
clc; clear; close all;
x=0:0.05:8; % Interval [0, 8] divided
% into 160 subintervals (Delta-x=0.05)
y=x.^2; %Function values at points x
% Compute integral using 'trapz'
area = trapz(x, y);
% Display the result
disp(['Integral = ', num2str(area)]);
% End of M - file
```

MATLAB Output:

```
Integral = 170.67
```

Example 5.16

Python code. This Python script, named `CH05Pcode8.py`, computes the integral of the function $y = x^2$ over the interval $[0, 8]$, using a step size of $\Delta x = 0.05$.

Python Input:

```
#This code estimates integral
#of f(x)=x^2 at interval [0,8]
import numpy as np
# Define interval and step size
x = np.arange(0, 8.05, 0.05)
y = x**2 #Function values at points x
# Compute the integral
area = np.trapz(y, x)
# Display the result
print(f"Integral = {area}")
#End of Python script
```

Python Output:

```
Integral = 170.67000000000002
```

Engineering application: This example is intended to estimate flow rate (Q) in an irregular channel or river. In hydraulic engineering, calculating the flow rate in open channels and rivers with irregular cross-sections is a common task. Analytical methods may not always be feasible due to the variability in the depth of flow along the channel width (Figure 5.10b). Numerical integration techniques, such as the trapezoidal rule, provide an effective approach to solve these problems.

The flow rate (Q) is determined as the product of the flow velocity (v) and the cross-sectional area (A). The cross-sectional area is calculated by integrating the depth of the channel, $y(x)$, over its width (x).

Consider an open channel (e.g., Figure 5.10b) with a width of 10 m which has a depth profile measured at intervals of 2 m across the width, as shown in Table 5.5. The flow velocity is uniform at $v = 2$ m/s. We calculate the flow rate Q using the trapezoidal rule with the data given in Table 5.5.

Solution:

***Step 1**: Calculate the cross-sectional area, A, using the trapezoidal rule.*

The trapezoidal rule formula for numerical integration is given as:

$$A = \frac{\Delta x}{2}\left[y_0 + 2(y_1 + y_2 + \cdots + y_{n-1}) + y_n\right], \tag{5.60}$$

TABLE 5.5
Depth measurements across the channel width

x (m)	$y(x)$ (m)
0	0.5
2	0.7
4	1.2
6	1.5
8	1.1
10	0.8

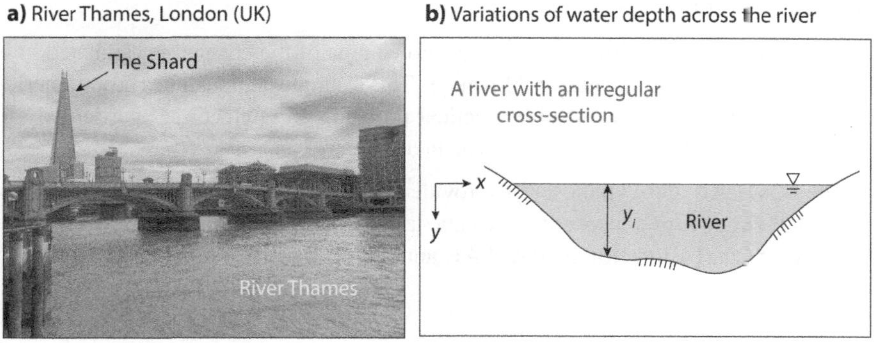

FIGURE 5.10 a) River Thames in London, UK. b) Sketch showing a river with an irregular cross section.

where:

- Δx is the uniform interval between points,
- $y_0, y_1, \ldots, y_n$ are the depth values at corresponding x-coordinates.

For this problem:

$$\Delta x = 2\,\text{m},$$

$$y_0 = 0.5,\ y_1 = 0.7,\ y_2 = 1.2,\ y_3 = 1.5,\ y_4 = 1.1,\ y_5 = 0.8.$$

Substitute the values into Equation 5.60:

$$A = \frac{2}{2}\big[0.5 + 2(0.7 + 1.2 + 1.5 + 1.1) + 0.8\big]. \tag{5.61}$$

Simplify the calculations:

$$A = 1\big[0.5 + 2(0.7 + 1.2 + 1.5 + 1.1) + 0.8\big], \tag{5.62}$$

$$A = 1\big[0.5 + 2(4.5) + 0.8\big], \tag{5.63}$$

$$A = 1\left[0.5 + 9 + 0.8\right], \tag{5.64}$$

$$A = 10.3\,\mathrm{m}^2 \tag{5.65}$$

The cross-sectional area is $A = 10.3\,\mathrm{m}^2$.

***Step 2**: Calculate the flow rate Q*

The flow rate Q is calculated as:

$$Q = v \cdot A \tag{5.66}$$

Substitute the values:

$$Q = 2 \times 10.3 = 20.6\,\mathrm{m}^3/\mathrm{s} \tag{5.67}$$

The flow rate is $Q = 20.6\,\mathrm{m}^3/\mathrm{s}$.

Using the trapezoidal rule, the cross-sectional area of the channel is estimated to be $10.3\,\mathrm{m}^2$, resulting in a flow rate of $20.6\,\mathrm{m}^3/\mathrm{s}$. This example demonstrates how numerical integration can be applied to solve practical problems in engineering, particularly for estimating flow rates in irregular open channels.

The MATLAB and Python scripts provided in Examples 5.17 (`CH05Mcode9.m`) and 5.18 (`CH05Pcode9.py`), respectively, can be used to solve the problem described above, where the `trapz()` function in MATLAB and the `np.trapz()` function in Python are applied.

Example 5.17

MATLAB code. This MATLAB script, named `CH05Mcode9.m`, computes the flow rate (Q) of a river channel with irregular cross section using numerical integration.

MATLAB Input:

```
%This code omputes the flow rate
% (Q) of an irregular river
%using numerical integration.
clc; clear; close all;
% Input data
x=[0,2,4,6,8,10]; % x (m)
y=[0.5,0.7,1.2,1.5,1.1,0.8];%Depths(m)
v = 2; % Flow velocity (m/s)
% Calculate cross-section area (A)
% using the trapezoidal rule
A = trapz(x, y);
% Calculate the flow rate
Q = v * A;
% Display results
fprintf('A = %.2f m^2\n',A);
```

```
fprintf('Q = %.2f m^3/s\n', Q);
% End of M - file
```

MATLAB Output:

```
A = 10.30 m^2
Q = 20.60 m^3/s
```

Example 5.18

Python code. This Python script, named CH05Pcode9.py, computes the flow rate (Q) of a river channel with irregular cross section using numerical integration.

Python Input:

```
#This code omputes the flow rate
# (Q) of an irregular river
#using numerical integration.
import numpy as np
# Input data
x=np.array([0,2,4,6,8,10]) # x(m)
y=np.array([0.5,0.7,1.2,1.5,1.1,0.8])
v = 2 # Flow velocity (m/s)
# Calculate cross-sectional area (A)
# using the trapezoidal rule
A = np.trapz(y, x)
# Calculate the flow rate
Q = v * A
# Display results
print(f"A = {A:.2f} m^2")
print(f"Q = {Q:.2f} m^3/s")
#End of Python script
```

Python Output:

```
A = 10.30 m^2
Q = 20.60 m^3/s
```

5.4 PROBLEMS

Problem 5.1: Manually apply the Bisection method to find the roots of $f(x) = x^2 + 3\sin(x) - 7$. Then, develop MATLAB and Python scripts, following the approach outlined in Examples 5.1 and 5.2, to solve the same problem.

Answer: The function has two roots; the first root is $x_1 = -2.8195$ and the second root is $x_2 = 2.1001$.

Problem 5.2: Apply the Newton-Raphson method to find the root of $f(x) = -x^3 + \cos^2(x) + 2.3$ both manually and using MATLAB and Python scripts, following the approach outlined in Examples 5.5 and 5.6.

Answer: The function has only one root, which is $x = 1.331$.

Problem 5.3: Use the built-in root-finding functions in MATLAB and Python to find the root of $f(x) = \frac{1}{\sqrt{2x}} - 2.5x$ by developing appropriate MATLAB and Python scripts.

Answer: The function has one root which is $x = 0.431$.

Problem 5.4: Table 5.6 provides the observed settlement (s, in mm) of a foundation under various applied forces (F, in kN). As the force increases, the settlement of the foundation increases linearly. Your task is to fit a straight-line equation of the form $s = aF + b$, where a and b are constants representing the slope and the intercept of the line, respectively.

i. Use the least squares method to determine the values of a and b manually.
ii. Write MATLAB or Python scripts to verify your results.
iii. Create a plot showing the measured data points and the fitted line on the same graph, similar to Figure 5.6b.
iv. Discuss how the fitted equation can be used to estimate settlement for unmeasured force values and how it may influence foundation design.

TABLE 5.6

Settlement of a foundation under various applied forces

Force, F (kN)	Settlement, s (mm)
0	1.5
50	4.2
100	6.8
150	9.1
200	11.5
250	14.0

Answer: The equation for the fitted line is $s = 0.0495F + 1.657$.

Problem 5.5: Table 5.7 provides data on the deflection of a beam (δ, in mm) at its mid-point under various applied forces (F, in kN). The deflection is known to follow a cubic relationship, and your task is to fit the data to the model $\delta = aF^3$, where a is a constant.

i. Determine the best-fit parameter a using nonlinear curve fitting in MATLAB or Python.
ii. Plot the observed data points and the fitted curve on the same graph.
iii. Use the fitted model to estimate the deflection at $F = 30$ kN.
iv. Explain how the parameter a relates to the beam's properties and how the model could assist in predicting deflections under varying loads.

TABLE 5.7

Deflection of a beam under various applied forces

Force, F (kN)	Deflection, δ (mm)
0	0.0
5	1.2
10	9.8
15	33.7
20	80.3
25	156.7

Answer: The equation for the fitted curve is $\delta = 0.010028F^3$. At $F = 30$ kN, deflection is $\delta = 270.76$ mm.

Problem 5.6: The relationship between traffic flow (q, vehicles/hour) and traffic density (k, vehicles/km) is often analyzed to optimize roadway performance. Table 5.8 shows traffic data collected from a highway segment.

TABLE 5.8

Traffic density vs. traffic flow data

Traffic density, k (vehicles/km)	Traffic flow, q (vehicles/hour)
10	200
20	380
30	520
40	600
50	580

i. Assume that the relationship between traffic flow (q) and density (k) can be modeled by a quadratic equation of the form:

$$q = ak^2 + bk + c$$

where a, b, and c are constants.

ii. Use the **least squares method** to find the values of a, b, and c that best fit the given data.
iii. Write the equation of the best-fit curve.
iv. Plot the given data points and the best-fit curve on the same graph. Clearly label the axes and include a legend.
v. Using your fitted equation, predict the traffic flow (q) when the density is $k = 35$ vehicles/km.

Answer: The fitted equation is: $q = -0.3286k^2 + 29.5143k - 68.0000$. The predicted traffic flow for k = 35 vehicles/km is: q = 562.50 vehicles/hour.

Problem 5.7: The elevation (y, in meters) of a road is measured at specific distances (x, in meters), as shown in Table 5.9. The slope of the road is the first derivative of the elevation profile, which represents the rate of change of elevation with respect to distance. Using numerical differentiation, calculate the slope ($\frac{dy}{dx}$) at $x = 500$ m, $x = 1500$ m, and $x = 2500$ m using the forward, central, and backward difference methods, respectively.

TABLE 5.9
Elevation of the road at different distances

Distance, x (m)	**Elevation, y (m)**
0	45
500	50
1000	60
1500	72
2000	90
2500	115
3000	140

Answer:

i. Slope (or first derivative) at $x = 500$ m using forward difference method: 0.02 m/m.
ii. Slope at $x = 1500$ m using central difference method: 0.03 m/m.
iii. Slope at $x = 2500$ m using backward difference method: 0.05 m/m.

Problem 5.8: Referring to the solved problem in the chapter on numerical integration of the function $y = x^2$ over the interval $[0, 8]$ (Figure 5.9), repeat the integration using different interval widths (Δx) such as $\Delta x = 1.0$, 0.5, and 0.25 using the trapezoidal rule. Critically analyze and discuss the results obtained.

Answer: The results are 172.000 (for $\Delta x = 1.0$), 171.000 (for $\Delta x = 0.5$), and 170.750 (for $\Delta x = 0.25$).

Problem 5.9: Apply the built-in numerical integration commands in MATLAB and Python to compute the integral of the function $f(x) = e^{-x^2} - 2.5x$ over the domain $[0.5, 2]$. Perform the integration using $n = 5$, $n = 10$, $n = 50$, and $n = 100$ intervals to observe the impact of the number of intervals on the accuracy of the result.

Answer: The results are -4.261355 (for $n = 5$), -4.265373 (for $n = 10$), -4.266647 (for $n = 50$), and -4.266686 (for $n = 100$).

Problem 5.10: A dam reservoir experiences inflow discharge from a catchment area during a torrential rainfall event lasting 8 hours. The input discharge data is given at hourly intervals, as shown in Table 5.10. The total inflow volume is the integral of the discharge time history, which is essentially the area beneath the discharge-time plot. First, calculate the total inflow volume into the reservoir manually using the trapezoidal rule. Then, apply MATLAB and Python built-in functions to compute the same total inflow volume using numerical integration. Finally, compare the results from the manual calculation with those obtained from MATLAB and Python, and discuss any variations.

TABLE 5.10
Inflow discharge data over the 8-hour rainfall period

Time (hours)	Inflow Discharge (m^3/s)
0	5.2
1	7.8
2	10.1
3	15.6
4	18.9
5	21.4
6	22.2
7	24.5
8	19.8

Answer: The total inflow volume during the flood event is 478,800 m^3.

REFERENCES

[1] Nicholas J Garber and Lester A Hoel. *Traffic & highway engineering*. 2009.

[2] Mohammad Heidarzadeh and Siamak Feizi. "A cascading risk model for the failure of the concrete spillway of the Toddbrook dam, England during the August 2019 flooding." In: *International Journal of Disaster Risk Reduction* 80 (2022), p. 103214.

[3] Mohammad Heidarzadeh, Ali A Mirghasemi, and Habib Niroomand. "Construction of relief wells under artesian flow conditions at dam toes: engineering experiences from Karkheh earth dam, Iran." In: *International Journal of Civil Engineering* 13(1) (2015). `https://doi.org/10.22068/IJCE.13.1.73`, p. 73–80.

[4] Mohammad Heidarzadeh et al. "Construction and performance of the Karkheh dam complementary cut-off wall: an innovative engineering solution." In: *International Journal of Civil Engineering* 17 (2019), p. 859–869.

[5] Mohammad Heidarzadeh et al. "Field surveys of tsunami runup and damage following the January 2024 Mw 7.5 Noto (Japan sea) tsunamigenic earthquake." In: *Ocean Engineering* 307 (2024), p. 118140.

[6] Erwin Kreyszig. *Advanced engineering mathematics 9th edition with Wiley plus set*. Vol. 334. John Wiley & Sons US, 2007.

[7] Adolf Darlington May. *Traffic flow fundamentals*. 1990.

[8] Stephen P Timoshenko. "Theory of suspension bridges." In: *Journal of the Franklin Institute* 235(3) (1943), p. 213–238.

6 Fourier Series and Transforms

Fourier series and transforms are fundamental mathematical tools used to analyze and represent periodic and non-periodic signals. In engineering, these techniques are indispensable for addressing problems related to structural vibrations, traffic flow analysis, wave propagation in materials, and ocean wave run-up in coastal areas [2, 4].

Fourier series allows complex periodic functions to be expressed as sums of simple sine and cosine waves, making it invaluable for studying structural dynamics and earthquake engineering (Figure 6.1a). The Fourier transform extends this concept to non-periodic functions, providing insights into the frequency components of signals, which is crucial in applications such as signal processing for structural health monitoring. Mastering these methods enables engineers to design more resilient structures, optimize material properties, and effectively analyze dynamic loads, such as the impacts of oceanic waves on coastal infrastructures [9] (Figure 6.1b).

This chapter provides an introduction to Fourier analysis, a central mathematical approach for studying signals and wave phenomena. The discussion begins with Fourier series, decomposing periodic functions into sums of `sine` and `cosine` components to reveal their frequency content. It explores the concepts of orthogonality and series convergence conditions in detail. Subsequently, the chapter extends these principles to non-periodic functions using the continuous Fourier transform and its inverse, facilitating the transition between the time and frequency domains. Additionally, essential signal-processing techniques, such as convolution and correlation, are introduced, highlighting their roles in analyzing interactions between signals [5–8].

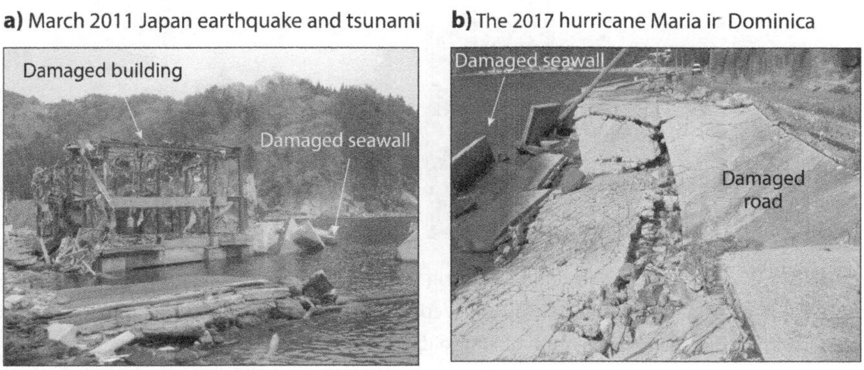

FIGURE 6.1 Examples of structural damage due to dynamic loading. **a)** Aftermath of the March 2011 earthquake and tsunami in Japan [10]. **b)** Damage resulting from Hurricane Maria (2017) in Dominica [3].

DOI: 10.1201/9781003511854-6

6.1 FOURIER SERIES: ORTHOGONAL FUNCTIONS AND CONVERGENCE

A function $f(x)$ is said to be 2π-periodic if it satisfies $f(x+2\pi) = f(x)$ for all values of x. Joseph Fourier [1] demonstrated that any sufficiently well-behaved 2π-periodic function can be represented as a sum of `sines` and `cosines` with different frequencies, expressed as:

$$f(x) = \frac{a_0}{2} + \sum_{n=1}^{\infty} \left[a_n \cos(nx) + b_n \sin(nx)\right] \tag{6.1}$$

where the constant term a_0 and the Fourier coefficients a_n and b_n are defined by the following integrals:

$$\begin{aligned} a_0 &= \frac{1}{\pi}\int_{-\pi}^{\pi} f(x)\,dx \\ a_n &= \frac{1}{\pi}\int_{-\pi}^{\pi} f(x)\cos(nx)\,dx \\ b_n &= \frac{1}{\pi}\int_{-\pi}^{\pi} f(x)\sin(nx)\,dx \end{aligned} \tag{6.2}$$

Fourier series rely fundamentally on the orthogonality of `sine` and `cosine` functions over intervals of length 2π. Orthogonality means that each frequency component can be isolated uniquely by integrating the function against corresponding `sine` or `cosine` waves. Figure 6.2 illustrates this concept by showing how a complex audio signal can be decomposed into simpler sinusoidal components.

A set of functions $\{f_n(x)\}$ is orthogonal on an interval $[a, b]$ if:

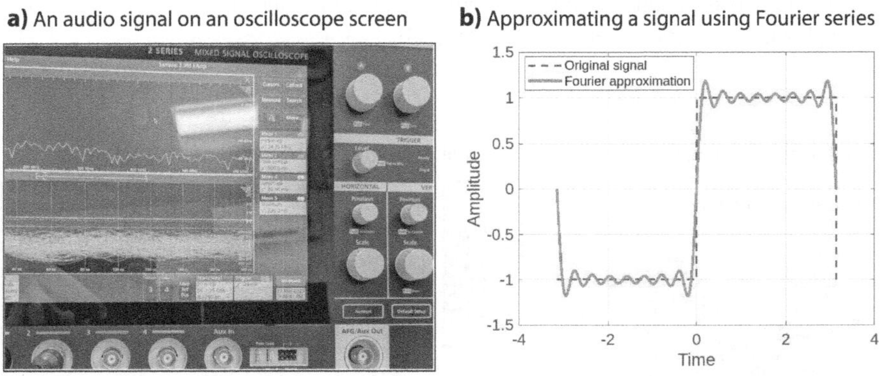

FIGURE 6.2 Decomposing an audio waveform with Fourier series. **a)** A photo of an audio signal displayed on an oscilloscope screen. **b)** A conceptual plot illustrating the approximation of a step signal using Fourier analysis as a sum of sinusoidal components.

$$\int_a^b f_m(x) f_n(x)\,dx = \begin{cases} 0, & m \neq n \\ \text{non-zero}, & m = n \end{cases} \tag{6.3}$$

The set is orthonormal if the integral equals 1 when $m = n$, and 0 when $m \neq n$. In particular, the `sine` and `cosine` functions satisfy these orthogonality conditions over the interval $[-\pi, \pi]$, as demonstrated below:

$$\int_{-\pi}^{\pi} \cos(mx)\,\cos(nx)\,\mathrm{d}x = \begin{cases} 0, & m \neq n \\ \pi, & m = n \neq 0 \\ 2\pi, & m = n = 0 \end{cases} \tag{6.4}$$

$$\int_{-\pi}^{\pi} \sin(mx)\,\sin(nx)\,\mathrm{d}x = \begin{cases} 0, & m \neq n \\ \pi, & m = n \end{cases} \tag{6.5}$$

$$\int_{-\pi}^{\pi} \sin(mx)\,\cos(nx)\,\mathrm{d}x = 0 \quad \text{for all integers } m \text{ and } n \tag{6.6}$$

Fourier series are also expressed in complex exponential form using Euler's identity ($\mathrm{e}^{ix} = \cos x + i \sin x$) as follows:

$$f(x) = \sum_{n=-\infty}^{\infty} c_n \mathrm{e}^{inx} \tag{6.7}$$

where c_n are complex Fourier coefficients.

When we say a Fourier series converges to a function $f(x)$, it means that as we include more and more terms in the Fourier series (i.e., as $N \to \infty$), the partial sums get closer and closer to the actual values of $f(x)$. There are different ways to measure how "close" two functions are, and thus there are different modes of convergence. Here are three common types:

- **Pointwise convergence:** Partial sums approach the function at individual points.
- **Uniform convergence:** The approximation uniformly approaches the function across the entire interval.
- **Mean square ($\mathbf{L^2}$) convergence:** Average squared differences become arbitrarily small as the number of terms increases:

$$\int_{-\pi}^{\pi} \left|f(x) - S_N(x)\right|^2 \mathrm{d}x \quad \longrightarrow \quad 0 \quad \text{as} \quad N \to \infty \tag{6.8}$$

  where $S_N(x)$ denotes the N-th partial sum. At a discontinuity, the series converges to the midpoint of the jump.

By denoting the N-th partial sum of the Fourier series of f by $S_N(f; x)$, we have:

$$S_N(f;x) = \frac{a_0}{2} + \sum_{n=1}^{N} \Big[a_n \cos(nx) + b_n \sin(nx)\Big] \tag{6.9}$$

We say the series converges pointwise to f at a point x if:

$$\lim_{N\to\infty} S_N(f;x) = \begin{cases} \frac{f(x^+)+f(x^-)}{2}, & \text{if } f \text{ is discontinuous at } x \\ f(x), & \text{if } f \text{ is continuous at } x \end{cases} \tag{6.10}$$

where $f(x^+)$ and $f(x^-)$ refer to the right-hand and left-hand limits of f at x, respectively.

Example: Calculate the Fourier series of the following function, and plot the original function alongside its approximations using 1, 3, 7, and 50 terms of the Fourier series.

$$f(x) = \begin{cases} 1, & 0 \le x < \pi \\ -1, & -\pi \le x < 0 \end{cases} \tag{6.11}$$

which is a periodic function with period 2π.

Solution:

The goal is to derive an equation of the following form for the function $f(x)$:

$$f(x) = \frac{a_0}{2} + \sum_{n=1}^{\infty} [a_n \cos(nx) + b_n \sin(nx)] \tag{6.12}$$

Next, we calculate the Fourier coefficients a_0, a_n, and b_n using the following procedures.

Computing a_0:

$$a_0 = \frac{1}{\pi} \int_{-\pi}^{\pi} f(x)\, dx \tag{6.13}$$

$$= \frac{1}{\pi} \left[\int_0^{\pi} 1\, dx + \int_{-\pi}^{0} (-1)\, dx \right] \tag{6.14}$$

$$= \frac{1}{\pi} [\pi + (-\pi)] = 0 \tag{6.15}$$

Computing a_n:

$$a_n = \frac{1}{\pi} \int_{-\pi}^{\pi} f(x) \cos(nx)\, dx \tag{6.16}$$

Since $f(x)$ is an odd function, the integral of $f(x)\cos(nx)$ over symmetric limits is zero, implying $a_n = 0$.

Computing b_n:

$$b_n = \frac{1}{\pi} \int_{-\pi}^{\pi} f(x) \sin(nx)\, dx \tag{6.17}$$

$$= \frac{1}{\pi}\left[\int_0^{\pi} \sin(nx)\,dx + \int_{-\pi}^{0} (-\sin(nx))\,dx\right] \tag{6.18}$$

We know that:

$$\int \sin(nx)\,dx = -\frac{\cos(nx)}{n} \tag{6.19}$$

Since the two integrals in Equation 6.18 are equal in magnitude, we can simplify the expression by computing only one of them and multiplying by 2. Therefore, we have:

$$b_n = \frac{2}{\pi}\int_0^{\pi} \sin(nx)\,dx = \frac{2}{\pi}\left[-\frac{\cos(n\pi)}{n} + \frac{\cos(0)}{n}\right] \tag{6.20}$$

$$= \frac{2}{\pi}\left[-\frac{(-1)^n}{n} + \frac{1}{n}\right] \tag{6.21}$$

$$= \frac{2}{\pi}\left[\frac{1-(-1)^n}{n}\right] \tag{6.22}$$

For odd n, $1-(-1)^n = 2$, and for even n, $1-(-1)^n = 0$, so:

$$b_n = \begin{cases} \frac{4}{n\pi}, & n \text{ odd} \\ 0, & n \text{ even} \end{cases} \tag{6.23}$$

Thus, the Fourier series representation of the function is:

$$f(x) = \begin{cases} 1, & 0 \le x < \pi \\ -1, & -\pi \le x < 0 \end{cases} = \sum_{\substack{n=1 \\ n \text{ odd}}}^{\infty} \frac{4}{n\pi}\sin(nx) \tag{6.24}$$

To illustrate the convergence of the derived Fourier series, Figure 6.3 shows the original function $f(x)$ alongside its approximations using 1, 3, 7, and 50 terms from the series. For example, in the case of $N = 3$, only the odd values $n = 1$ and $n = 3$ are included in the summation, since the even terms vanish. This results in the following two-term approximation:

$$f_3(x) = \frac{4}{\pi}\left[\sin(x) + \frac{1}{3}\sin(3x)\right] \tag{6.25}$$

Similarly, for $N = 7$, the approximation includes the odd harmonics $n = 1, 3, 5, 7$, resulting in a four-term Fourier series:

$$f_7(x) = \frac{4}{\pi}\left[\sin(x) + \frac{1}{3}\sin(3x) + \frac{1}{5}\sin(5x) + \frac{1}{7}\sin(7x)\right] \tag{6.26}$$

It can be observed from Figure 6.3 that the approximation is quite poor when only one term is used ($N = 1$), but it improves as more Fourier terms are included in the approximation of $f(x)$. With just seven terms ($N = 7$), the approximation becomes reasonably accurate. A high level of accuracy is achieved when 50 terms ($N = 50$) are used.

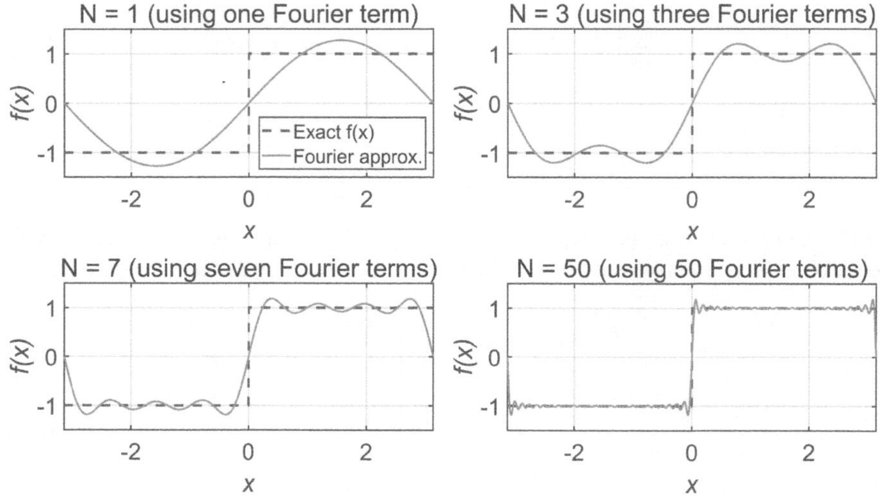

FIGURE 6.3 The original function $f(x)$, defined as $f(x) = \begin{cases} 1, & 0 \le x < \pi \\ -1, & -\pi \le x < 0 \end{cases}$, is shown alongside its Fourier series approximation. Each subplot overlays the exact function (blue dashed line) with the n-term partial sum (pink line) for $n = 1, 3, 7$, and 50.

Example: Consider the following function defined on the interval $[-\pi, \pi]$. Derive its Fourier series representation:

$$g(x) = x, \qquad -\pi \le x \le \pi \tag{6.27}$$

which is periodic with period 2π.

Solution:

The goal is to derive an equation of the following form for the function $f(x)$:

$$f(x) = \frac{a_0}{2} + \sum_{n=1}^{\infty} [a_n \cos(nx) + b_n \sin(nx)] \tag{6.28}$$

Next, we calculate the Fourier coefficients a_0, a_n, and b_n using the following procedures.

Computing a_0:

$$a_0 = \frac{1}{\pi} \int_{-\pi}^{\pi} g(x)\, dx \tag{6.29}$$

Since the integrand is odd, it integrates to zero:

$$a_0 = 0 \tag{6.30}$$

Computing a_n:

$$a_n = \frac{1}{\pi} \int_{-\pi}^{\pi} g(x) \cos(nx)\, dx \tag{6.31}$$

Since x is an odd function and $\cos(nx)$ is even, the integral evaluates to zero:

$$a_n = 0 \tag{6.32}$$

Computing b_n:

$$b_n = \frac{1}{\pi} \int_{-\pi}^{\pi} g(x) \sin(nx)\, dx \tag{6.33}$$

Since $g(x)$ is an odd function and $\sin(nx)$ is also odd, their product is even. Therefore, we can simplify the integral by evaluating it over the interval $[0, \pi]$ and then doubling the result:

$$b_n = \frac{2}{\pi} \int_{0}^{\pi} x \sin(nx)\, dx \tag{6.34}$$

Using integration by parts, where:

$$\int x \sin(nx)\, dx = -\frac{x \cos(nx)}{n} + \frac{\sin(nx)}{n^2} \tag{6.35}$$

we compute:

$$b_n = \frac{2}{\pi} \left[-\frac{\pi \cos(n\pi)}{n} + \frac{\sin(n\pi)}{n^2} + \frac{0}{n} - \frac{\sin(0)}{n^2} \right]. \tag{6.36}$$

Since $\sin(n\pi) = 0$, this simplifies to:

$$b_n = \frac{2}{\pi} \left[-\frac{\pi \cos(n\pi)}{n} \right] \tag{6.37}$$

$$b_n = -\frac{2\cos(n\pi)}{n} \tag{6.38}$$

Since $\cos(n\pi) = (-1)^n$, we get:

$$b_n = -\frac{2(-1)^n}{n} \tag{6.39}$$

Thus, the Fourier series for $g(x)$ is:

$$g(x) = x = \sum_{n=1}^{\infty} -\frac{2(-1)^n}{n} \sin(nx) \tag{6.40}$$

By the Dirichlet conditions, the Fourier series converges to $g(x)$ where $g(x)$ is continuous and to the average of the left and right limits at discontinuities.

Figure 6.4 presents the original function alongside its approximation using Fourier series with 1, 5, 10, and 50 terms. As shown in Figure 6.4, the convergence is non-uniform near the discontinuity. Similar to the previous example, the approximation is poor when only one term ($n = 1$) is used, but it improves significantly with 5 terms ($n = 5$). An excellent approximation is achieved when $n = 50$. When we refer to the approximation of $g(x)$ using five terms ($n = 5$), this corresponds to the following equation:

$$\begin{aligned} x = &-\frac{2(-1)^1}{1}\sin(x) - \frac{2(-1)^2}{2}\sin(2x) \\ &-\frac{2(-1)^3}{3}\sin(3x) - \frac{2(-1)^4}{4}\sin(4x) - \frac{2(-1)^5}{5}\sin(5x) \end{aligned} \tag{6.41}$$

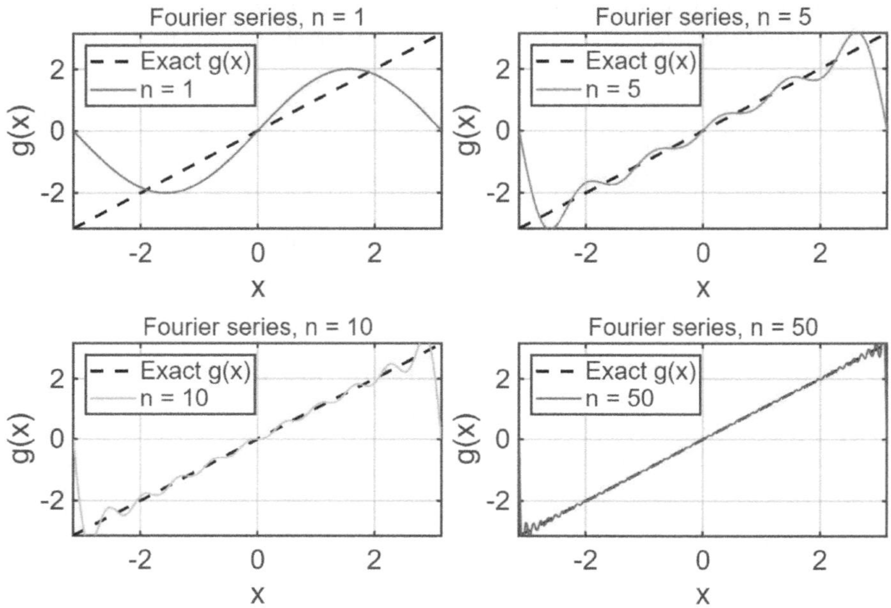

FIGURE 6.4 Fourier series approximation of an odd sawtooth function: $g(x) = x$, on $[-\pi, \pi]$. Each subplot overlays the exact function (black dashed lines) with the n-term partial sum of the Fourier series (colored solid curves) for n=1, 5, 10, and 50.

In this section, we apply both MATLAB and Python to solve the given example. Example 6.1 demonstrates the use of MATLAB to compute and simplify the first three terms of the Fourier series for $f(x) = x$ using the script `CH06Mcode1.m`. Example 6.2 replicates this calculation in Python with `Sympy`, using the script `CH06Pcode1.py` and retaining symbolic coefficients for clarity. While MATLAB's `simplify` function typically attempts to express floating-point quantities as ratios of large integers, `Sympy` in Python preserves π symbolically and directly cancels it with the $\frac{1}{\pi}$ in the definition of b_n, resulting in clean, rational coefficients.

The MATLAB script `CH06Mcode1.m` in Example 6.1 symbolically computes the first few terms of the Fourier series for a given function using the following key symbolic commands:

- `syms x n real`: Declares symbolic variables x and n, representing the continuous variable and the Fourier index, respectively.

- `int( ... )`: Computes definite integrals symbolically to calculate the Fourier coefficients.
- `subs( ... )`: Substitutes specific integer values for n into the general expressions for the coefficients.
- `simplify( ... )`: Simplifies symbolic expressions to make the result more interpretable.
- A `for` loop is used to build the partial Fourier series sum up to N terms, using both cosine and sine components.

The Python script in Example 6.2 uses the `SymPy` library to compute the Fourier series of a function symbolically. The main commands and their purposes are as follows:

- `symbols('x n', real=True)`: Declares x and n as real symbolic variables used in the definition of the function and its Fourier coefficients.
- `integrate( ... )`: Computes definite integrals symbolically, which are required to find the Fourier coefficients.
- `subs( ... )`: Substitutes specific values of n into the general expressions for a_n and b_n.
- `simplify( ... )`: Simplifies the symbolic expressions for clearer output.
- A `for` loop iterates over $k = 1$ to N, constructing the partial sum of the Fourier series by adding the cosine and sine terms.

Example 6.1

MATLAB code. This MATLAB script, named `CH06Mcode1.m`, computes the first three terms of the Fourier series for $f(x) = x$ in domain $[-\pi, \pi]$. This MATLAB script, along with all other scripts presented in this book, is available at: `https://github.com/prog871/CRC-book2025`

MATLAB Input:

```
syms x n real
f = x; % Define the function
%
% Compute a0
a0 = (1/pi) * int(f, x, -pi, pi);
%
% Define a_n and b_n as symbolic expressions
a_n = (1/pi) * int(f .* cos(n*x), x, -pi, pi);
b_n = (1/pi) * int(f .* sin(n*x), x, -pi, pi);
%
% Build the partial sum (N=3)
N = 3;
S = a0/2; % Start with a0/2
```

```
%
for k = 1:N
   ak = simplify(subs(a_n, n, k));
   bk = simplify(subs(b_n, n, k));
   S = S + ak*cos(k*x) + bk*sin(k*x);
end
%
disp(['The ', num2str(N), ...
'-term Fourier series partial sum is:']);
S_simplified = simplify(S);
disp(S_simplified);
```

MATLAB Output:

```
The 3-term Fourier series partial sum is:
(1911387046407553*pi*(2*sin(3*x) - 3*sin(2*x) + 6*sin(x)))
    /18014398509481984
```

Example 6.2

Python code. This Python code, named `CH06Pcode1.py`, computes the first three terms of the Fourier series for $f(x) = x$ on interval $[-\pi, \pi]$ using `sympy`.

Python Input:

```
from sympy import symbols, sin, cos, pi, integrate, simplify
#
# Define symbols
x, n = symbols('x n', real=True)
#
# Define the function
f = x
#
# 3. Compute a0
a0 = (1/pi) * integrate(f, (x, -pi, pi))
#
# 4. Define the Fourier coefficients
a_n = (1/pi) * integrate(f * cos(n*x), (x, -pi, pi))
b_n = (1/pi) * integrate(f * sin(n*x), (x, -pi, pi))
#
# 5. Build the N-term partial sum
N = 3
```

```
S = simplify(a0/2) # start with the constant term
for k in range(1, N+1):
    ak = simplify(a_n.subs(n, k))
    bk = simplify(b_n.subs(n, k))
    S += ak * cos(k*x) + bk * sin(k*x)
#
print(f"The {N}-term Fourier series partial sum is:")
print(simplify(S))
```

Python Output:

```
The 3-term Fourier series partial sum is:
2*sin(x) - sin(2*x) + 2*sin(3*x)/3
```

Example: Derive the Fourier series for the following function on the interval $[-\pi, \pi]$ using MATLAB and Python:

$$f(x) = x^2 \tag{6.42}$$

Solution:

This time, we directly use MATLAB and Python to address this problem. Example 6.3 applies MATLAB via the script `CH06Mcode2.m` to demonstrate that all `sine` coefficients vanish due to the even symmetry of x^2. Example 6.4 performs the same computation in Python using the script `CH06Pcode2.py`, producing the same result and neatly identifying the constant term as $\pi^2/3$.

Example 6.3

MATLAB code. This MATLAB script, named `CH06Mcode2.m`, computes the first three terms of the Fourier series for $f(x) = x^2$ on domain $[-\pi, \pi]$.

MATLAB Input:

```
% Fourier series of x^2 on [-pi,pi], first 3 terms
syms x
%
% Define the function
f = x^2;
%
% Compute the a0 coefficient
a0 = (1/pi)*int(f, x, -pi, pi); % a0 = 2*pi^2/3
%
% Number of harmonics
```

```
N = 3;
%
% Preallocate arrays for a_n, b_n
an = sym(zeros(1,N));
bn = sym(zeros(1,N));
%
% Compute a_n and b_n for n=1..3
for n = 1:N
   an(n) = (1/pi)*int( f*cos(n*x), x, -pi, pi );
   bn(n) = (1/pi)*int( f*sin(n*x), x, -pi, pi );
end
%
% Simplify coefficients
a0 = simplify(a0);
an = simplify(an);
bn = simplify(bn);
%
% Build the 3-term Fourier series
f3 = a0/2;
for n = 1:N
   f3 = f3 + an(n)*cos(n*x) + bn(n)*sin(n*x);
end
f3 = simplify(f3);
%
% Show the approximation
disp('3-term Fourier approximation:');
disp(f3)
```

MATLAB Output:

```
3-term Fourier approximation:
-(1911387046407553*pi*(24*cos(x) - 18*cos(x)^2 + 16*cos(x)^3 -
    3*pi^2 + 9))/54043195528445952
```

Example 6.4

Python code. This Python code, named CH06Pcode2.py, computes the first three terms of the Fourier series for $f(x) = x^2$ on interval $[-\pi, \pi]$.

Python Input:

```
from sympy import symbols, sin, cos, pi, integrate, simplify
```

```
#
# Define symbols
x, n = symbols('x n', real=True, integer=True)
#
# Define the function
f = x**2
#
# Compute a0
a0 = (1/pi) * integrate(f, (x, -pi, pi))
#
# Define a_n and b_n
a_n = (1/pi) * integrate(f * cos(n*x), (x, -pi, pi))
b_n = (1/pi) * integrate(f * sin(n*x), (x, -pi, pi))
#
# Build the partial sum (N = 3)
N = 3
S = a0/2
for k in range(1, N+1):
    ak = simplify(a_n.subs(n, k))
    # this will turn out to be 0 every time
    bk = simplify(b_n.subs(n, k))
    S = S + ak*cos(k*x) + bk*sin(k*x)
print(f"{N}-term Fourier series partial sum =", simplify(S))
```

Python Output:

```
3-term Fourier series partial sum = -4*cos(x) + cos(2*x) - 4*
    cos(3*x)/9 + pi**2/3
```

6.2 CONTINUOUS FOURIER TRANSFORM AND INVERSE TRANSFORM

In practical engineering applications, many signals such as sound waves, electromagnetic radiation, or mechanical vibrations vary continuously over time. Analyzing such signals directly in the time domain can be challenging, especially when multiple frequencies are present. By transforming these signals into the frequency domain using the Fourier series or Fourier transform, we can more easily identify the constituent frequencies and their respective amplitudes. In this context, t typically denotes time, while ω (the Greek letter omega) represents angular frequency in radians per second.

The Fourier transform extends the idea of the Fourier series to non-periodic signals, making it an essential tool in a wide range of applications, including radar systems, noise cancellation, and real-time signal processing. By examining a signal's frequency content, engineers can design filters and detect meaningful patterns in complex environments—for example, within the aviation industry and airport systems (see Figure 6.5).

Consider a recording of a person speaking. In the time domain, this signal corresponds to variations in air pressure over time. However, the Fourier transform of this signal reveals its frequency-domain structure, highlighting the pitch (fundamental frequency) and harmonics (overtones). Such insights are valuable for applications like speech recognition, noise reduction, auto-tuning, and music analysis.

Another key example arises in communications. Wireless transmission involves modulating signals onto carrier frequencies. Using Fourier analysis, we can evaluate the signal's spectrum (i.e., how its energy is distributed across different frequencies). This is critical for understanding bandwidth, optimizing channel capacity, and designing efficient communication systems.

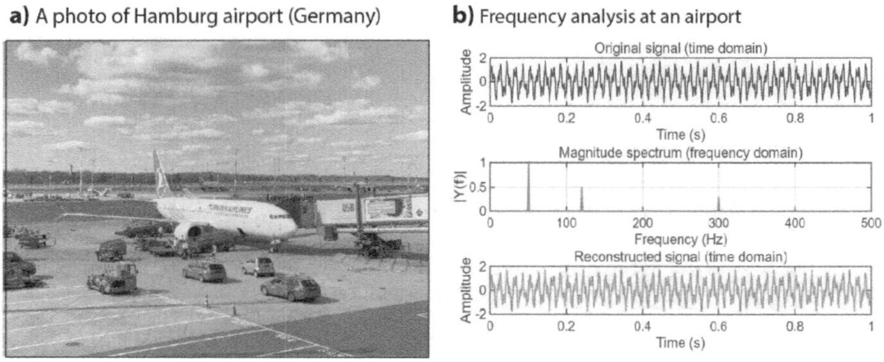

FIGURE 6.5 Frequency analysis applied in airports for aviation control purposes. **a)** A photograph of Hamburg Airport in Germany, where multiple overlapping sounds—such as engine noise and public announcements—are present. **b)** A diagram showing how the continuous Fourier transform represents these signals in the frequency domain and the inverse transform reconstructs them in time.

A commonly used definition of the continuous Fourier transform of a function $f(x)$ is given below, where the transform is denoted by $\hat{F}(\omega)$:

$$\hat{F}(\omega) = \mathcal{F}\{f(x)\} = \int_{-\infty}^{\infty} f(x)\, \mathrm{e}^{-i\omega x}\, dx \tag{6.43}$$

where x is typically time or space (depending on context), ω is the angular frequency, and the exponential $\mathrm{e}^{-i\omega x}$ represents a pure complex sinusoid. Once we have obtained the frequency-domain representation $\hat{F}(\omega)$, we can recover our original function $f(x)$ using the Inverse Fourier Transform as follows:

$$f(x) = \frac{1}{2\pi} \int_{-\infty}^{\infty} \hat{F}(\omega)\, \mathrm{e}^{i\omega x}\, d\omega \tag{6.44}$$

Textbooks sometimes place the 2π factor in different places (forward transform, inverse transform, or split evenly), but the core idea remains the same. The process of transforming a signal's time series into the frequency domain using the Fourier transform, and subsequently reconstructing it through the Inverse Fourier Transform, is illustrated in Figure 6.5b for noise signals in an airport.

Example: Find the Fourier transform $\hat{F}(\omega)$ of the following function $f(t)$ and verify that the inverse transform reconstructs $f(t)$:

$$f(x) \;=\; \begin{cases} 1, & \text{if } -\frac{1}{2} \leq x \leq +\frac{1}{2} \\ 0, & \text{otherwise} \end{cases} \tag{6.45}$$

Solution:

This function looks like a "box" of height 1, centered at $x = 0$, and extending from $-\frac{1}{2}$ to $+\frac{1}{2}$. It is referred to as a rectangular pulse.

Fourier transform: For the forward transform of the rectangular pulse, we compute:

$$\hat{F}(\omega) \;=\; \int_{-\infty}^{\infty} f(x)\, \mathrm{e}^{-\,i\,\omega x}\, dx \tag{6.46}$$

Since $f(x) = 0$ everywhere except between $-\frac{1}{2}$ and $+\frac{1}{2}$, the integral reduces to:

$$\hat{F}(\omega) \;=\; \int_{-\frac{1}{2}}^{+\frac{1}{2}} 1 \cdot \mathrm{e}^{-\,i\,\omega x}\, dx \tag{6.47}$$

This integral is straightforward to evaluate:

$$\hat{F}(\omega) = \left[\frac{\mathrm{e}^{-\,i\,\omega x}}{-\,i\,\omega} \right]_{x=-1/2}^{x=+1/2} = \frac{\mathrm{e}^{-\,i\,\omega/2} - \mathrm{e}^{\,i\,\omega/2}}{-\,i\,\omega} \tag{6.48}$$

Recall the identity for the `sine` function in terms of exponentials as follows:

$$\sin(\alpha) \;=\; \frac{\mathrm{e}^{\,i\,\alpha} - \mathrm{e}^{-\,i\,\alpha}}{2i} \tag{6.49}$$

By applying this identity equation, we obtain:

$$\hat{F}(\omega) = \frac{\sin\left(\frac{\omega}{2}\right)}{\frac{\omega}{2}} = 2 \cdot \frac{\sin\left(\frac{\omega}{2}\right)}{\omega} \tag{6.50}$$

Depending on the convention, this is often written in $\text{sinc}(x)$ form. One common definition of $\text{sinc}(x)$ is:

$$\text{sinc}(x) = \frac{\sin(\pi x)}{\pi x} \quad \text{or} \quad \text{sinc}(x) = \frac{\sin(x)}{x} \tag{6.51}$$

Either way, the important takeaway is that the Fourier Transform of a rectangular pulse is a sinc-like function.

Inverse transform: If we only had $\hat{F}(\omega)$, we could reconstruct the original function $f(x)$ by the inverse transform as follows:

$$f(x) \;=\; \frac{1}{2\pi}\int_{-\infty}^{\infty} \hat{F}(\omega)\,\mathrm{e}^{\,i\,\omega x}\,d\omega \tag{6.52}$$

Substituting our expression for $\hat{F}(\omega)$, the $\operatorname{sinc}(x)$ form, into that integral, we would indeed recover the rectangular function:

$$f(x) = \begin{cases} 1, & -\frac{1}{2} \le x \le +\frac{1}{2} \\ 0, & \text{otherwise} \end{cases} \tag{6.53}$$

Conceptually, this inverse transform adds up (integrates) all possible frequencies, each weighted by how strongly it appears in $\hat{F}(\omega)$.

Example: Find the Fourier Transform $\hat{F}(\omega)$ of the following function $f(t)$ and verify that the inverse transform reconstructs $f(t)$:

$$f(x) = \mathrm{e}^{-x^2} \tag{6.54}$$

Solution:

In this section, we utilize MATLAB (Example 6.5, via script `CH06Mcode3.m`) and Python (Example 6.6, via script `CH06Pcode3.py`) to solve the problem, bypassing manual calculations. In MATLAB, the Fourier transform is computed using the function `fourier(f, x)`, while in Python, the corresponding function is `sympy.fourier_transform(f, x, w)`. For the inverse transform, MATLAB uses `ifourier(F, x)`, while Python employs `sympy.inverse_fourier_transform(F, w, x)`.

From the outcome of the MATLAB code (Example 6.5), it is confirmed that the Fourier transform of e^{-x^2} results in another Gaussian function. Depending on the version of MATLAB, the output may vary slightly in its formatting, for instance, presenting $\sqrt{\pi}\,\mathrm{exp}\left(-\frac{w^2}{4}\right)$ instead. Example 6.6 illustrates the same Gaussian pair using SymPy's default 2π convention.

Example 6.5

MATLAB code. This MATLAB script, named `CH06Mcode3.m`, computes the Fourier transform of $f(x) = \mathrm{e}^{-x^2}$ and then computes the Inverse Fourier Transform of the result.

MATLAB Input:

```
% Define symbolic variable and function
syms x;
f = exp(-x^2);
%
```

```
% Compute the Fourier Transform F(w) of f(x)
% By default, transforms with respect to e^(-i*x*w)
F = fourier(f, x);
disp('Fourier Transform of exp(-x^2):');
disp(F);
%
% Compute the Inverse Fourier Transform of F(w)
f_inverse = ifourier(F, x);
disp('Inverse Fourier Transform of the above result:');
disp(f_inverse);
```

MATLAB Output:

```
Fourier Transform of exp(-x^2):
pi^(1/2)*exp(-x^2/4)

Inverse Fourier Transform of the above result:
exp(-x^2)
```

Example 6.6

Python code. This Python code, named CH06Pcode3.py, computes the Fourier transform of $f(x) = \mathrm{e}^{-x^2}$ and then computes the Inverse Fourier Transform of the result.

Python Input:

```
#
import sympy as sp
#
# Define the symbolic variable and function
x = sp.Symbol('x', real=True)
f = sp.exp(-x**2)
#
# Define the transform variable (commonly k or omega)
w = sp.Symbol('w', real=True)
#
# Compute the Fourier Transform of f(x)
# By default, sympy.fourier_transform uses the definition:
# F(k) = int f(x)*exp(-2*pi*i*x*k) dx
F = sp.fourier_transform(f, x, w)
#
```

```
# Compute the Inverse Fourier Transform
f_inverse = sp.inverse_fourier_transform(F, w, x)
#
# Print results
print("Fourier Transform of exp(-x^2):")
print(F)
print("\nInverse Fourier Transform of the above result:")
print(f_inverse)
```

Python Output:

```
Fourier Transform of exp(-x^2):
sqrt(pi)*exp(-pi**2*w**2)

Inverse Fourier Transform of the above result:
exp(-x**2)
```

6.3 FAST FOURIER TRANSFORM (FFT)

In signal processing, the Fast Fourier Transform (FFT) is an efficient algorithm for computing the Discrete Fourier Transform (DFT) of a signal, which transforms a time-domain signal into its frequency-domain representation. In both MATLAB and Python, the FFT is widely used for analyzing the frequency content of signals. The FFT decomposes a signal into a series of sinusoidal components, providing insights into its frequency characteristics. By applying FFT to a sampled time-domain signal, one can identify the dominant frequencies, their amplitudes, and phases, which are crucial for applications such as signal analysis, filtering, and communications.

The Inverse Fast Fourier Transform (IFFT) reverses this process, converting a frequency-domain signal back into the time-domain representation. In MATLAB, the functions `fft()` and `ifft()` are used to compute the FFT and IFFT, respectively. Similarly, in Python, the `numpy.fft.fft()` and `numpy.fft.ifft()` functions are employed. By using IFFT, the original time-domain signal can be reconstructed from its frequency components. This dual capability of FFT and IFFT is essential for a wide range of engineering applications, including signal processing, audio analysis, and communications, where both frequency-domain and time-domain information are critical.

Example: Calculate Fast Fourier Transform (FFT) and Inverse Fast Fourier Transform (IFFT) for the following time-domain function using MATLAB and Python:

$$f(t) = \cos(2\pi f_0 t) \tag{6.55}$$

Assume that $f_0 = 50$.

Solution:

Example 6.7 (script `CH06Mcode4.m`) demonstrates the use of MATLAB's FFT and IFFT to transform the signal into the frequency domain, revealing a sharp peak at 50 Hz. The time-domain signal is then perfectly recovered using IFFT. While the exact numerical values of the FFT magnitude may vary due to floating-point precision, a distinct peak will appear at 50 Hz. Example 6.8 (script `CH06Pcode4.py`) replicates this FFT procedure in Python. As shown in Figure 6.6, the left panel displays the discrete-time `cosine` waveform $f(t)$, while the right panel shows its amplitude spectrum $|X(f)|$, clearly highlighting the spectral peak at 50 Hz.

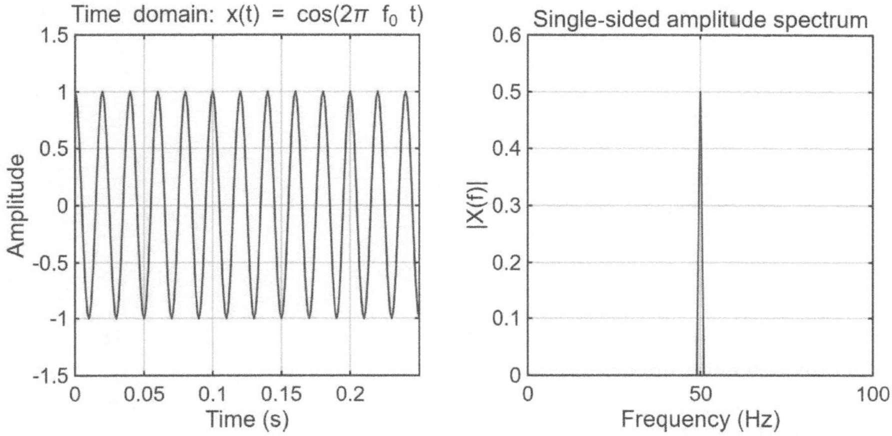

FIGURE 6.6 Time-domain signal $f(t) = \cos(2\pi 50\, t)$ (left) and its FFT magnitude $|X(f)|$ (right), sampled at $F_s = 1000$ Hz over a signal length of $L = 1000$ points, showing the spectral peak at 50 Hz.

Example 6.7

MATLAB code. This MATLAB script, named `CH06Mcode4.m`, computes the Discrete-Time Fourier Transform approximation of $f(t) = \cos(2\pi f_0 t)$ with a chosen sampling frequency (Fs) via FFT (Fast Fourier Transform) which converts time-domain signals into frequency-domain signals. It then performs an inverse transform (IFFT: Inverse Fast Fourier Transform) which does the reverse, converting frequency-domain signals back into time-domain signals to recover the original time-domain signal.

MATLAB Input:

```
% Parameters
Fs = 1000; % Sampling frequency (Hz)
T = 1/Fs; % Sampling period
L = 1000; % Length of signal
t = (0:L-1)*T; % Time vector
%
```

```
f0 = 50; % Frequency of the cosine (Hz)
x = cos(2*pi*f0*t); % Our discrete signal
%
% Compute the FFT
X = fft(x);
%
% Frequency axis for plotting
fAxis = (0:L-1)*(Fs/L);
%
% Compute the IFFT
x_recovered = ifft(X);
%
% Display results
disp('First few samples of the original signal x(t):');
disp(x(1:5));
disp('First few samples of the magnitude of FFT(X):');
disp(abs(X(1:5)));
disp('First few samples of the recovered signal (IFFT):');
disp(x_recovered(1:5));
```

MATLAB Output:

```
First few samples of the original signal x(t):
    1.0000 0.9511 0.8090 0.5878 0.3090

First few samples of the magnitude of FFT(X):
   1.0e-12 *

    0.1724 0.1342 0.3976 0.0938 0.3520

First few samples of the recovered signal (IFFT):
    1.0000 0.9511 0.8090 0.5878 0.3090
```

Example 6.8

Python code. This Python code, named `CH06Pcode4.py`, computes the Discrete-Time Fourier Transform approximation of $f(t) = \cos(2\pi f_0 t)$ with a chosen sampling frequency via FFT (Fast Fourier Transform) which converts time-domain signals into frequency-domain signals and $NumPy$ in Python. It then performs an inverse transform (IFFT: Inverse Fast Fourier Transform) which does the reverse, converting frequency-domain signals back into time-domain signals to recover the original time-domain signal.

Python Input:

```
import numpy as np
#
# Parameters
Fs = 1000.0 # Sampling frequency (Hz)
T = 1.0 / Fs # Sampling period
L = 1000 # Number of sample points
t = np.arange(L) * T # Time vector
#
f0 = 50.0
x = np.cos(2.0 * np.pi * f0 * t)
#
# Compute the FFT
X = np.fft.fft(x)
#
# Frequency axis
fAxis = np.fft.fftfreq(L, d=T)
#
# Compute the IFFT
x_recovered = np.fft.ifft(X)
#
# Display partial results
print("First few samples of x(t):", x[:5])
print("First few samples of |X|:", np.abs(X[:5]))
print("First few samples of recovered x(t) from IFFT:", \
x_recovered[:5].real)
```

Python Output:

```
First few samples of x(t): [1. 0.95105652 0.80901699 0.58778525
     0.30901699]
First few samples of |X|: [1.76619476e-13 1.34059013e-13
    4.03408971e-13 9.35880035e-14
 3.51218104e-13]
First few samples of recovered x(t) from IFFT: [1. 0.95105652
    0.80901699 0.58778525 0.30901699]
```

6.4 CONVOLUTION AND CORRELATION IN SIGNAL PROCESSING

In signal processing, convolution and correlation are two fundamental operations that help us understand the interaction between signals. **Convolution** describes how one signal modifies or influences another, while **Correlation** quantifies the similarity between signals. These operations are essential in areas such as filtering, noise reduction, and characterizing system responses in electronics and communication systems. Figure 6.7

illustrates both a real-world application (an electronic circuit used in a racing car) and a schematic representation of how convolution and correlation are used to filter signals and recognize patterns.

This section focuses on the discrete-time definitions of these operations, which are commonly used in signal processing courses and implemented in MATLAB and Python. The examples, figures, and accompanying MATLAB and Python codes in this section provide practical insights on how these concepts are applied in engineering applications.

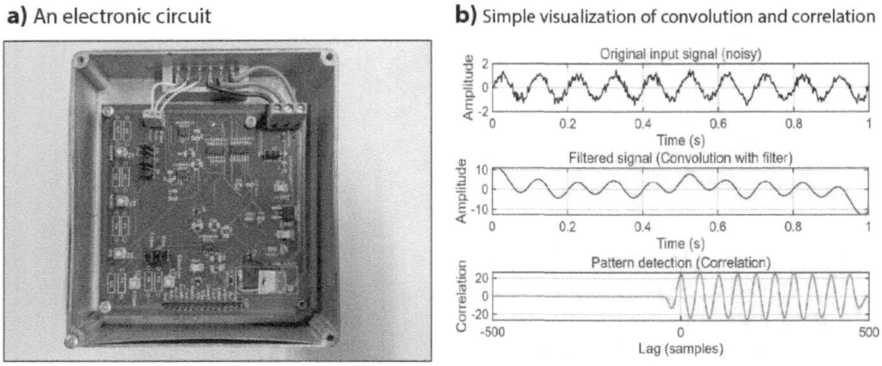

FIGURE 6.7 Filtering and pattern recognition using convolution and correlation. **a)** A photo of an electronic circuit used in a racing car. **b)** Simplified visualization of signal processing operations: convolution shows how the input signal is smoothed by the filter, and correlation illustrates the detection of a known pattern within the signal.

6.4.1 CONVOLUTION

For two discrete-time signals $x[n]$ and $h[n]$, the convolution $y[n] = x[n] * h[n]$ is defined as:

$$y[n] = \sum_{k=-\infty}^{\infty} x[k]\, h[n-k] \tag{6.56}$$

In practice, signals often have a finite length. For example, if

$$x[n] = \begin{cases} x[n], & 0 \le n \le L_x - 1 \\ 0, & \text{otherwise} \end{cases} \quad \text{and} \quad h[n] = \begin{cases} h[n], & 0 \le n \le L_h - 1 \\ 0, & \text{otherwise,} \end{cases}$$

then the convolution result $y[n]$ typically has length $L_x + L_h - 1$. To visualize the process, think of convolution as sliding one signal across the other. At each step (shift), you multiply the overlapping samples and add them up. The result of each shift provides one value of the output signal $y[n]$.

6.4.2 CORRELATION

For **cross-correlation**, the discrete-time definition between two signals $x[n]$ and $y[n]$ is:

$$r_{xy}[m] = \sum_{n=-\infty}^{\infty} x[n]\, y[n+m] \tag{6.57}$$

This operation measures the similarity between one signal and a time-shifted version of another. A high correlation value indicates that the signals are more similar at that particular shift. While correlation also involves sliding one signal across another, unlike convolution which reverses one signal, correlation directly assesses how "aligned" or similar the signals are at each shift.

Example: Calculate convolution and correlation for the following two short signals:

$$x[n] = [1, 2, 3], \quad h[n] = [1, 1, 1] \tag{6.58}$$

Solution:

Index them starting at $n = 0$. Then:

- $x[0] = 1,\ x[1] = 2,\ x[2] = 3$, and $x[n] = 0$ for $n < 0$ or $n > 2$.
- $h[0] = 1,\ h[1] = 1,\ h[2] = 1$, and $h[n] = 0$ for $n < 0$ or $n > 2$.

Convolution: To compute $y[n] = x[n] * h[n]$:

$$y[n] = \sum_{k=-\infty}^{\infty} x[k]\, h[n-k] \tag{6.59}$$

For finite indices:

$$y[0] = x[0]h[0] = 1 \cdot 1 = 1$$
$$y[1] = x[0]h[1] + x[1]h[0] = 1 \cdot 1 + 2 \cdot 1 = 3$$
$$y[2] = x[0]h[2] + x[1]h[1] + x[2]h[0] = 1 \cdot 1 + 2 \cdot 1 + 3 \cdot 1 = 6$$
$$y[3] = x[1]h[2] + x[2]h[1] = 2 \cdot 1 + 3 \cdot 1 = 5$$
$$y[4] = x[2]h[2] = 3 \cdot 1 = 3$$

Hence,

$$y[n] = [1,\ 3,\ 6,\ 5,\ 3] \tag{6.60}$$

Correlation: For cross-correlation of $x[n]$ with $h[n]$, we compute:

$$r_{xh}[m] = \sum_{n=-\infty}^{\infty} x[n]\, h[n+m] \tag{6.61}$$

You can compute each shift m similarly. If you prefer *auto-correlation*, you would just use the same signal for both x and h.

6.4.3 USING MATLAB AND PYTHON FOR CONVOLUTION AND CORRELATION

In MATLAB, the convolution of two discrete-time signals can be computed using the built-in function `conv(x, h)`, where `x` and `h` are the input signals. This function directly applies the convolution operation and returns the resulting signal `y[n]`. Similarly, for correlation, MATLAB provides the function `xcorr(x, h)`, which computes the cross-correlation between two signals `x[n]` and `h[n]`. The result is a vector that represents the similarity between the two signals at different time shifts.

In Python, using the `NumPy` library, you can compute convolution with the function `np.convolve(x, h, mode='full')`, where `x` and `h` are the input signals. The `mode='full'` option ensures that the full output of the convolution is returned. For correlation, Python's `NumPy` library provides `np.correlate(x, h, mode='full')`, which computes the cross-correlation of `x` and `h`. Both operations in Python work in a similar manner to MATLAB.

Example: Apply MATLAB and Python to calculate convolution and correlation for the following two short signals:

$$x[n] = [1, 2, 3], \quad h[n] = [1, 1, 1] \tag{6.62}$$

Solution:

Example 6.9 (script `CH06Mcode5.m`) demonstrates the use of MATLAB's `conv` and `xcorr` functions to compute the convolution and cross-correlation of the input signals $x = [1,\ 2,\ 3]$ and $h = [1,\ 1,\ 1]$. The convolution yields $[1,\ 3,\ 6,\ 5,\ 3]$. The resulting cross-correlation sequence also reflects the expected alignment behavior. Note that the specific indexing and padding behavior of `xcorr` may vary slightly depending on the MATLAB version or signal lengths, but the core output—central values such as $[3,\ 5,\ 6,\ 5,\ 3]$—remains consistent.

Example 6.10 (script `CH06Pcode5.py`) performs the equivalent operations using Python's `numpy.convolve` and `numpy.correlate` functions. Depending on the version of Python and the `NumPy` library, the axis orientation for cross-correlation may differ (e.g., reversed lags), but the computed values $[3,\ 5,\ 6,\ 5,\ 3]$ will be the same.

Plots of convolution and cross-correlation results for the two discrete-time signals $x[n](x = [1, 2, 3])$ and $h[n](h = [1, 1, 1])$ are shown in Figure 6.8.

Example 6.9

MATLAB code. This MATLAB script, named `CH06Mcode5.m`, computes convolution and cross-correlation for two discrete-time signals $x[n](x = [1, 2, 3])$ and $h[n](h = [1, 1, 1])$. Convolution (y_conv) measures how one discrete-time signal (h) modifies or "filters" another (x) by summing pairwise products as one signal slides over the other in time, while cross-correlation ($r_x h$) measures the similarity between two discrete-time signals (x and h) as one is shifted relative to the other. A larger value indicates higher similarity at that particular time shift.

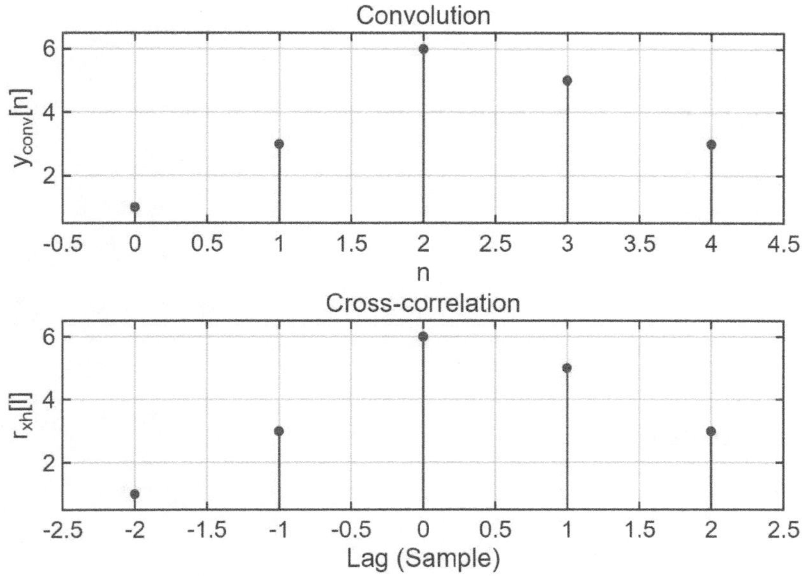

FIGURE 6.8 Convolution and cross-correlation results for two short signals $x = [1, 2, 3]$ and $h = [1, 1, 1]$.

MATLAB Input:

```
% We use:
% x = [1, 2, 3]
% h = [1, 1, 1]
%
% Then we compute:
% - y_conv = conv(x, h)
% - r_xh = xcorr(x, h)
%
% Finally, we plot both results in subplots with captions.
%
% 1. Define the input signals
x = [1, 2, 3];
h = [1, 1, 1];
%
% 2. Compute convolution of x and h
y_conv = conv(x, h); % Convolution
%
% 3. Compute cross-correlation between x and h (r_xh)
r_xh = xcorr(x, h); % Cross-correlation
```

```
%
% 4. Display the results
disp('Convolution y_conv = ');
disp(y_conv); % Expected: [1 3 6 5 3]
%
disp('Cross-correlation r_xh = ');
disp(r_xh); % Expected (7 points):
% For x=[1,2,3], h=[1,1,1], the theoretical
% cross-correlation (for all lags) should yield:
% [3, 5, 6, 5, 3] at the central lags,
% plus extra zeros/padding depending on xcorr's
% shift convention.
```

MATLAB Output:

```
Convolution y_conv =
   1 3 6 5 3

Cross-correlation r_xh =
  1.0000 3.0000 6.0000 5.0000 3.0000
```

Example 6.10

Python code. This Python code, named CH06Pcode5.py, computes convolution and cross-correlation for two discrete-time signals $x[n](x = [1, 2, 3])$ and $h[n](h = [1, 1, 1])$.

Python Input:

```
import numpy as np
#
# 1. Define the input signals
x = np.array([1, 2, 3])
h = np.array([1, 1, 1])
#
# 2. Compute full convolution y[n] = x[n] * h[n]
y_conv = np.convolve(x, h, mode='full')
#
# 3. Compute full cross-correlation r_xh[l]
r_xh = np.correlate(x, h, mode='full')
#
# 4. Print results
```

```
print("Convolution y_conv =", y_conv) # [1 3 6 5 3]
print("Cross-correlation r_xh =", r_xh) # [3 5 6 5 3]
```

Python Output:

```
Convolution y_conv = [1 3 6 5 3]
Cross-correlation r_xh = [1 3 6 5 3]
```

6.5 PROBLEMS

Problem 6.1: In an electrical circuit, the voltage across a capacitor can exhibit a ramp (sawtooth) behavior. Consider the following function:

$$f(x) = x, \quad -\pi < x < \pi,$$

which is assumed to be extended periodically with period 2π. Find the Fourier series of $f(x)$ using both MATLAB and Python.

Answer:

$$f(x) = 2\sum_{n=1}^{\infty} \frac{(-1)^{n+1}}{n} \sin(nx)$$

Problem 6.2: A simplified on-and-off traffic light can be modeled by the following square-wave function:

$$g(x) = \begin{cases} 1, & 0 \le x < \pi, \\ -1, & \pi \le x < 2\pi, \end{cases}$$

which is repeated with period 2π. Compute the Fourier series of $g(x)$ using both MATLAB and Python.

Answer:

$$g(x) = \frac{4}{\pi}\sum_{k=0}^{\infty} \frac{1}{2k+1} \sin\big[(2k+1)x\big]$$

Problem 6.3: A building's climate control system cycles between two temperature settings over one period, defined as:

$$h(x) = \begin{cases} 5, & 0 < x < \pi, \\ 0, & -\pi < x < 0, \end{cases}$$

and extend $h(x)$ periodically with period 2π. Find the Fourier series of $h(x)$ using both MATLAB and Python.

Answer:

$$a_0 = \frac{1}{\pi}\int_{-\pi}^{\pi} h(x)\,dx = 5, \quad \text{(so the average term is } \tfrac{a_0}{2} = \tfrac{5}{2}\text{)},$$

$$a_n = 0 \quad \text{for all } n,$$

$$b_n = \frac{5}{\pi}\int_{0}^{\pi} \sin(nx)\,dx = \frac{5}{\pi n}\left[1-\cos(n\pi)\right] = \begin{cases} \frac{10}{\pi n}, & n \text{ odd}, \\ 0, & n \text{ even}. \end{cases}$$

Thus,

$$h(x) = \frac{5}{2} + \sum_{\substack{n=1 \\ n \text{ odd}}}^{\infty} \frac{10}{\pi n}\sin(nx)$$

Problem 6.4: In a simplified mechanical vibration setting, a triangular waveform is used to model a displacement. Let,

$$p(x) = A\left(1 - \frac{2|x|}{\pi}\right), \qquad -\pi < x < \pi,$$

extended periodically with period 2π. Compute the Fourier series of $p(x)$ using both MATLAB and Python.

Answer:

$$p(x) = \frac{A\pi}{2} - \frac{4A}{\pi}\sum_{k=1}^{\infty} \frac{\cos\left[(2k-1)x\right]}{(2k-1)^2}$$

Problem 6.5: In digital communications, a rectangular pulse $f(t)$ is often used to represent a bit of information as follows:

$$f(t) = \begin{cases} 1, & |t| < 1, \\ 0, & |t| \geq 1. \end{cases}$$

Compute the Fourier transform of $f(t)$ using the convention $\hat{F}(\omega) = \int_{-\infty}^{\infty} f(t)\, e^{-i\omega t}\, dt$.

Answer:

$$\hat{F}(\omega) = \int_{-1}^{1} \mathrm{e}^{-i\,\omega\,t}\, dt = \frac{2\sin(\omega)}{\omega}$$

Problem 6.6: A decaying exponential is used to model the discharge of a capacitor:

$$g(t) = \mathrm{e}^{-at}\, u(t), \quad a > 0,$$

where $u(t)$ is the unit step function. Find the Fourier transform of $g(t)$.

Answer:

$$\hat{G}(\omega) = \int_{0}^{\infty} e^{-at}\mathrm{e}^{-i\omega t}\, dt = \frac{1}{a + i\omega}$$

Problem 6.7: In certain diffusion or smoothing processes, the frequency-comain representation may look like the following:

$$H(\omega) = \mathrm{e}^{-\alpha\,|\omega|}, \quad \alpha > 0.$$

Using the inverse Fourier transform $h(t) = \dfrac{1}{2\pi}\displaystyle\int_{-\infty}^{\infty} H(\omega)\,\mathrm{e}^{i\omega t}\, d\omega$, find $h(t)$.

Answer:

$$h(t) = \frac{1}{2\pi}\int_{-\infty}^{\infty} \mathrm{e}^{-\alpha|\omega|}\mathrm{e}^{i\omega t}\, d\omega = \frac{\alpha}{\pi\left(\alpha^2 + t^2\right)}$$

Problem 6.8: Consider the following rectangular function:

$$F(\omega) = \mathrm{rect}\left(\frac{\omega}{2}\right) = \begin{cases} 1, & |\omega| < 1, \\ \frac{1}{2}, & |\omega| = 1, \\ 0, & |\omega| > 1. \end{cases}$$

(Note: some definitions set it strictly 1 for $|\omega| \le 1$, but this does not affect the main result.) Using the inverse Fourier transform $f(t) = \frac{1}{2\pi}\displaystyle\int_{-\infty}^{\infty} F(\omega)\,\mathrm{e}^{i\omega t}\, d\omega$, compute $f(t)$.

Answer:

$$f(t) = \frac{1}{2\pi}\int_{-1}^{1} \mathrm{e}^{\,i\omega t}\, d\omega = \frac{\sin t}{\pi\, t}$$

Problem 6.9: In a simple communication system, the follwoing two pulses $f(t)$ and $g(t)$ are used:

$$f(t) = \begin{cases} 1, & 0 \leq t \leq 1, \\ 0, & \text{otherwise}, \end{cases} \qquad g(t) = \begin{cases} 1, & 0 \leq t \leq 2, \\ 0, & \text{otherwise}. \end{cases}$$

Find the convolution $(f * g)(t)$.

Answer: For $t < 0$ or $t > 3$, $(f * g)(t) = 0$. For $0 \leq t \leq 1$,

$$(f * g)(t) = \int_0^t 1 \cdot 1 \, d\tau = t$$

For $1 < t \leq 2$,

$$(f * g)(t) = \int_0^1 1 \cdot 1 \, d\tau = 1$$

For $2 < t \leq 3$,

$$(f * g)(t) = \int_{t-2}^1 1 \cdot 1 \, d\tau = 3 - t$$

Hence,

$$(f * g)(t) = \begin{cases} 0, & t < 0, \\ t, & 0 \leq t \leq 1, \\ 1, & 1 \leq t \leq 2, \\ 3 - t, & 2 \leq t \leq 3, \\ 0, & t \geq 3. \end{cases}$$

Problem 6.10: A rectangular pulse $h(t)$ of duration 2s can be written as:

$$h(t) = \begin{cases} 1, & 0 \leq t \leq 2\text{s}, \\ 0, & \text{otherwise}. \end{cases}$$

Its auto-correlation is defined (for real signals) by:

$$R_h(\tau) = (h \star h)(\tau) = \int_{-\infty}^{\infty} h(t) \, h(t + \tau) \, dt$$

Determine $R_h(\tau)$.

Answer:

$$R_h(\tau) = \begin{cases} 2\text{s} - |\tau|, & |\tau| \leq 2\text{s}, \\ 0, & |\tau| > 2\text{s}. \end{cases}$$

This is a triangular function centered at $\tau = 0$ with maximum height $R_h(0) = 2\text{s}$.

REFERENCES

[1] Joseph Fourier. *The analytical theory of heat.* Cambridge university press, 1878.

[2] Mohammad Heidarzadeh et al. *A Practical Approach to Advanced Mathematical Modelling in Civil Engineering*. Oxford University Press, 2025.

[3] Mohammad Heidarzadeh et al. "Storm wave runups and sea level variations for the September 2017 Hurricane Maria along the coast of Dominica, eastern Caribbean sea: Evidence from field surveys and sea-level data analysis." In: *Coastal Engineering Journal* 60(3) (2018), p. 371–384.

[4] Mohammad Heidarzadeh et al. "Tsunami hazards and risks from the Philippine Trench: The cases of 2012 and 2023 Mw 7.6 tsunamigenic earthquakes." In: *Ocean Engineering* 329 (2025), p. 120985.

[5] Payam Khazaeinejad and Asif S. Usmani. "On thermo-mechanical nonlinear behaviour of shallow shells." In: *International Journal of Non-Linear Mechanics* 82 (2016), p. 114–123.

[6] Payam Khazaeinejad and Asif S. Usmani. "Temperature-dependent nonlinear analysis of shallow shells: A theoretical approach." In: *Composite Structures* 141 (2016), p. 1–13.

[7] Payam Khazaeinejad, Asif S. Usmani, and O. Laghrouche. "Temperature-dependent nonlinear behaviour of thin rectangular plates exposed to through-depth thermal gradients." In: *Composite Structures* 132 (2015), p. 652–664.

[8] Payam Khazaeinejad, Asif S. Usmani, and Omar Laghrouche. "An analytical study of the nonlinear thermo-mechanical behaviour of thin isotropic rectangular plates." In: *Computers and Structures* 141 (2014), p. 1–8.

[9] Henry Moss, Mohammad Heidarzadeh, and Ramtin Sabeti. "Innovative coastal defence using rock bag revetments: preliminary physical modelling." In: *Results in Engineering* (2025), p. 105151.

[10] Yoshinobu Tsuji et al. "Field surveys of tsunami heights from the 2011 off the Pacific Coast of Tohoku, Japan Earthquake." In: *Bulletin of Earthquake Research Institute of University of Tokyo* 86 (2011), p. 29–279.

7 Laplace Transforms and Applications

Laplace transform is a powerful analytical tool for solving ordinary differential equations and is therefore fundamental in engineering disciplines concerned with dynamic behavior. By converting time-domain equations into algebraic expressions in the Laplace domain (s-domain), they greatly simplify problems in structural dynamics, fluid mechanics and control. In structural engineering, for example, Laplace transforms are used to model building and bridge vibrations under dynamic loads such as earthquakes [1] and wind (Figure 7.1a). In hydraulics, they facilitate the analysis of unsteady flow in pipelines, rivers and canals (Figure 7.1b).

This chapter introduces the definition, properties and common engineering applications of the Laplace transform, providing a systematic approach to transient- and dynamic-response problems. The discussion begins with the fundamental definition of the Laplace transform and its role in transforming complex time-domain problems into a more manageable form. Key properties and theorems will be explored. The inverse Laplace transform will also be examined as a method for recovering original functions from their transformed counterparts. Finally, practical applications in engineering, particularly in control systems and dynamic analysis, will be discussed. Mastering the Laplace transform provides a valuable analytical tool for solving a wide range of engineering problems.

FIGURE 7.1 Examples of damage from extreme natural hazards. **a)** Roof damage to a school caused by Hurricane Maria (2017), Dominica. **b)** High river flow near Batheaston, UK.

DOI: 10.1201/9781003511854-7

7.1 INTRODUCTION TO LAPLACE TRANSFORMS

Consider a simple model for a car's suspension, which typically includes a spring (that resists compression and extension), a damper or shock absorber (that dissipates energy), and a mass (the car body or part of it) supported by the spring and damper. In engineering terms, this system can be represented by a second-order differential equation. When the car hits a bump (an input force or displacement), the suspension moves up and down, trying to settle back to equilibrium (see Figure 7.2). A simplified version of the suspension's motion can be written as:

$$m\frac{d^2x(t)}{dt^2} + c\frac{dx(t)}{dt} + kx(t) = F_{\text{in}}(t) \tag{7.1}$$

where m is the mass, $x(t)$ is the displacement of the mass (car body), c is the damping coefficient (damper), k is the spring's stiffness constant, and $F_{\text{in}}(t)$ is the external force or input (e.g., force from a bump).

Solving this second-order differential equation directly in the time domain can be done, but it sometimes involves tedious steps, including guessing solutions, matching initial conditions and handling forcing functions. The **Laplace transform** provides a systematic shortcut by transforming the time-domain differential equation into an algebraic equation in what is called the "Laplace domain" or "s-domain."

For a time-domain function like $f(t)$, the Laplace transform $\mathcal{L}\{f(t)\}$ is defined as follows:

$$\mathcal{L}\{f(t)\} = F(s) = \int_0^\infty \mathrm{e}^{-st} f(t)dt \tag{7.2}$$

where s is a complex variable. It can be viewed as a method of compressing all the information about the behavior of $f(t)$ over time into a single function $F(s)$ in the s-domain. Although a mechanical suspension was used as the example, the same procedure applies to electrical circuits and other physical systems governed by linear differential equations. For example, Figure 7.3 illustrates how Laplace transforms support transportation control by converting differential equations into simpler algebraic forms, enabling the design of stable and efficient systems like cruise control and adaptive braking.

Example: Find the Laplace transform of the following constant function:

$$f(t) = 1 \tag{7.3}$$

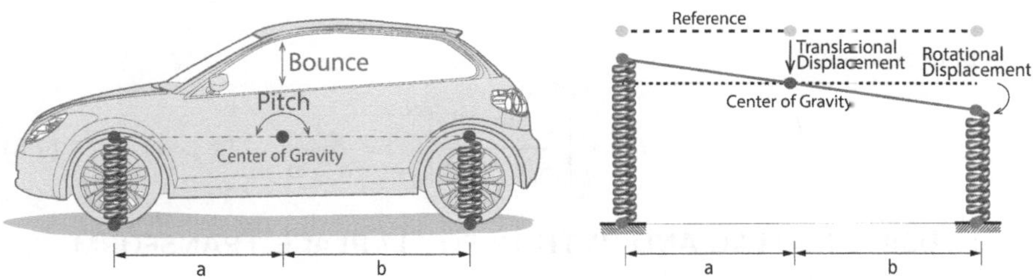

FIGURE 7.2 A simplified model of an automobile capable of pitch and bounce motions.

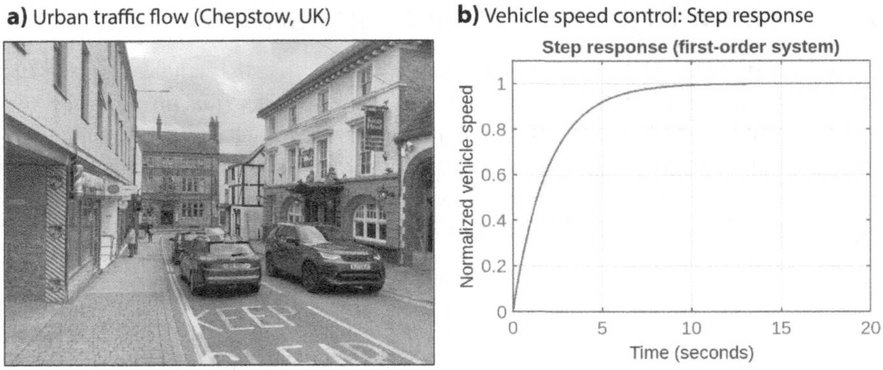

FIGURE 7.3 Laplace transforms in transportation control. **a)** City street traffic illustrating real-world speed control mechanisms. **b)** Graph showing how the Laplace transform converts a differential equation into an algebraic relation, thereby simplifying the design of stable automotive control systems.

Solution:

$$\mathcal{L}\{1\} = \int_0^\infty e^{-st}\, dt \tag{7.4}$$

Evaluating the integral:

$$\mathcal{L}\{1\} = \left[\frac{e^{-st}}{-s}\right]_0^\infty = \frac{1}{s}, \quad (s > 0) \tag{7.5}$$

Example: Find the Laplace transform of the following exponential function:

$$f(t) = e^{at} \tag{7.6}$$

Solution:

$$\mathcal{L}\{e^{at}\} = \int_0^\infty e^{-st} e^{at}\, dt. \tag{7.7}$$

Combine exponents:

$$\mathcal{L}\{e^{at}\} = \int_0^\infty e^{-(s-a)t}\, dt = \left[\frac{e^{-(s-a)t}}{-(s-a)}\right]_0^\infty = \frac{1}{s-a}, \quad (s > a) \tag{7.8}$$

7.1.1 USING MATLAB AND PYTHON FOR LAPLACE TRANSFORM

In both MATLAB and Python, the Laplace transform is performed using symbolic computation. The process begins by defining symbolic variables using `syms` in MATLAB or appropriate symbolic declarations in Python. In MATLAB, the command `laplace(f, t, s)` is used to compute the Laplace transform of a function `f`, where `t` is the time variable and `s` is the complex frequency variable.

In Python, using the `sympy` library, the equivalent command is `laplace_transform(f, t, s)`. These functions automatically convert a time-domain expression into its Laplace-domain representation, facilitating the analysis of mathematical and engineering problems.

Example: Find the Laplace transform of the following constant function using MATLAB and Python:

$$f(t) = 1 \tag{7.9}$$

Solution:

Example 7.1 (script `CH07Mcode1.m`) demonstrates how MATLAB's `laplace(f, t, s)` command converts the constant function $f(t) = 1$ into its representation in the s-domain. Example 7.2 repeats the calculation in Python using the script `CH07Pcode1.py`, which applies the `laplace_transform(f, t, s)` command from the `SymPy` library.

Example 7.1

MATLAB code. This MATLAB script, named "CH07Mcode1.m," computes the Laplace transform of $f(t) = 1$. This MATLAB script, along with all other scripts presented in this book, is available at:

https://github.com/prog871/CRC-book2025

MATLAB Input:

```
syms t s
f = 1; % Define the function f(t) = 1
F = laplace(f, t, s) % Compute the Laplace transform
```

MATLAB Output:

```
F =
1/s
```

Example 7.2

Python code. This Python code, named "CH07Pcode1.py," computes the Laplace transform of $f(t) = 1$.

Python Input:

```
from sympy import symbols, laplace_transform
#
t, s = symbols('t s', real=True, positive=True)
f = 1 # Define f(t) = 1
#
F = laplace_transform(f, t, s)
print("Laplace Transform of 1:", F)
```

Python Output:

```
Laplace Transform of 1: (1/s, 0, True)
```

Example: Find the Laplace transform of the following exponential function using MATLAB and Python:

$$f(t) = e^{2t} \tag{7.10}$$

Solution:

Example 7.3 demonstrates the Laplace transform of $f(t) = e^{2t}$ in MATLAB using the script `CH07Mcode2.m`. Example 7.4, implemented in Python through the script `CH07Pcode2.py`, confirms the same analytical result using the `SymPy` library.

Example 7.3

MATLAB code. This MATLAB script, named "CH07Mcode2.m," computes the Laplace transform of $f(t) = e^{2t}$.

MATLAB Input:

```
syms t s
f = exp(2*t);
F = laplace(f, t, s) % Compute Laplace transform of exp(2*t)
```

MATLAB Output:

```
F =
1/(s - 2)
```

Example 7.4

Python code. This Python code, named "CH07Pcode2.py," computes the Laplace transform of $f(t) = e^{2t}$.

Python Input:

```
from sympy import symbols, laplace_transform, exp
#
t, s = symbols('t s', real=True, positive=True)
f = exp(2*t) # f(t) = e^{2t}
#
F = laplace_transform(f, t, s)
print("Laplace Transform of exp(2*t):", F)
```

Python Output:

```
Laplace Transform of exp(2*t): (1/(s - 2), 2, True)
```

Figure 7.4 compares the original time domain signal with its frequency domain (Laplace) representation. The left panel in Figure 7.4 shows the exponential function $f(t) = e^{2t}$, illustrating the rapid growth that is characteristic of a positive real exponent. In contrast, the right panel depicts the Laplace transform $F(s) = 1/(s-2)$. Because $F(s)$ has a simple pole at $s = 2$, the plot is restricted to $s > 2$, making the vertical asymptote explicit. As the complex frequency approaches the pole from the right, the magnitude of the transform increases without bound.

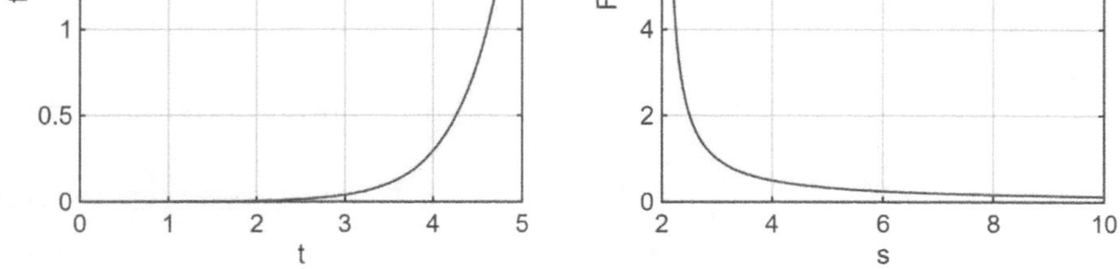

FIGURE 7.4 Visualization of the exponential signal and its Laplace transform. *Left*: the time-domain function $f(t) = e^{2t}$ plotted for $0 \leq t \leq 5$, highlighting its rapid growth. *Right*: the corresponding Laplace transform $F(s) = \frac{1}{s-2}$, shown for $s > 2$. The vertical asymptote at the pole $s = 2$ is avoided by starting the horizontal axis just to the right of that pole.

7.2 LAPLACE TRANSFORM PROPERTIES AND THEOREMS

Properties like time shifting, frequency shifting, and convolution in the Laplace domain help engineers systematically solve complex problems, ranging from abrupt starts in heavy machinery to on-and-off control actions in industrial processes. Table 7.1 provides some basic Laplace rules for derivatives and other functions.

TABLE 7.1

Common Laplace transform rules for derivatives and standard functions. Here, $x(0)$ is the initial value of the function and $\dot{x}(0)$ is its initial derivative

Function	Laplace Transform
$x(t)$	$\mathcal{L}\{x(t)\} = X(s)$
$\frac{dx(t)}{dt}$	$\mathcal{L}\left\{\frac{dx(t)}{dt}\right\} = sX(s) - x(0)$
$\frac{d^2x(t)}{dt^2}$	$\mathcal{L}\left\{\frac{d^2x(t)}{dt^2}\right\} = s^2X(s) - sx(0) - \dot{x}(0)$
$\frac{d^nx(t)}{dt^n}$	$\mathcal{L}\left\{\frac{d^nx(t)}{dt^n}\right\} = s^nX(s) - s^{n-1}x(0) - \cdots - \frac{d^{n-1}x}{dt^{n-1}}(0)$
1	$\frac{1}{s}$
t	$\frac{1}{s^2}$
t^n	$\frac{n!}{s^{n+1}}$
e^{at}	$\frac{1}{s-a}$
$\sin(at)$	$\frac{a}{s^2+a^2}$
$\cos(at)$	$\frac{s}{s^2+a^2}$
$e^{at}\sin(bt)$	$\frac{b}{(s-a)^2+b^2}$
$e^{at}\cos(bt)$	$\frac{s-a}{(s-a)^2+b^2}$

Example: Take the Laplace transform of each term in the differential equation presented in Equation 7.1:

Solution:

$$\mathcal{L}\left\{\frac{d^2x(t)}{dt^2}\right\} = s^2X(s) - s\,x(0) - \dot{x}(0) \tag{7.11}$$

$$\mathcal{L}\left\{\frac{dx(t)}{dt}\right\} = s\,X(s) - x(0) \tag{7.12}$$

$$\mathcal{L}\{x(t)\} = X(s) \tag{7.13}$$

$$\mathcal{L}\{F_{\text{in}}(t)\} = F_{\text{in}}(s). \tag{7.14}$$

Here, $X(s)$ is the Laplace transform of $x(t)$, and $x(0)$, $\dot{x}(0)$ are the initial position and velocity of the mass. Substituting these transforms into the differential equation results in:

$$m\big[s^2X(s) - s\,x(0) - \dot{x}(0)\big] + c\big[s\,X(s) - x(0)\big] + k\,X(s) = F_{\text{in}}(s) \tag{7.15}$$

Rearrange the equation to:

$$\left(ms^2 + cs + k\right) X(s) \; - \; m\left[s\,x(0) + \dot{x}(0)\right] \; - \; c\,x(0) = F_{\text{in}}(s) \tag{7.16}$$

The solution for $X(s)$, leads to the following:

$$X(s) = \frac{F_{\text{in}}(s) + m\left[s\,x(0) + \dot{x}(0)\right] + c\,x(0)}{m\,s^2 + c\,s + k} \tag{7.17}$$

7.2.1 LINEARITY PROPERTY OF THE LAPLACE TRANSFORM

The Laplace transform is a linear operator. For constants a and b, and functions $f(t)$ and $g(t)$, we have:

$$\mathcal{L}\{af(t) + bg(t)\} = a\mathcal{L}\{f(t)\} + b\mathcal{L}\{g(t)\}. \tag{7.18}$$

This is called the linearity of the Laplace transform.

Example: Calculate the Laplace transform for the following function:

$$h(t) = 2t + 3\sin(t) \tag{7.19}$$

Solution:

Let $f(t) = t$ and $g(t) = \sin(t)$. Then, referring to Table 7.1, we have:

$$\mathcal{L}\{t\} = \frac{1}{s^2}, \qquad \mathcal{L}\{\sin t\} = \frac{1}{s^2+1}. \tag{7.20}$$

Using linearity,

$$\mathcal{L}\{2t + 3\sin t\} = 2\,\mathcal{L}\{t\} + 3\,\mathcal{L}\{\sin t\} = 2 \cdot \frac{1}{s^2} + 3 \cdot \frac{1}{s^2+1} \tag{7.21}$$

These calculations can be carried out using both MATLAB and Python. Example 7.5 (script `CH07Mcode3.m`) illustrates MATLAB's ability to apply the linearity property by transforming t, $\sin t$, and their weighted sum within a single script. Example 7.6 (script `CH07Pcode3.py`) replicates this process using Python.

Example 7.5

MATLAB code. This MATLAB script, named "CH07Mcode3.m," computes $\mathcal{L}\{t\}, \mathcal{L}\{\sin t\}$, and $\mathcal{L}\{2t + 3\sin t\}$ by demonstrating the linearity property with $f(t) = t$ and $g(t) = \sin(t)$.

MATLAB Input:

```
syms t s
f = t;
g = sin(t);
%
Ff = laplace(f, t, s);
```

```
Fg = laplace(g, t, s);
F_combined = laplace(2*f + 3*g, t, s);
%
disp("Laplace of t = " + string(Ff))
disp("Laplace of sin(t) = " + string(Fg))
disp("Laplace of 2*t + 3*sin(t) = " + string(F_combined))
```

MATLAB Output:

```
Laplace of t = 1/s^2
Laplace of sin(t) = 1/(s^2 + 1)
Laplace of 2*t + 3*sin(t) = 3/(s^2 + 1) + 2/s^2
```

Example 7.6

Python code. This Python code, named "CH07Pcode3.py," computes $\mathcal{L}\{t\}$, $\mathcal{L}\{\sin t\}$, and $\mathcal{L}\{2t + 3\sin t\}$ by demonstrating the linearity property with $f(t) = t$ and $g(t) = \sin t$.

Python Input:

```
from sympy import symbols, laplace_transform, sin
#
t, s = symbols('t s', real=True, positive=True)
#
f = t
g = sin(t)
#
Ff = laplace_transform(f, t, s)
Fg = laplace_transform(g, t, s)
F_combined = laplace_transform(2*f + 3*g, t, s)
#
print("Laplace of t:", Ff)
print("Laplace of sin(t):", Fg)
print("Laplace of 2*t + 3*sin(t):", F_combined)
```

Python Output:

```
Laplace of t: (s**(-2), 0, True)
Laplace of sin(t): (1/(s**2 + 1), 0, True)
Laplace of 2*t + 3*sin(t): (3/(s**2 + 1) + 2/s**2, 0, True)
```

7.2.2 FIRST SHIFTING THEOREM (LAPLACE DOMAIN SHIFT)

If $\mathcal{L}\{f(t)\} = F(s)$, then using First Shifting Theorem:

$$\mathcal{L}\{e^{at} f(t)\} = F(s-a), \tag{7.22}$$

provided the Laplace transform exists for $s > \text{Re}(a)$.

Example: Calculate the Laplace Transform for the following function:

$$e^{2t} \sin(t) \tag{7.23}$$

Solution:

Let $f(t) = \sin(t)$. We know:

$$\mathcal{L}\{\sin(t)\} = \frac{1}{s^2+1} \tag{7.24}$$

By the first shift theorem,

$$\mathcal{L}\{e^{2t} \sin(t)\} = \frac{1}{(s-2)^2+1} \tag{7.25}$$

Example 7.7 (script `CH07Mcode4.m`) demonstrates the application of the First Shifting Theorem to $e^{at} \sin t$ and illustrates the resulting pole shift. Example 7.8 (script `CH07Pcode4.py`) verifies the shifted denominator $(s-2)^2+1$ using Python.

Example 7.7

MATLAB code. This MATLAB script, named "CH07Mcode4.m," demonstrates the first shift theorem with $e^{at} f(t)$ when $f(t) = \sin t$ and $a = 2$.

MATLAB Input:

```
syms t s
f = sin(t);
%
F_shifted = laplace(exp(2*t)*f, t, s);
disp("Laplace of e^(2*t)*sin(t) = " + string(F_shifted))
```

MATLAB Output:

```
Laplace of e^(2*t)*sin(t) = 1/((s - 2)^2 + 1)
```

Example 7.8

Python code. This Python code, named "CH07Pcode4.py," demonstrates the first shift theorem with $e^{at} f(t)$ when $f(t) = \sin t$ and $a = 2$.

Python Input:

```
from sympy import symbols, laplace_transform, sin, exp
#
t, s = symbols('t s', real=True, positive=True)
#
f = sin(t)
F_shifted = laplace_transform(exp(2*t)*f, t, s)
print("Laplace of e^(2*t)*sin(t):", F_shifted)
```

Python Output:

```
Laplace of e^(2*t)*sin(t): (1/((s - 2)**2 + 1), 2, True)
```

7.3 INVERSE LAPLACE TRANSFORMS

After solving a differential equation using the Laplace transform, the inverse Laplace transform is used to return the solution to the time domain. This step is essential for interpreting the physical behavior of a system over time. Figure 7.5 visually links the mathematical operation to practice by pairing a photograph of an actual feedback-control circuit with a schematic plot, showing how partial-fraction expansion of $X(s)$ unfolds into the transient and steady-state response $x(t)$. Once $X(s)$ is obtained, the goal is typically to recover $x(t)$, the time-domain response. This is achieved by applying the inverse Laplace transform as follows:

$$x(t) = \mathcal{L}^{-1}\{X(s)\} \tag{7.26}$$

Example: Find Inverse Laplace of the following rational function:

$$F(s) = \frac{1}{s} \tag{7.27}$$

Solution:

We recognize that $\mathcal{L}\{1\} = \frac{1}{s}$. Hence,

$$\mathcal{L}^{-1}\left\{\frac{1}{s}\right\} = 1 \tag{7.28}$$

Example: Find Inverse Laplace of the following exponential form:

$$F(s) = \frac{1}{s-a} \tag{7.29}$$

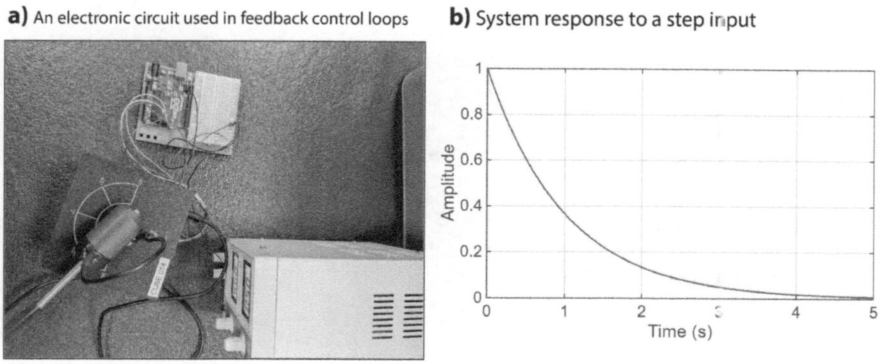

FIGURE 7.5 Reconstructing time-domain behavior in electronics. **a)** A photo of electronic circuits used in feedback control loops. **b)** A conceptual graph highlighting the inverse Laplace transform and partial-fraction expansions, enabling engineers to assess stability and transient responses in the time domain.

Solution:

We know that $\mathcal{L}\{e^{at}\} = \frac{1}{s-a}$. Therefore,

$$\mathcal{L}^{-1}\left\{\frac{1}{s-a}\right\} = e^{at} \tag{7.30}$$

7.3.1 USING MATLAB AND PYTHON FOR INVERSE LAPLACE TRANSFORM

Both MATLAB and Python can symbolically compute inverse Laplace transforms. MATLAB uses the `ilaplace(F, s, t)` command, while Python employs `inverse_laplace_transform(F, s, t)` from the `sympy` library.

Example: Determine the inverse Laplace transform of the following function using MATLAB and Python:

$$F(s) = \frac{1}{s} \tag{7.31}$$

Solution:

Example 7.9 (script `CH07Mcode5.m`) computes the inverse Laplace transform of the simple rational function $1/s$, yielding the constant 1 in MATLAB. Example 7.10 (script `CH07Pcode5.py`) confirms the same result using Python.

Example 7.9

MATLAB code. This MATLAB script, named "CH07Mcode5.m," computes the inverse Laplace transform of $F(s) = \frac{1}{s}$.

MATLAB Input:

```
syms s t
F = 1/s;
f = ilaplace(F, s, t);
disp("Inverse Laplace of 1/s = " + string(f))
```

MATLAB Output:

```
Inverse Laplace of 1/s = 1
```

Example 7.10

Python code. This Python code, named "CH07Pcode5.py," computes the inverse Laplace transform of $F(s) = \frac{1}{s}$.

Python Input:

```
from sympy import symbols, inverse_laplace_transform
#
t, s = symbols('t s', real=True, positive=True)
F = 1/s
#
f = inverse_laplace_transform(F, s, t)
print("Inverse Laplace of 1/s:", f)
```

Python Output:

```
Inverse Laplace of 1/s: 1
```

Example: Determine the inverse Laplace transform of the following function using MATLAB and Python:

$$F(s) = \frac{1}{s-3} \tag{7.32}$$

Solution:

The solutions are provided in Examples 7.11 and 7.12 using MATLAB and Python, respectively. Example 7.11 (script `CH07Mcode6.m`) retrieves e^{3t} from the shifted pole at $s = 3$ in MATLAB. Example 7.12 (script `CH07Pcode6.py`) reproduces the same exponential time response using Python.

Example 7.11

MATLAB code. This MATLAB script, named "CH07Mcode6.m," computes the inverse Laplace transform of $F(s) = \frac{1}{s-3}$.

MATLAB Input:

```
syms s t
F = 1/(s-3);
f = ilaplace(F, s, t);
disp("Inverse Laplace of 1/(s-3) = " + string(f))
```

MATLAB Output:

```
Inverse Laplace of 1/(s-3) = exp(3*t)
```

Example 7.12

Python code. This Python code, named "CH07Pcode6.py," computes the inverse Laplace transform of $F(s) = \frac{1}{s-3}$.

Python Input:

```
from sympy import symbols, inverse_laplace_transform
#
t, s = symbols('t s', real=True, positive=True)
F = 1/(s-3)
#
f = inverse_laplace_transform(F, s, t)
print("Inverse Laplace of 1/(s-3):", f)
```

Python Output:

```
Inverse Laplace of 1/(s-3) = exp(3*t)
```

7.4 APPLICATIONS IN CONTROL SYSTEMS

Laplace transforms are essential tools in control systems because they turn difficult differential equations (from the time domain) into simpler algebraic equations (in the s-domain). This makes it much easier to study how a system behaves—such as whether it is stable, how it responds to changes, and what it does in the long run. Below are some real-world examples that show how Laplace transforms help us understand and design control systems.

7.4.1 AUTOMOTIVE CRUISE CONTROL

Modern cars can automatically keep a steady speed. To do this, the system increases or decreases engine power to adjust the speed, even when road slopes, wind, or other conditions change. The Laplace transform helps by:

- **System modeling:** The car's behavior (e.g., mass, drag, and engine response) is described by equations involving rates of change. The Laplace transform simplifies these into algebraic equations, enabling the creation of the car's transfer function.

- **Controller design:** Common controllers are often designed and tuned via transfer functions in the s-domain. The Laplace transform helps engineers quickly analyze the open-loop and closed-loop responses (e.g., how fast the car's speed converges to the set point).

- **Stability analysis:** By examining the poles of the transfer function (the roots of the denominator in the s-domain), control engineers can determine if the system will maintain stable speed, oscillate or drift.

7.4.2 AEROSPACE FLIGHT CONTROL SYSTEMS

Aircraft autopilots adjust flight controls to maintain altitude and heading, particularly under various aerodynamic and environmental disturbances such as wind gusts and turbulence. The Laplace transform helps in:

- **Linearized aircraft models:** Around a steady-state flight condition, engineers linearize the complicated nonlinear dynamics of the aircraft. The linearized differential equations are then transformed using Laplace transforms, creating state-space or transfer-function-based models.

- **Controller synthesis:** Using the transformed model (e.g., pitch angle to elevator deflection transfer function), classical control approaches are applied in the s-domain to ensure the aircraft responds smoothly and stably.

- **Frequency-domain analysis:** The Laplace transform underpins frequency-response methods which are essential in robust control design to handle uncertainties and disturbances.

7.4.3 STRUCTURAL VIBRATION CONTROL

In high-precision environments such as semiconductor manufacturing equipment, tall buildings subjected to winds or seismic events, and aerospace structures in flight, vibration control is key. The Laplace transform helps in:

- **Vibration modeling:** Structural vibrations are typically governed by second-order differential equations (mass-spring-damper systems). In the Laplace domain, these become algebraic expressions where natural frequencies and damping ratios are easily extracted.

- **Active damping systems:** Systems that employ active actuators (e.g., dynamic vibration absorbers and active damping in tall buildings) are designed using feedback loops. Control laws, often developed in the s-domain, ensure the vibration is reduced to acceptable levels.

- **Modal analysis integration:** Although modal analysis is typically performed in the frequency domain, the Laplace transform provides the theoretical foundation for describing system dynamics and bridging time-domain simulations with frequency-domain insights.

7.5 PROBLEMS

Problem 7.1: An electrical circuit has a current $i(t) = 2e^{-3t}$ amperes for $t > 0$. Find the Laplace transform of $i(t)$ using both MATLAB and Python.

Answer:

$$\mathcal{L}\{2e^{-3t}\}(s) = \frac{2}{s+3}$$

Problem 7.2: A certain bacterial population is growing according to the model $p(t) = 200e^{0.5t}$ (in thousands) for $t \geq 0$. Compute the Laplace transform of $p(t)$ using both MATLAB and Python.

Answer:

$$\mathcal{L}\{200e^{0.5t}\}(s) = \frac{200}{s-0.5}, \quad \text{valid for } s > 0.5$$

Problem 7.3: A mass-spring system with no damping undergoes simple harmonic motion. The displacement is given by:

$$y(t) = 3\sin(4t)$$

Find its Laplace transform using both MATLAB and Python.

Answer:

$$\mathcal{L}\{3\sin(4t)\}(s) = \frac{3 \cdot 4}{s^2+16} = \frac{12}{s^2+16}$$

Problem 7.4: The daily temperature deviation (in °C) from some average in a certain region is modeled by:

$$T(t) = 10\cos(2t)$$

Find the Laplace transform of $T(t)$ using both MATLAB and Python.

Answer:

$$\mathcal{L}\{10\cos(2t)\}(s) = \frac{10\,s}{s^2+4}$$

Problem 7.5: A lightly damped oscillator has displacement given by the following equation:

$$d(t) = e^{-t}\sin(3t) \quad (t \geq 0)$$

Compute its Laplace transform using both MATLAB and Python.

Answer:

$$\mathcal{L}\{e^{-t}\sin(3t)\}(s) = \frac{3}{(s+1)^2+3^2} = \frac{3}{(s+1)^2+9}$$

Problem 7.6: In a chemical process, the Laplace transform of the concentration $c(t)$ is:

$$C(s) = \frac{6}{s+2}.$$

Find $c(t)$ by taking the inverse Laplace transform using both MATLAB and Python.

Answer:

$$c(t) = \mathcal{L}^{-1}\left\{\frac{6}{s+2}\right\} = 6\,e^{-2t}$$

Problem 7.7: A population in a wildlife preserve is described in the s-domain by:

$$P(s) = \frac{10\,s}{s^2+4}$$

Determine $p(t)$ in the time domain using both MATLAB and Python.

Answer:

$$p(t) = \mathcal{L}^{-1}\left\{\frac{10\,s}{s^2+4}\right\} = 10\,\cos(2t)$$

Problem 7.8: An engineering system has velocity $v(t)$ whose Laplace transform is:

$$V(s) = \frac{5}{s^2}$$

Find $v(t)$ using both MATLAB and Python.

Answer:

$$v(t) = \mathcal{L}^{-1}\left\{\frac{5}{s^2}\right\} = 5\,t$$

Problem 7.9: A damped harmonic oscillator has displacement $Y(s)$ in the s-domain given by the following equation:

$$Y(s) = \frac{2}{(s+1)^2+9}$$

Find $y(t)$ using both MATLAB and Python.

Answer:

$$y(t) = \mathcal{L}^{-1}\left\{\frac{2}{(s+1)^2+9}\right\} = \frac{2}{3}\,e^{-t}\,\sin(3t)$$

Problem 7.10: The Laplace transform of a signal $x(t)$ is given by:

$$X(s) = \frac{s-2}{(s-1)^2+4}$$

Find $x(t)$ using both MATLAB and Python.

Answer:

$$x(t) = \mathcal{L}^{-1}\left\{\frac{s-2}{(s-1)^2+4}\right\} = \mathrm{e}^t\cos(2t) \;-\; \tfrac{1}{2}\,\mathrm{e}^t\sin(2t)$$

REFERENCES

[1] Mohammad Heidarzadeh, Abdul Muhari, and Antonius B Wijanarto. “Insights on the source of the 28 September 2018 Sulawesi tsunami, Indonesia based on spectral analyses and numerical simulations.” In: *Pure and Applied Geophysics* 176 (2019), p. 25–43.

8 Statistics and Probabilities

Statistics and probability play a fundamental role in the design, analysis, and decision-making processes in engineering. Many infrastructure systems rely on statistical methods to assess uncertainties, forecast future events, and optimize performance based on observed data. Measurements collected from various sensors, such as rainfall gauges, river discharge stations, wave recorders, and traffic counters require statistical analysis to extract meaningful insights and ensure their applicability in engineering design.

For example, in flood engineering, historical rainfall and river discharge data are analyzed using statistical techniques to estimate the probability of extreme flood events (e.g., Figure 8.1b). Engineers determine the design flood discharge for different return periods, such as 50 or 100 years, and calculate the probability of exceedance of critical water levels. This information is essential for designing flood protection measures such as levees, dams, and drainage systems.

In coastal engineering, wave records collected during storms are statistically analyzed to determine the maximum expected wave height and the probability of wave exceedance beyond a critical threshold. These analyses help engineers design coastal defense structures, such as breakwaters and seawalls, to withstand extreme wave forces and minimize coastal erosion (e.g., Figure 8.1a) [1].

In traffic engineering, vehicle count data collected from roads and highways are used for traffic flow analysis, capacity estimation, and accident risk assessment. Engineers employ statistical models to predict peak-hour traffic volumes, assess congestion levels, and optimize traffic signal timings to improve road safety and efficiency.

Beyond these applications, statistical and probabilistic methods are widely used in structural engineering to assess material strength variability, in geotechnical engineering to evaluate soil property distributions, and in environmental engineering to analyze pollutant concentrations and predict contamination risks. By applying statistical tools, engineers can make data-driven decisions, improve the reliability of infrastructure systems, and enhance resilience against extreme events. Understanding the principles of probability and statistics is therefore essential for solving real-world engineering problems and designing safe, efficient, and sustainable infrastructure.

8.1 DATA

Data plays a fundamental role in engineering, providing the basis for informed decision-making in design, construction, and maintenance of infrastructure systems. Engineers rely on data to assess structural performance, analyze environmental conditions, and optimize construction materials. Understanding data and its types is essential for interpreting engineering problems and applying appropriate mathematical and statistical methods.

DOI: 10.1201/9781003511854-8

FIGURE 8.1 Examples of damage from extreme natural hazards. **a)** Completely destroyed houses and leveled ground in a coastal town following the March 2011 Japan tsunami. **b)** Flooding of River Avon in Bath, UK.

8.1.1 TYPES OF DATA

Data in engineering can be broadly classified into two categories: **qualitative data** and **quantitative data**.

i) Qualitative Data

Qualitative data represents non-numerical information that categorizes objects or characteristics. Examples include:

- Soil classification (e.g., sand, clay, silt).
- Bridge types (e.g., suspension, beam, arch).
- Pavement conditions (e.g., good, fair, poor).

ii) Quantitative Data

Quantitative data consists of numerical values that can be measured or counted. This category is further divided into:

- **Discrete data** (countable values): Examples include the number of cracks in a concrete beam or the number of vehicles passing a bridge per hour.

- **Continuous data** (measurable values): Examples include the compressive strength of concrete (measured in megapascals, MPa) or the settlement of a foundation (measured in millimeters).

8.1.2 DATA COLLECTION IN ENGINEERING

Data in engineering is obtained through various methods, including field measurements, laboratory experiments, and computer simulations.

- **Field measurements**: Engineers collect real-world data using instruments and sensors. For instance, land surveyors measure terrain elevations before designing highways, while hydrologists monitor river flow rates to assess flood risks. Coastal engineers measure tsunami run-up heights along the affected coasts following tsunami incidents through field surveys (see Figure 8.2a).
- **Laboratory experiments**: Testing materials in controlled environments allows engineers to determine properties such as concrete compressive strength, soil permeability, and asphalt durability. Engineers and scientists study wave dynamics by making water waves in hydraulic laboratories and measuring them for various research activities (see Figure 8.2b,c).
- **Computer simulations**: Computational models help predict structural behavior under different conditions. For example, finite element analysis (FEA) is used to simulate the response of bridges to dynamic loads.

Engineering application: When designing a new road, civil engineers collect data on traffic volume, road surface conditions, and historical accident records. This data is used to determine optimal road alignment, material selection, and safety measures, ensuring efficient and durable infrastructure. As another example, coastal engineers conduct field surveys following tsunami and storms to measure wave runup heights and document damage (see Figure 8.2a).

8.1.3 REPRESENTING DATA

Once collected, data must be organized and presented effectively for analysis. The most common methods include:

- **Tables:** Raw data is structured into tables for easier interpretation (e.g., Figure 8.3a).
- **Graphs and charts:** Engineers often use bar charts, histograms, and scatter plots to visualize data trends (e.g., Figure 8.3b).
- **Descriptive statistics:** Measures such as mean, median, and standard deviation summarize key aspects of the data, providing insight into variations and distributions.

Engineering application: A civil engineer conducts a traffic study to analyze vehicle flow at a four-way intersection. During peak hours (8:00 AM to 9:00 AM), they record the number of cars passing in each direction (North, South, East, and West) at 15-min intervals. The collected data are presented in both a table and a bar chart, as shown in Figure 8.3.

8.2 FREQUENCY DISTRIBUTION AND HISTOGRAMS

Frequency distribution is an essential tool in engineering, used to summarize how often different values occur within a dataset. This method allows engineers to efficiently interpret large volumes of data by organizing it into intervals or "bins." It is particularly useful in applications like signal processing, quality control, and data analysis.

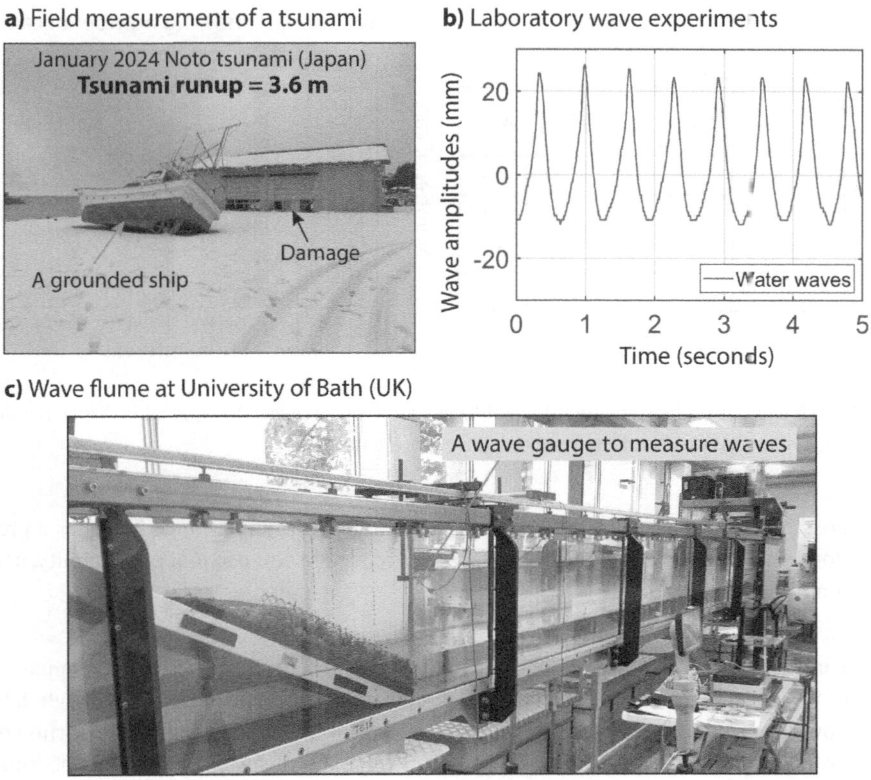

FIGURE 8.2 Examples of field and laboratory measurements of tsunami runup and water waves. **a)** Field measurement of tsunami runup by the author MH following the Noto tsunami (Japan) in January 2024 [1]. **b)** Time series showing laboratory measurement of water waves. **c)** The wave flume at the University of Bath (UK) which was used to measure the wave presented in panel "b."

Key Components of a frequency distribution are listed in the following

- **Class intervals:** Class intervals represent the predefined ranges into which data is categorized for systematic analysis. In civil engineering, for example, when evaluating the compressive strength of concrete samples, the data may be grouped into intervals such as 20–25 MPa, 25–30 MPa, and so forth. The selection of appropriate intervals is critical, as it ensures logical data grouping, facilitating the identification of underlying patterns and trends.

- **Frequency:** Frequency denotes the number of occurrences of a particular value within a specified class interval. It serves as a fundamental measure in statistical analysis, enabling the identification of common and rare values within a dataset. For instance, in civil engineering, when assessing the distribution of soil particle sizes in a sample, the frequency of particles within specific size ranges—such as 0.075–0.150 mm or 0.150–0.300 mm—helps engineers determine the soil gradation and its suitability for

a) Table of data for number of cars

Time interval	North	South	Esat	West
8:00 - 8:15	25	30	20	15
8:15 - 8:30	40	35	25	20
8:30 - 8:45	35	28	22	18
8:45 - 9:00	30	33	28	22

b) Bar chart showing number of cars

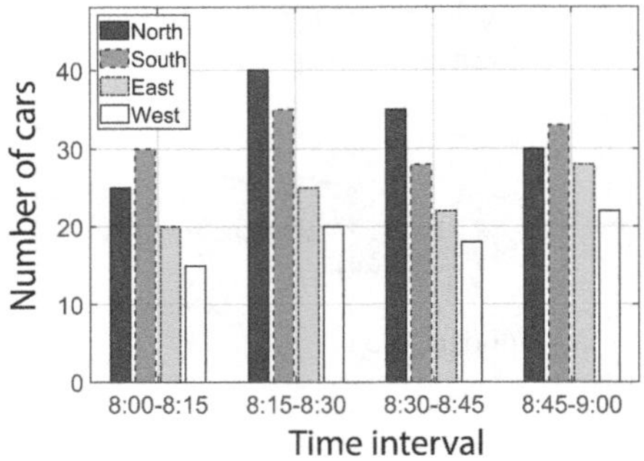

FIGURE 8.3 A traffic study on vehicle flow at a four-way intersection. **a)** A table displaying the collected data. **b)** A bar chart representing the collected data.

construction purposes. A higher frequency within a particular range indicates a predominant particle size, influencing decisions related to compaction, permeability, and load-bearing capacity.

- **Cumulative frequency:** Cumulative frequency represents the progressive summation of frequencies up to a given class interval. This statistical measure provides insight into how data accumulates, making it particularly useful for determining the proportion of observations that fall below or at a specified value. For instance, in civil engineering, when analyzing the compressive strength of concrete samples, cumulative frequency can be employed to assess the proportion of samples that achieve a strength below a critical threshold. If the compressive strength data is grouped into intervals such as 20–25 MPa, 25–30 MPa, and so forth, the cumulative frequency at 30 MPa indicates the total number of samples with strengths at or below this value.

- **Relative frequency:** Relative frequency represents the proportion of total observations that fall within a specific class interval. It is often expressed as a percentage or decimal and provides a clearer understanding of the distribution by highlighting the concentration of data points across different intervals. By normalizing raw frequency values, relative frequency facilitates comparisons between datasets of varying sizes and helps visualize the overall shape of the distribution. For example, in civil engineering, when analyzing the grain size distribution of soil samples, relative frequency can indicate the proportion of particles within a given size range. If a sieve analysis reveals that 30 out of 100 soil particles fall within the 0.075–0.150 mm range, the relative frequency for this interval would be 0.30 (or 30%).

Frequency distributions can be graphically represented using **histograms**, where each bar corresponds to a class interval, making it easier to see the distribution at a glance. An example of a histogram is presented in Figure 8.3b.

Engineering application: In coastal engineering, waves are classified based on their height. Suppose a coastal engineer records the heights (in meters) of 30 waves at a beach during a storm: 1.2 m, 1.5, 2.0, 1.8, 1.7, 2.2, 2.5, 2.8, 1.9, 1.6, 2.4, 2.5, 3.0, 2.7, 2.1, 1.4, 2.3, 2.0, 1.3, 2.9, 1.1, 3.1, 1.9, 2.2, 2.7, 2.4, 1.8, 2.5, 1.6, and 2.3 m. It is desired to construct a frequency distribution table by grouping the wave heights into equal class intervals: 1.0–1.5 m, 1.5–2.0 m, 2.0–2.5 m, 2.5–3.0 m, and 3.0–3.5 m, and determining the number of occurrences in each class. The upper limit of each class interval is not included in that class. For example, the class of 1.0–1.5 m excludes wave heights of 1.5 m, which instead belong to the next interval (1.5–2.0 m). A histogram should then be plotted to visualize the distribution of wave heights. Based on the histogram, the most frequent wave height range should be identified, and an analysis should be conducted to interpret the wave conditions at this beach.

a) Frequency distribution of wave data

Wave height range (m)	Frequency (number of waves)
1.0 – 1.5	4
1.5 – 2.0	8
2.0 – 2.5	9
2.5 – 3.0	7
3.0 – 3.5	2

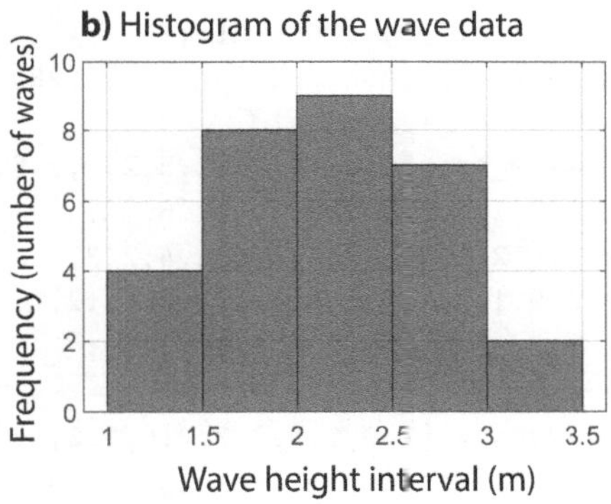

FIGURE 8.4 **a)** Frequency distribution table for the 30 waves of the storm presented in the preceding example. **b)** The corresponding histogram.

The frequency distribution table and the corresponding histogram of wave heights for the 30 waves of the storm are shown in Figure 8.4. The MATLAB and Python scripts presented in Examples 8.1 (script `CH08Mcode1.m`) and 8.2 (`CH08Pcode1.py`) can be used to plot the histogram for this example (Figure 8.4).

The MATLAB `histogram` function generates a histogram plot from data by automatically determining bin ranges or by using user-specified bins. In the MATLAB script presented in Example 8.1, the vector named as `edges` is used to specify the bin boundaries, ensuring that the wave height data are grouped into the desired equal-width intervals.

The Python script presented in Example 8.2 (script `CH08Pcode1.py`) uses the `hist` function from the `matplotlib` library to generate a histogram plot of wave height data. The bin boundaries are defined by a list named `edges`, which allows the data to be grouped into equal-width intervals as desired. The script also customizes the plot appearance by adjusting the grid, spine thickness, and color settings, providing a clear and professional visualization of the distribution of 30 storm wave heights.

The above MATLAB scripts, together with all other scripts presented in this book, are available at: `https://github.com/prog871/CRC-book2025`

Example 8.1

MATLAB code. This MATLAB script, named `CH08Mcode1.m`, plots a histogram for the heights of 30 waves of a storm. This MATLAB script, along with all other scripts presented in this book, is available at:

`https://github.com/prog871/CRC-book2025`

MATLAB Input:

```
% This code plots a histogram
% for heights of 30 storm waves
clc; clear; close all;
set(0,'defaultaxesfontsize',28);
clc; clear; close all;
set(0,'defaultaxesfontsize',28);
% Wave height data
wave_heights = [1.2,1.5,2.0,1.8, ...
1.7,2.2,2.5,2.8,1.9,1.6,  ...
2.4,2.6,3.0,2.7,2.1,1.4,2.3, ...
2.0,1.3,2.9,1.1,3.1,1.9, 2.2, ...
2.7, 2.4,1.8,2.5,1.6, 2.3];
% Define class intervals
edges = [1.0,1.5,2.0,2.5,3.0,3.5];
%Plot histogram using specified bins
subplot('position',[0.2 0.2 0.5 0.7]);
histogram(wave_heights,'BinEdges',edges, ...
'FaceColor','b','EdgeColor','k', ...
'LineWidth',1.5);
set(gca,'linewidth',1.5); grid on;
% end of M - file
```

MATLAB Output:

```
The output of this code is the histogram of wave heights for
    the storm waves (see Figure 8.4).
```

Example 8.2

Python code. This Python script, named `CH08Pcode1.py`, plots a histogram for the heights of 30 waves of a storm.

Python Input:

```
#This code plots a histogram
#for heights of 30 storm waves
import numpy as np
import matplotlib.pyplot as plt
# Wave height data
wave_heights =[1.2,1.5,2.0,1.8,
1.7, 2.2, 2.5, 2.8, 1.9, 1.6,
2.4,2.6,3.0,2.7,2.1,1.4,2.3,
2.0,1.3, 2.9,1.1,3.1,1.9,2.2,
2.7, 2.4,1.8,2.5,1.6,2.3]
# Define class intervals
edges=[1.0,1.5,2.0,2.5,3.0,3.5]
# Create figure and axis
fig,ax=plt.subplots(figsize=(6,5))
# Plot histogram
ax.hist(wave_heights,bins=edges,\
color='b',edgecolor='k',linewidth=1.5)
# Customize plot
ax.grid(True,linestyle='--',alpha=0.6)
ax.spines['top'].set_linewidth(1.5)
ax.spines['right'].set_linewidth(1.5)
ax.spines['left'].set_linewidth(1.5)
ax.spines['bottom'].set_linewidth(1.5)
# Show plot
plt.show()
# End of Python script
```

Python Output:

```
The output of this code is the histogram of wave heights for
    the storm waves (see Figure 8.4).
```

8.3 STANDARD DEVIATION AND ERROR ANALYSIS

Engineering relies heavily on measurements, whether assessing the strength of a material, the speed of a motor, or the voltage in a circuit. However, no measurement is ever perfectly precise. There is always some degree of uncertainty or variation, and understanding this variation is essential for designing reliable systems. This is where standard deviation and error analysis play a crucial role.

Standard deviation quantifies the spread of data around the mean, providing insight into the consistency of measurements. For instance, consider measuring the length of a metal rod multiple times, obtaining values such as 10.1 cm, 9.9 cm, and 10.0 cm. While the average length may be 10 cm, the standard deviation indicates how closely the individual measurements cluster around this mean. A small standard deviation suggests high precision, whereas a larger standard deviation signifies greater variability.

Error analysis helps identify the sources of variation in measurements. Common types of measurement errors include:

- **Instrument limitations**—for example, a ruler with slight inaccuracies or limited precision.
- **Environmental factors**—such as temperature changes affecting material dimensions.
- **Human error**—like misreading a scale or recording incorrect values.

By analyzing these errors, engineers can assess the reliability of their measurements and improve experimental accuracy. Applying standard deviation and error analysis allows them to quantify uncertainty, refine experimental methods, and make informed decisions based on reliable data.

8.3.1 STANDARD DEVIATION

The standard deviation (σ) for a dataset is defined as:

$$\sigma = \sqrt{\frac{\sum(x_i - \bar{x})^2}{n}} \tag{8.1}$$

where:

- x_i represents each individual measurement (e.g., wave height in meters),
- $\bar{x}$ is the mean (average) of all measurements,
- $\sum$ denotes summation over all data points,
- n is the total number of measurements.

Although this equation may initially seem complex, its underlying principle is straightforward: it measures how spread out the data points are by computing the differences from the mean ($\bar{x}$), squaring them, averaging the squared differences, and then taking the square root to return to the original units of measurement.

Note that σ^2 (before taking the square root) is known as the **variance**:

$$\sigma^2 = \frac{\sum(x_i - \bar{x})^2}{n} \tag{8.2}$$

Sometimes, a slightly different formula is used where the denominator is $n-1$ instead of n. This adjustment, known as Bessel's correction, is commonly applied when estimating the standard deviation from a sample rather than the entire population. However, in this book, we consistently use the version divided by n, assuming the data represents the full set of measurements.

Engineering application: Building on the wave height data analyzed in the previous section, where a coastal engineer recorded the heights (in meters) of 30 waves during a storm, we now compute the standard deviation to assess the variability of wave heights. The recorded data is as follows: 1.2 m, 1.5, 2.0, 1.8, 1.7, 2.2, 2.5, 2.8, 1.9, 1.6, 2.4, 2.6,

3.0, 2.7, 2.1, 1.4, 2.3, 2.0, 1.3, 2.9, 1.1, 3.1, 1.9, 2.2, 2.7, 2.4, 1.8, 2.5, 1.6, and 2.3 m. In the histogram analysis, it was observed that the most frequently occurring wave height range was 2.0–2.5 m. We now proceed with calculating the standard deviation to quantify the spread of the wave heights around the mean.

Step 1: Compute the mean wave height.

The mean wave height ($\bar{x}$) is obtained by summing all recorded wave heights and dividing by the total number of observations:

$$\bar{x} = \frac{\sum x_i}{n} = \frac{1.2 + 1.5 + 2.0 + \cdots + 2.3}{30} = \frac{63.5}{30} = 2.12 \text{ m} \tag{8.3}$$

Thus, the average wave height is 2.12 m.

Step 2: Compute squared deviations.

For each recorded wave height, we subtract the mean and square the result. A few calculations are shown below:

$$(1.2 - 2.12)^2 = (-0.92)^2 = 0.8464 \tag{8.4}$$

$$(1.5 - 2.12)^2 = (-0.62)^2 = 0.3844 \tag{8.5}$$

$$(2.0 - 2.12)^2 = (-0.12)^2 = 0.0144 \tag{8.6}$$

This process is repeated for all 30 values.

Step 3: Compute the variance (σ^2).

Summing the squared differences yields:

$$\sum (x_i - \bar{x})^2 = 8.742 \tag{8.7}$$

The variance (σ^2) is then computed by dividing this sum by the number of measurements:

$$\sigma^2 = \frac{\sum (x_i - \bar{x})^2}{n} = \frac{8.742}{30} = 0.2914 \tag{8.8}$$

Step 4: Compute the standard deviation.

Finally, the standard deviation is obtained by taking the square root of the variance:

$$\sigma = \sqrt{\sigma^2} = \sqrt{0.2914} = 0.54 \text{ m} \tag{8.9}$$

Interpretation of the standard deviation: A standard deviation of 0.54 m indicates that most wave heights (approximately **68%** of the data, assuming a normal distribution) fall within $\bar{x} \pm \sigma$, or between:

$$2.12 - 0.54 = 1.58 \text{ m}, \quad 2.12 + 0.54 = 2.66 \text{ m} \tag{8.10}$$

This aligns with the histogram analysis, where the majority of wave heights were in the range of 1.5–2.5 m (see Figure 8.4).

8.3.2 USING MATLAB AND PYTHON TO COMPUTE MEAN AND STANDARD DEVIATION

The above calculations of the mean value and standard deviation of a dataset can be efficiently performed using built-in functions in MATLAB and Python. In MATLAB, the functions `mean` and `std` are used to compute the mean and standard deviation, respectively, as shown in script `CH08Mcode2.m` (see Example 8.3). In Python, the `NumPy` library provides the functions `numpy.mean()` and `numpy.std()` to perform the same calculations (see script `CH08Pcode2.py` in Example 8.4).

Example 8.3

MATLAB code. This MATLAB script, named `CH08Mcode2.m`, calculates the mean wave height and standard deviation for the heights of 30 waves of a storm.

MATLAB Input:

```
%This code mean wave height
%and standard deviation for
%heights of 30 storm waves
clc; clear; close all;
%Wave height data (in meters)
wave_heights=[1.2,1.5,2.0,1.8, ...
1.7,2.2,2.5,2.8,1.9,1.6,  ...
2.4,2.6,3.0,2.7,2.1,1.4,2.3, ...
2.0,1.3,2.9,1.1,3.1,1.9, 2.2, ...
2.7, 2.4,1.8,2.5,1.6, 2.3];
% Compute mean wave height
mean_height=mean(wave_heights);
% Compute standard deviation
std_dev = std(wave_heights,1);
% Display results
fprintf('Mean height: %.2f m\n', ...
mean_height);
fprintf('Std Deviation: %.2f m\n', ...
std_dev);
% end of M - file
```

MATLAB Output:

```
Mean height: 2.12 m
Std Deviation: 0.54 m
```

Example 8.4

Python code. This Python script, named CH08Pcode2.py, calculates the mean wave height and standard deviation for the heights of 30 waves of a storm.

Python Input:

```
#This code mean wave height
#and standard deviation for
#heights of 30 storm waves
import numpy as np
# Wave height data (in meters)
wave_heights=[1.2,1.5,2.0,1.8,
 1.7,2.2,2.5,2.8,1.9,1.6,
 2.4,2.6,3.0,2.7,2.1,1.4, 2.3,
 2.0,1.3,2.9,1.1,3.1,1.9, 2.2,
 2.7,2.4,1.8,2.5,1.6,2.3]
# Compute mean wave height
mean_height=np.mean(wave_heights)
# Compute standard deviation
# ddof=0 population std deviation
std_dev=np.std(wave_heights, ddof=0)
# Display results
print(f'Mean height: {mean_height:.2f} m')
print(f'Std Deviation: {std_dev:.2f} m')
# End of Python script
```

Python Output:

```
Mean height: 2.12 m
Std Deviation: 0.54 m
```

8.3.3 ERROR ANALYSIS

In engineering, precise measurements are crucial, yet they are rarely identical across multiple trials. This variability is particularly evident in fields like structural engineering, where material properties must be tested rigorously to ensure safety and reliability. A classic example is the compressive strength of concrete, a critical factor in infrastructure design.

Despite using the same batch of concrete, we often obtain a range of values rather than a single, fixed number. To quantify and understand this variation, statistical tools such as mean and standard deviation are employed. The mean provides a central estimate, while the standard deviation quantifies the spread of data around that mean. However, why does this spread exist, and how much confidence can we place in our measurements? This is where error analysis becomes essential.

Engineering application: Measuring compressive strength of concrete. Consider a civil engineer testing the compressive strength of 12 concrete samples intended for a sea wall structure (Figure 8.5a). The recorded strengths (in megapascals, MPa) are: 32.1, 33.4, 31.8, 32.9, 34.2, 33.0, 31.5, 32.7, 33.8, 32.3, 34.0, and 31.9 (Figure 8.5b). Each value represents the maximum stress the sample can withstand before failure.

Step 1: Calculating the mean ($\bar{x}$).

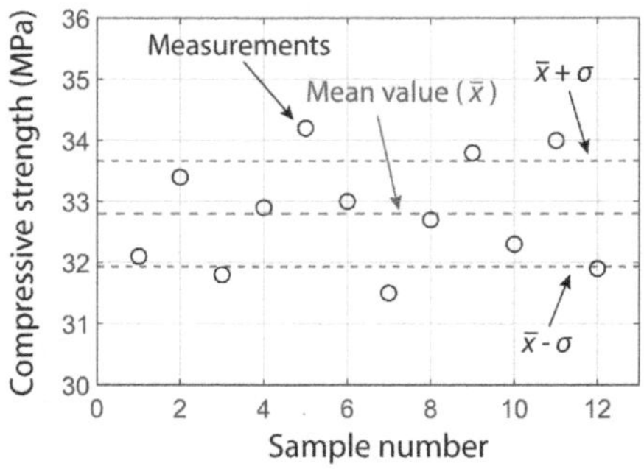

FIGURE 8.5 **a)** Concrete seawall in Dawlish, UK. **b)** Measurements of the compressive strength of concrete (the circles), the mean value, $\bar{x}$, (the dashed blue line), and the interval $\bar{x} \pm \sigma$ (the dashed black lines).

The mean compressive strength is given by:

$$\bar{x} = \frac{\sum x_i}{n} = \frac{32.1 + 33.4 + 31.8 + 32.9 + ... + 31.9}{12} \tag{8.11}$$

$$\bar{x} = \frac{393.6}{12} = 32.8 \text{ MPa} \tag{8.12}$$

Thus, the average compressive strength is 32.8 MPa (see Figure 8.5b).

Step 2: Calculating the standard deviation (σ).

The standard deviation is computed as:

$$\sigma = \sqrt{\frac{\sum (x_i - \bar{x})^2}{n}} \tag{8.13}$$

By computing the squared deviations from the mean for each value and summing them, we get:

$$\sum (x_i - \bar{x})^2 = 9.06 \tag{8.14}$$

$$\sigma = \sqrt{\frac{9.06}{12}} = 0.87 \text{ MPa} \tag{8.15}$$

The standard deviation of 0.87 MPa indicates that most of the recorded strengths (approximately 68% of values, assuming a normal distribution) lie between $\bar{x} \pm \sigma$:

$$32.8 - 0.87 = 31.93 \text{ MPa} \quad \text{and} \quad 32.8 + 0.87 = 33.67 \text{ MPa} \tag{8.16}$$

This range, 31.93 - 33.67 MPa, provides insight into the typical variation in compressive strength for this batch (see Figure 8.5b).

The MATLAB and Python scripts provided in Examples 8.5 (script `CH08Mcode3.m`) and 8.6 (script `CH08Pcode3.py`) can be used to perform the calculations for the previous problem and to plot Figure 8.5b.

Example 8.5

MATLAB code. This MATLAB script, named `CH08Mcode3.m`, calculates the mean and standard deviation for a dataset of 12 measurements of compressive strength of concrete and plots them.

MATLAB Input:

```
%This code gives mean
%and %standard deviation for
%12 records and plots them
clc; clear; close all;
set(0,'defaultaxesfontsize',27)
%Data:Compressive Strength(MPa)
A = [32.1, 33.4,31.8, 32.9,34.2, ...
33.0,31.5,32.7,33.8,32.3,34.0,31.9];
mean_A = mean(A);
std_A = std(A, 1);
% Plot the values as markers
subplot('position',[.2 .2 .5 .7]);
plot(1:12,A,'ko','MarkerSize',16, ...
   'linewidth',2);
hold on;
yline(mean_A, '--b','LineWidth',2.5);
% Plot (Mean +/- Sigma)
yline(mean_A+std_A,'--k','LineWidth',2);
yline(mean_A-std_A,'--k','LineWidth',2);
ylim([30 36]); xlim([0 13]);
fprintf('Mean = %.2f MPa\n',mean_A);
fprintf('Std Deviation= %.2f MPa\n',std_A);
grid on;set(gca,'linewidth',1.5)
% end of M - file
```

MATLAB Output:

```
Mean = 32.80 MPa
Std Deviation= 0.87 MPa
This code also plots Figure 8.5b.
```

Example 8.6

Python code. This Python script, named `CH08Pcode3.py`, calculates the mean and standard deviation for a dataset of 12 measurements of compressive strength of concrete and plots them.

Python Input:

```
#This code gives mean
#and %standard deviation for
#12 records and plots them
import numpy as np
import matplotlib.pyplot as plt
#Data:Compressive Strength(MPa)
A=np.array([32.1,33.4,31.8,32.9,34.2,
33.0,31.5,32.7,33.8,32.3,34,31.9])
mean_A = np.mean(A)
std_A = np.std(A, ddof=0)
plt.figure(figsize=(7, 8))
plt.plot(range(1, 13), A, 'ko',\
markersize=10, linewidth=2)
plt.axhline(mean_A, linestyle='--',\
color='b', linewidth=2.5)
plt.axhline(mean_A+std_A,linestyle='--',\
color='k', linewidth=2)
plt.axhline(mean_A-std_A,linestyle='--',\
color='k', linewidth=2)
plt.ylim([30, 36])
plt.xlim([0, 13])
plt.grid(True, linewidth=1.5)
print(f'Mean = {mean_A:.2f} MPa')
print(f'Std Deviation = {std_A:.2f} MPa')
plt.show()
# End of Python script
```

Python Output:

```
Mean = 32.80 MPa
Std Deviation = 0.87 MPa
This code also plots Figure 8.5b.
```

8.3.4 SOURCES OF MEASUREMENT VARIATION AND THEIR MITIGATION

Although the standard deviation quantifies the variability, it does not explain why the variation occurs. Error analysis helps engineers assess the accuracy and reliability of measurements by identifying the sources of error.

Types of errors in engineering measurements. There are generally two main types of errors in engineering measurements, as listed below:

(i) Random errors:

- These arise due to small, unpredictable variations in the measurement process.
- In our concrete strength example, random errors could result from slight inconsistencies in the cement mix, curing conditions, or testing machine sensitivity.
- Random errors increase data scatter, but they do not systematically bias the results in one direction.

(ii) Systematic errors:

- These are persistent biases in measurement, often due to faulty equipment or improper methodology.
- In this case, if the testing machine consistently overestimates strengths by 0.5 MPa due to improper calibration, all values will be shifted upwards, misleading engineers about the true strength of the concrete.
- Systematic errors affect the accuracy of results and can lead to unsafe engineering decisions.

Impact of systematic errors on engineering decisions: For the previous example on the compressive strength of concrete samples, suppose the engineer discovers that the testing machine has a consistent overestimation bias of 0.5 MPa. This means the true mean strength is:

$$32.8 - 0.5 = 32.3 \text{ MPa} \tag{8.17}$$

If the design requirements specify a minimum compressive strength of 32.5 MPa, this systematic error could falsely indicate that the concrete meets the safety standards, when in reality, it falls short. This could have serious consequences for bridge safety and structural integrity.

Correcting and mitigating errors: To ensure accurate engineering measurements, the following steps can be taken:

i) Minimizing random errors:

- Increase the number of samples to reduce statistical uncertainty.
- Improve material consistency (e.g., better cement mixing techniques).
- Maintain controlled curing conditions for all samples.

ii) Eliminating systematic errors:

- Regular calibration of testing equipment to avoid measurement bias.
- Cross-checking results with alternative testing methods.
- Using control samples with known properties to verify accuracy.

8.4 PROBABILITY ANALYSIS

Probability analysis is a fundamental tool in engineering, allowing us to quantify uncertainty and make informed decisions. Whether designing a bridge, assessing structural reliability, or predicting equipment failure, engineers rely on probability to evaluate risks and optimize performance. Previously, we explored histograms, which visually represent how data is distributed. Now, we will delve into the rules of probability, which help us analyze data systematically and make predictions based on observed trends. These concepts will also serve as a foundation for our next topic, the normal distribution, a key model for engineering measurements.

What is probability?

Probability is a measure of how likely an event is to occur. It is expressed as a number between **0** (impossible) and **1** (certain). For example:

- The probability that a structural component fails under excessive load.
- The probability that a sensor gives an incorrect reading.

Engineering connection:

In engineering, probability helps us predict outcomes when exact results cannot be guaranteed. For instance, when designing a machine, we cannot say with certainty when a part will wear out, but we can estimate the likelihood that it will fail before its warranty period ends. This helps in planning maintenance schedules and ensuring safety and reliability.

8.4.1 BASIC PROBABILITY RULES FOR ENGINEERS

Probability follows specific rules that allow us to calculate the likelihood of events. Think of these rules as an instruction manual for working with uncertainty in engineering problems.

(a) The addition rule (for mutually exclusive events)

If two events cannot happen at the same time (i.e., they are mutually exclusive), the probability of either one occurring is the sum of their individual probabilities. The addition rule is given by the following formula:

$$P(A \text{ or } B) = P(A) + P(B) \tag{8.18}$$

where:

- $P(A)$ = probability of event A occurring,
- $P(B)$ = probability of event B occurring,
- $P(A \text{ or } B)$ = probability that either event A or event B occurs. The symbol $P(A \text{ or } B)$ is also written as $P(A \cup B)$, indicating the union of events A and B:

$$P(A \cup B) = P(A) + P(B) \tag{8.19}$$

Example: Suppose you are testing two types of bolts:

- Bolt A fails 10% of the time: $P(A) = 0.1$
- Bolt B fails 15% of the time: $P(B) = 0.15$

Calculate the probability that either Bolt A or Bolt B fails, assuming only one bolt is used at a time.

Solution:

If only one bolt is used at a time, the probability that either A or B fails is calculated as:

$$P(A \cup B) = 0.1 + 0.15 = 0.25 \quad (25\%) \tag{8.20}$$

Note that if the events are not mutually exclusive (i.e., both could happen together), we adjust for double-counting:

$$P(A \cup B) = P(A) + P(B) - P(A \cap B) \tag{8.21}$$

where:

- $P(A)$ = probability of event A occurring,
- $P(B)$ = probability of event B occurring,
- $P(A \cap B)$ = probability that both events A and B occur at the same time, which is also written as $P(A \text{ and } B)$,
- $P(A \cup B)$ = probability that either event A or event B (or both) occurs.

(b) The multiplication rule (for independent events)

If two events occur independently (i.e., one does not affect the other), the probability of both happening together is the product of their individual probabilities, which is given by the following formula:

$$P(A \text{ and } B) = P(A) \times P(B) \tag{8.22}$$

It is also written as:

$$P(A \cap B) = P(A) \times P(B) \tag{8.23}$$

which represents the probability that both events A and B occur simultaneously.

Example: Imagine two sensors monitoring a structure:

- Sensor 1 fails 5% of the time: $P(A) = 0.05$
- Sensor 2 fails 2% of the time: $P(B) = 0.02$

By assuming that they work independently, calculate the probability that both sensors fail at the same time.

Solution:

If they work independently, the probability that both sensors fail at the same time is:

$$P(A \text{ and } B) = 0.05 \times 0.02 = 0.001 \quad (0.1\%) \tag{8.24}$$

This rule is important in safety system design. Engineers often use multiple sensors to reduce the risk of complete system failure. Even if one sensor fails, the system may still function properly.

(c) The complement rule

The probability that an event does not happen is 1 minus the probability that it does happen. This rule is represented by the following formula:

$$P(\text{not } A) = 1 - P(A) \tag{8.25}$$

Example: A water pump has a 20% chance of failing during a 12-hour shift: $P(failure) = 0.2$. Calculate the probability that the pump does not fail.

Solution:

The probability that the pump does not fail is:

$$P(\text{no failure}) = 1 - 0.2 = 0.8 \quad (80\%) \tag{8.26}$$

Reliability analysis in engineering often relies on the complement rule. For example, if a critical machine in a construction project has a high failure probability, engineers may introduce backup systems to minimize downtime.

Engineering application: To see how probability applies in real engineering scenarios, let us consider the stress on a steel beam. A steel beam in a structure is designed to withstand a maximum allowable stress of 250 MPa. Based on previous tests, the actual stress levels follow the probability distribution listed in Table 8.1. What is the probability that the stress exceeds the allowable limit of 250 MPa?

TABLE 8.1

Probability distribution of stress levels in a steel beam

Stress, σ (MPa)	Probability
200	0.30
220	0.40
250	0.20
270	0.10

Solution:

The stress exceeds 250 MPa only when it reaches 270 MPa. We have:

$$P(\text{Stress} > 250) = P(270) = 0.10 \tag{8.27}$$

8.4.2 RETURN PERIOD AND DESIGN EVENTS

In civil engineering, extreme events such as floods, earthquakes, and hurricanes are of particular concern. Engineers must design infrastructure to withstand events that occur rarely but have significant consequences. One way to quantify the likelihood of such events is through the return period, which is the average interval of time between occurrences of a given event.

Definition of return period:

The return period, T, of an event with probability P in any given year is defined as:

$$T = \frac{1}{P} \tag{8.28}$$

For example:

- A 100-year flood has a 1 in 100 chance ($P = 0.01$) of occurring in any given year.
- A 500-year earthquake has a 1 in 500 chance ($P = 0.002$) of occurring in any given year.

Probability of occurrence over multiple years:

While the return period provides a long-term average, it does not mean that a 100-year flood will occur exactly once every 100 years. Instead, we use probability theory to determine the likelihood of at least one such event occurring within a given time span.

The probability that an event does not occur in a single year is:

$$1 - P \tag{8.29}$$

The probability that it does not occur over n years is:

$$(1 - P)^n \tag{8.30}$$

Thus, the probability that the event occurs at least once over n years is:

$$P(\text{at least once}) = 1 - (1 - P)^n \tag{8.31}$$

Example: Consider a 100-year flood event. Calculate the probability that it does not occur over a 30-year period, as well as the probability that at least one such flood occurs within the same timeframe.

Solution:

As it is a 100-year flood, we know that: $P = 0.01$ per year. The probability that it does not occur in a 30-year period is:

$$(1 - 0.01)^{30} = (0.99)^{30} = 0.74 \tag{8.32}$$

Thus, the probability that at least one 100-year flood occurs in 30 years is:

$$P(\text{at least once}) = 1 - 0.74 = 0.26 \quad (26\%) \tag{8.33}$$

This means that a structure designed for a 100-year flood has a 26% chance of experiencing such a flood within 30 years.

Engineering application: In civil engineering, culverts are designed to manage flood flows and prevent road damage. Consider a culvert that faces two independent risks: high rainfall, which exceeds the culvert's capacity 8% of the time ($P(R) = 0.08$), and debris blockage, which restricts flow 5% of the time ($P(D) = 0.05$). We are tasked with determining the probability that the culvert will fail in a given year, as well as the probability that it will survive for 10 years.

Solution:

- **Annual failure probability:** Since the events are not mutually exclusive, we use the general addition rule (Equation 8.21) to calculate the total probability of failure:

$$P(\text{failure}) = P(R) + P(D) - P(R \cap D) \tag{8.34}$$

  For independent events, the probability of both events occurring together is given by the product $P(R \cap D) = P(R) \times P(D)$ (Equation 8.22):

$$P(R \cap D) = 0.08 \times 0.05 = 0.004 \tag{8.35}$$

  Substituting into Equation 8.34:

$$P(\text{failure}) = 0.08 + 0.05 - 0.004 = 0.126 \quad (12.6\%) \tag{8.36}$$

- **No failure in one year:** Using the complement rule (Equation 8.25), we calculate the probability of no failure in a single year:

$$P(\text{no failure}) = 1 - 0.126 = 0.874 \quad (87.4\%) \tag{8.37}$$

- **No failure in 10 years:** Assuming the events in each year are independent, we apply the formula for no failure over multiple years (Equation 8.30):

$$P(\text{no failure in 10 years}) = 0.874^{10} \tag{8.38}$$

$$P(\text{no failure in 10 years}) = 0.260 \quad (26.0\%) \tag{8.39}$$

- **At least one failure in 10 years:** Finally, we calculate the probability of at least one failure occurring within the 10-year period using the formula for at least one event (Equation 8.31):

$$P(\text{at least one failure}) = 1 - 0.260 = 0.74 \quad (74\%) \tag{8.40}$$

The return period, which represents the average time between failures, is given by:

$$T = \frac{1}{0.126} = 7.94 \text{ years} \tag{8.41}$$

As per Equation 8.28, the return period suggests that, on average, the culvert will fail approximately once every 7.9 years. With a 74% probability of failure over the next 10 years, this may prompt the need for a larger culvert design to reduce the risk of failure.

8.5 NORMAL DISTRIBUTION

So far, we have dealt with specific distributions, such as the uniform distribution. The normal distribution, also known as the Gaussian distribution, is one of the most important in statistics and is widely used in various fields [2]. It is defined by its bell-shaped curve and characterized by its mean (μ) and standard deviation (σ). In this section, we explore the normal distribution, the concept of z-scores, and how to use the cumulative distribution function (CDF) to calculate probabilities.

Normal distribution curve:

The normal distribution has a probability density function (PDF), introduced as [2]:

$$f(x) = \frac{1}{\sigma\sqrt{2\pi}} e^{-\frac{(x-\mu)^2}{2\sigma^2}} \tag{8.42}$$

where:

- $f(x)$ is the probability density function (PDF),
- x is the variable,
- μ is the mean (previously in this chapter, the mean was represented by $\bar{x}$),
- σ is the standard deviation, and
- π is the mathematical constant approximately equal to 3.1416.

Figure 8.6 presents examples of probability density functions (PDFs) for normal distributions with different values of μ (mean) and σ (standard deviation). As σ increases, the distribution becomes wider and flatter, reflecting greater variability in the data.

Note that Equation 8.42 gives the height of the curve at any point x, but it does not directly give probabilities over a range. To calculate the probability that a random variable X falls within a certain range, we use the cumulative distribution function (CDF) as follows.

The cumulative distribution function (CDF) and z-scores:

The cumulative distribution function (CDF), denoted as $\Phi(z)$, represents the probability that a random variable Z is less than or equal to a specific z-score, i.e., $P(Z \leq z)$. Mathematically, the CDF is the area under the PDF curve from $-\infty$ to z:

$$\Phi(z) = \int_{-\infty}^{z} \frac{1}{\sqrt{2\pi}} e^{-\frac{t^2}{2}}\, dt \tag{8.43}$$

This integral cannot be solved analytically, which is why software (such as MATLAB's `normcdf`, Excel's `NORM.DIST`, or Python's `scipy.stats.norm.cdf`) computes it numerically. The CDF provides the cumulative probability up to a specific z-score.

Figure 8.7 presents the probability density functions (PDFs) and cumulative distribution functions (CDFs) for two normal distributions. The area under each PDF curve is equal to one (see Figure 8.7, left), which ensures that the maximum value of the corresponding CDF reaches one (see Figure 8.7, right).

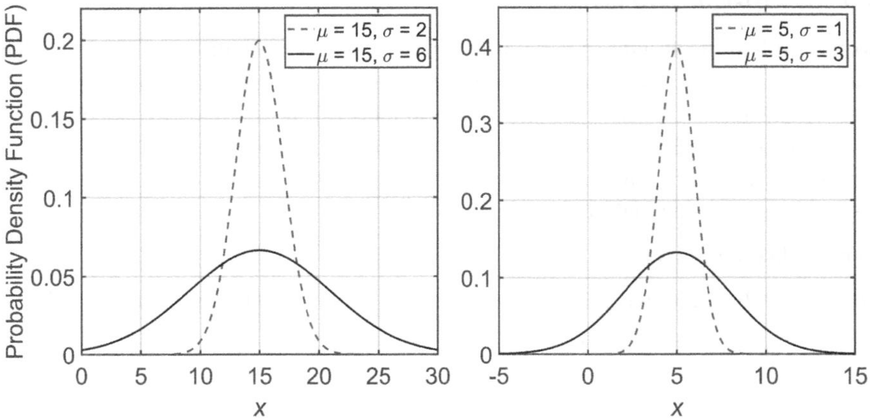

FIGURE 8.6 Examples of probability density functions (PDF) for normal distribution with varying μ (mean) and σ (standard deviation) values.

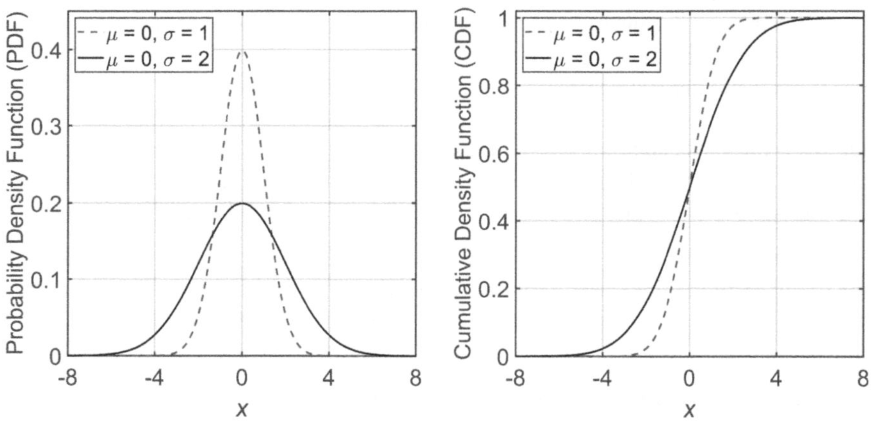

FIGURE 8.7 Examples of probability density functions (PDF) (left) for two normal distributions and corresponding cumulative distribution functions (CDF) (right).

Z-score calculation:

The z-score is a standardized value that represents the number of standard deviations a data point is from the mean. It is calculated using the formula:

$$z = \frac{x - \mu}{\sigma} \tag{8.44}$$

where:

- x is the value of the data point,
- μ is the mean,
- σ is the standard deviation.

The z-score tells us how far x is from the mean in terms of standard deviations. A key reason for using z-scores is that they allow us to compare values from different normal distributions by converting them into a common scale. This is particularly useful in statistical analysis and hypothesis testing. Additionally, many software packages such as MATLAB (`normpdf` and `normcdf`), Python's `scipy.stats.norm`, and Excel's `NORM.DIST` rely on z-scores to compute probabilities from the normal distribution. By standardizing the distribution to the standard normal form ($\mu = 0, \sigma = 1$), these tools simplify calculations and improve computational efficiency.

Engineering application: In the design of a concrete foundation, tests indicate that the compressive strength of concrete samples (f_c) follows a normal distribution with a mean (μ) of 30 MPa and a standard deviation (σ) of 3 MPa. The foundation must sustain a minimum compressive strength of 25 MPa (see Figure 8.8).

- Determine the probability that a randomly selected sample does not meet this requirement.
- Calculate the strength value that corresponds to the 95th percentile of the distribution.

Solution:

Part 1: Probability that a sample does not meet the requirement.

To find the probability that the strength is less than 25 MPa, we first calculate the z-score for 25 MPa using Equation (8.44):

$$z = \frac{25 - 30}{3} = \frac{-5}{3} = -1.67 \tag{8.45}$$

A z-score of -1.67 means 25 MPa is 1.67 standard deviations below the mean.

Find the probability

Now, we aim to calculate $P(f_c < 25)$, which is equivalent to $P(Z < -1.67) = \Phi(-1.67)$. This represents the cumulative distribution function (CDF) value at $z = -1.67$, which indicates the probability that a randomly selected concrete sample will have a compressive strength less than 25 MPa. Using software tools such as Excel (with the function `NORM.DIST`), MATLAB (with `normcdf`), or Python (with `scipy.stats.norm.cdf`), we input $\mu = 30$, $\sigma = 3$, and $z = -1.67$. For example, in MATLAB, you would simply enter the following command: `normcdf(-1.67)`. The software then computes the CDF directly, providing the result:

$$P(Z < -1.67) = 0.0475 \tag{8.46}$$

Or, it can be rewritten as:

$$P(f_c < 25) = 0.0475 \tag{8.47}$$

So, there is approximately a 4.75% chance the concrete will have a strength below 25 MPa (see Figure 8.8).

Complement rule:

To find the probability that the concrete meets the requirement of at least 25 MPa, we use the complement rule:

$$P(f_c \geq 25) = 1 - P(f_c < 25) = 1 - 0.0475 = 0.9525 \tag{8.48}$$

Thus, there is a 95.25% chance the concrete will be strong enough (see Figure 8.8).

Part 2: Strength value corresponding to the 95th percentile.

The 95th percentile means we need to find the strength value f_c such that:

$$P(f_c \leq X) = 0.95 \tag{8.49}$$

The z-score corresponding to the 95th percentile can be found using MATLAB, Python, or Excel as detailed in the following:

- MATLAB: Use the `norminv` function:

$$z = \texttt{norminv}(0.95, 0, 1) \tag{8.50}$$

- Python: Use the `scipy.stats` module:

$$z = \texttt{scipy.stats.norm.ppf}(0.95) \tag{8.51}$$

- Excel: Use the formula:

$$z = \texttt{NORM.S.INV}(0.95) \tag{8.52}$$

The calculated z-score value is:

$$z = 1.645 \tag{8.53}$$

We use the inverse z-score formula:

$$X = \mu + z\sigma \tag{8.54}$$

Substituting the given values ($\mu = 30$ MPa, $\sigma = 3$ MPa):

$$X = 30 + (1.645 \times 3) \tag{8.55}$$

$$X = 30 + 4.935 = 34.94\, MPa \tag{8.56}$$

Thus, the strength value corresponding to the 95th percentile is:

$$f_c = 34.94 \text{ MPa} \tag{8.57}$$

The MATLAB and Python scripts provided in Examples 8.7 (script `CH08Mcode4.m`) and 8.8 (script `CH08Pcode4.py`) can be employed to solve this example. As shown in the Probability Density Function (PDF) and Cumulative Distribution Function (CDF) plots (Figure 8.8), the probability of a randomly selected concrete sample having a compressive strength below 25 MPa can be determined using either representation. When using the PDF plot, the probability corresponds to the shaded area under the normal distribution curve to the left of 25 MPa. In contrast, when using the CDF plot, the probability is obtained directly from the cumulative probability value on the y-axis at $f_c = 25$ MPa.

The normal distribution is a powerful tool in statistics, and understanding z-scores and the CDF allows us to calculate probabilities effectively. By using software functions such as `normcdf` in MATLAB, `NORM.DIST` in Excel, or `scipy.stats.norm.cdf` in Python, we can easily compute the cumulative probability for any z-score. Similarly, to find the inverse (i.e., determine the z-score corresponding to a given probability), we can use `norminv` in MATLAB, `NORM.S.INV` in Excel, or `scipy.stats.norm.ppf` in Python, making statistical analysis more accessible.

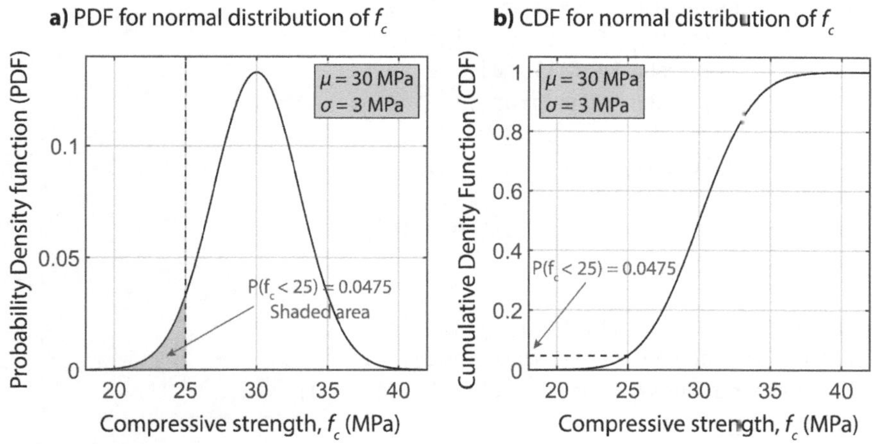

FIGURE 8.8 **a)** Probability Density Function (PDF) for normal distribution of compressive strength of concrete samples (f_c) showing the probability that a randomly selected sample does not meet a compressive strength of 25 MPa. **b)** The corresponding Cumulative Distribution Function (CDF) curve showing the same.

Example 8.7

MATLAB code. This MATLAB script, named `CH08Mcode4.m`, plots the PDF and CDF for normal distribution of compressive strength of concrete samples.

MATLAB Input:

```
%This code plots the CDF
%for normal distribution
clc; clear; close all;
set(0,'defaultaxesfontsize',27)
mu=30;sigma=3; %mean & St. dev.(MPa)
x=25;%Strength value to check (MPa)
z = (x - mu) / sigma; %z-score
probab= normcdf(z);% probability
% Display the probability
fprintf(['The probability that ' ...
'compressive strength is less' ...
'than 25 MPa = %.4f\n'],probab);
subplot('position',[0.07 .1 .37 .7]);
fplot(@(x) normpdf(x, mu, sigma), ...
[mu-4*sigma,mu+4*sigma],'k','LineWidth',2);
hold on; grid on;
% Shade area to the left of x = 25
```

```
x_fill=linspace(mu - 4*sigma, x,100);
y_fill = normpdf(x_fill, mu, sigma);
fill([x_fill, x], [y_fill, 0], 'r', ...
'FaceAlpha', 0.3, 'EdgeColor', 'none');
% Add a vertical line at x = 25
line([x x],[0 10],'Color','k', ...
'LineStyle','--','linewidth',2);
set(gca,'linewidth',2); ylim([0 0.14])
%%%%%%%%%%%%%%%%%%%%%%%%%%
% Add CDF plot to the right
subplot('position',[.55 .1 .37 .7]);
fplot(@(x) normcdf((x - mu)/sigma), ...
[mu-4*sigma,mu+4*sigma],'k','LineWidth',2);
hold on; grid on;
line([x x], [0 probab],'Color','k', ...
'LineStyle', '--', 'LineWidth', 2.0);
line([mu-4*sigma x], [probab probab], ...
'Color','k','LineStyle','--','LineWidth',2);
set(gca, 'linewidth', 2); ylim([0 1.05]);
% end of M - file
```

MATLAB Output:

```
The probability that compressive strength is lessthan 25 MPa =
    0.0478.
This code also plots Figure 8.8.
```

Example 8.8

Python code. This Python script, named `CH08Pcode4.py`, plots the PDF and CDF for normal distribution of compressive strength of concrete samples.

Python Input:

```
#This code plots the CDF
#for normal distribution
import numpy as np
import matplotlib.pyplot as plt
from scipy.stats import norm
# Parameters
mu=30 #Mean (MPa)
sigma=3 #Standard dev. (MPa)
```

```
x=25 #Strength to check (MPa)
z = (x - mu) / sigma
probab = norm.cdf(z)
# Display probability
print(f'The probability that \
compressive strength is less than\
25 MPa = {probab:.4f}')
# Create figure
fig,axes=plt.subplots(1,2,figsize=(12,6))
# Plot PDF with shaded region
x_vals=np.linspace(mu - 4 * sigma,\
   mu + 4 * sigma, 1000)
pdf_vals=norm.pdf(x_vals,mu,sigma)
axes[0].plot(x_vals, \
 pdf_vals,'k',linewidth=2)
axes[0].grid(True)
# Shade area to left of x = 25
x_fill=np.linspace(mu-4*sigma,x,100)
y_fill=norm.pdf(x_fill,mu,sigma)
axes[0].fill_between(x_fill,\
 y_fill,color='r',alpha=0.3)
# A vertical dashed line at x=25
axes[0].axvline(x,color='k',\
  linestyle='--',linewidth=2)
# Formatting
axes[0].set_ylim([0, 0.14])
axes[0].set_xlabel('Strength (MPa)')
axes[0].set_ylabel('Probab. Density')
axes[0].tick_params(axis='both',\
       labelsize=14)
# plot CDF with reference lines
cdf_vals=norm.cdf((x_vals-mu)/sigma)
axes[1].plot(x_vals,cdf_vals,\
   'k',linewidth=2)
axes[1].grid(True)
# Add dashed reference lines
axes[1].axvline(x, color='k',\
 linestyle='--', linewidth=2)
axes[1].axhline(probab,xmin=0,\
xmax=(x-(mu-4*sigma))/(8*sigma),\
color='k',linestyle='--',linewidth=2)
# Formatting
axes[1].set_ylim([0, 1.05])
```

```
axes[1].set_xlabel('Strength (MPa)')
axes[1].set_ylabel('Cumulative Probab.')
axes[1].tick_params(axis='both',labelsize=14)
plt.tight_layout()
plt.show()
% end of Python script.
```

Python Output:

```
The probability that compressive strength is less than 25 MPa =
    0.0478.
This code also plots Figure 8.8.
```

Engineering application: A hydroelectric dam relies on the daily river flow (Q) to generate electrical power. Engineers have compiled an extensive dataset of historical daily river flow measurements (in cubic meters per second, m^3/s) over a 30-year period, from March 1995 to March 2025, to evaluate the dam's operational reliability and safety under varying hydrological conditions. From this comprehensive dataset of approximately 10,950 daily observations, a representative sample of 40 randomly selected daily flow values is provided in Table 8.2. Statistical analysis assumes that the daily flow values conform to a normal distribution, with the sample used to estimate the population mean and standard deviation.

The dam requires a minimum daily flow of Q = 300 m^3/s to maintain stable hydropower generation, while a flow exceeding Q = 700 m^3/s activates a flood warning due to potential risks to the dam's structural integrity and downstream areas. Using the sample data in Table 8.2, calculate the following:

- The probability that the river flow exceeds 700 m^3/s in a single day (flood risk), assuming independence of daily flow events.
- The probability that the river flow falls below 300 m^3/s at least once in 2 months (power failure risk).

Solution:

Step 1: Calculate the mean (μ) and standard deviation (σ)

- Number of data points, $n = 40$
- Flow values (in m^3/s) range from 350 m^3/s to 730 m^3/s

Mean (μ):

$$\mu = \frac{1}{n}\sum_{i=1}^{n} x_i \tag{8.58}$$

TABLE 8.2

Representative sample of 40 daily river flow observations (m³/s) randomly selected from a 30-year dataset

Day	River flow, Q (m³/s)	Day	River flow, Q (m³/s)
1	450	21	540
2	620	22	390
3	700	23	460
4	350	24	590
5	480	25	710
6	520	26	370
7	550	27	540
8	600	28	500
9	730	29	620
10	400	30	670
11	530	31	510
12	610	32	680
13	580	33	420
14	490	34	570
15	720	35	630
16	510	36	460
17	650	37	690
18	470	38	580
19	380	39	410
20	560	40	530

where x_i represents the individual daily flow values. Using the flow values in the table, the sample mean is calculated as:

$$\mu = 546.75\,\text{m}^3/\text{s} \tag{8.59}$$

Variance and standard deviation (σ):

The formula for population standard deviation is:

$$\sigma = \sqrt{\frac{1}{n}\sum_{i=1}^{n}(x_i - \mu)^2} \tag{8.60}$$

The sum of squared differences from the mean is computed to be:

$$\sum_{i=1}^{40}(x_i - \mu)^2 = 438677.50 \tag{8.61}$$

Now, divide by $n = 40$ and take the square root:

$$\sigma = \sqrt{\frac{438677.50}{40}} = 104.72\,\text{m}^3/\text{s} \tag{8.62}$$

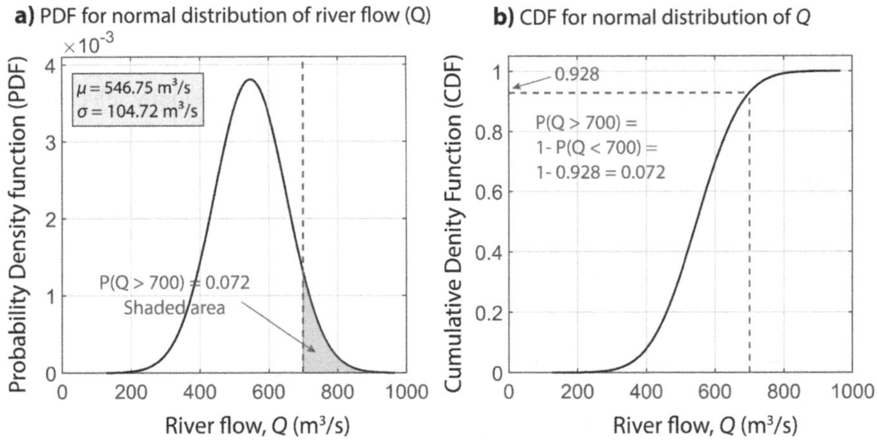

FIGURE 8.9 **a)** Probability Density Function (PDF) for normal distribution of river flow (Q) showing the probability that the river flow exceeds Q =700 m³/s in a single day. **b)** The corresponding Cumulative Distribution Function (CDF) curve showing the same.

Figure 8.9 shows the Probability Density Function (PDF) and Cumulative Distribution Function (CDF) of the normal distribution for river flow, using the mean and standard deviation values given above.

Step 2: Calculate the probabilities

(a) Flood risk: The probability that the river flow exceeds 700 m³/s in a single day.

We calculate the z-score for $Q = 700$ m³/s using the formula:

$$z = \frac{Q - \mu}{\sigma} \tag{8.63}$$

Substituting the known values:

$$z = \frac{700 - 546.75}{104.72} = 1.46 \tag{8.64}$$

Using software tools such as Excel (with the function `NORM.DIST`), MATLAB (with `normcdf`), or Python (with `scipy.stats.norm.cdf`), we find that the cumulative probability for $z = 1.46$ is approximately:

$$P(Q < 700) = P(Z < 1.46) = 0.9279 \tag{8.65}$$

Thus, the probability that the flow exceeds 700 m³/s is:

$$P(Q > 700) = 1 - P(Z < 1.46) = 1 - 0.9279 = 0.0721 \tag{8.66}$$

Flood risk: The probability that the river flow exceeds 700 m³/s is approximately 7.2%. This solution is illustrated in Figure 8.9.

(b) Power failure risk: The probability that the river flow falls below Q = 300 m³/s at least once in 60 days.

First, calculate the z-score for $Q = 300$ m³/s:

$$z = \frac{300 - 546.75}{104.72} = -2.36 \tag{8.67}$$

The cumulative probability for $Z = -2.36$ is:

$$P(Q < 300) = P(Z < -2.36) = 0.0091 \tag{8.68}$$

Thus, the probability that the flow falls below 300 m³/s on a single day is:

$$P(Q < 300) = 0.0091 \tag{8.69}$$

The probability that the river flow does not fall below Q = 300 m³/s on a given day is:

$$P(\text{not below 300}) = 1 - 0.0091 = 0.9909 \tag{8.70}$$

For 60 days, the probability that the river flow never falls below Q = 300 m³/s is:

$$P(\text{not below 300 for 60 days}) = 0.9909^{60} = 0.578 \tag{8.71}$$

Thus, the probability that the flow falls below 300 m³/s at least once in 60 days is:

$$P(\text{below 300 at least once}) = 1 - 0.578 = 0.422 \tag{8.72}$$

Power failure risk: The probability that the river flow falls below Q = 300 m³/s at least once in 60 days is approximately 42.2%.

Final answers:

- Flood risk (flow exceeds Q = 700 m³/s in a single day): 7.2%.
- Power failure risk (flow falls below Q = 300 m³/s at least once in 60 days): 42.2%.

Now, we use MATLAB and Python to solve the first part of this problem, which involves assessing flood risk by determining the probability of daily flow exceeding 700 m³/s. The scripts provided in Examples 8.9 (`CH08Mcode5.m`) and 8.10 (`CH08Pcode5.py`) can be used for this analysis.

Example 8.9

MATLAB code. This MATLAB script, named `CH08Mcode5.m`, plots the PDF and CDF for normal distribution of flows in a hydroelectric dam project.

MATLAB Input:

```
%This code plots the PDF
%and CDF for a dam problem
clc; clear; close all;
set(0,'defaultaxesfontsize',25)
```

```
% Step 1:Input the sample data
data = [450,620,700,350,480,520, ...
550,600,730,400,530,610,580,490, ...
720,510,650,470,380,560,640,390, ...
460,590,710,370,540,500,620,670,  ...
510,680,420,570,630, ...
460,690,580,410,530];
mu=mean(data); % Mean
sigma=std(data, 1);% std dev
% Step 3: Define flood threshold
flood1=700;% Flood threshold(m3/s)
% Step 4:flood risk probability
z_flood=(flood1-mu)/sigma;% z-score
p_flood =1-normcdf(z_flood);% P(X>700)
flood_prcnt = p_flood * 100;
cdf_at_700=normcdf(z_flood);% CDF@700
% Step 5: Display result
fprintf('Mean (mu) = %.2f m3/s\n',mu);
fprintf('St Dev (sigma)=%.2f m3/s\n',sigma);
fprintf(['Flood Risk (P(X>700))'  ...
    '= %.2f%%\n'], flood_prcnt);
% Subplot 1: plot PDF %%%%%%%%%%%%%%
subplot('position',[0.09 .1 .37 .7]);
x = (mu - 4*sigma):0.1:(mu + 4*sigma);
pdf = normpdf(x, mu, sigma); %PDF
plot(x,pdf,'k-','LineWidth',2.5);
hold on; ylim([0 0.0041]);
x_flood=x(x  ≥flood1);% x values>700
pdf_flood=normpdf(x_flood,mu,sigma);%pdf
fill([x_flood fliplr(x_flood)], ...
[pdf_flood zeros(size(pdf_flood))],'m', ...
'FaceAlpha', 0.3, 'DisplayName', ...
sprintf('P(X>700)=%.2f%%',flood_prcnt));
line([flood1 flood1],[0 100], ...
'Color','b','LineStyle','--','LineWidth',2);
grid on;set(gca,'linewidth',1.5);
% Subplot 2: plot CDF %%%%%%%%%%%%%%
subplot('position',[.57 .1 .37 .7]);
cdf = normcdf(x, mu, sigma); %CDF
plot(x,cdf,'k-','LineWidth',2.5);
hold on;ylim([0 1.05]);
line([flood1 flood1], [0 cdf_at_700], ...
'Color','b','LineStyle','--','LineWidth',2);
```

```
line([0 flood1], ...
[cdf_at_700 cdf_at_700],'Color','b', ...
'LineStyle', '--','LineWidth', 2);
grid on;set(gca,'linewidth',1.5);
% end of M - file
```

MATLAB Output:

```
Mean = 546.75 m^3/s
St Dev =104.72 m^3/s
Flood Risk (P(X>700))= 7.17%
This script also plots Figure 8.9.
```

Example 8.10

Python code. This Python script, named `CH08Pcode5.py`, plots the PDF and CDF for normal distribution of flows in a hydroelectric dam project.

Python Input:

```
#This code plots the PDF
#and CDF for a dam problem
import numpy as np
import matplotlib.pyplot as plt
from scipy.stats import norm
# Step 1: Input the sample data
data=np.array([450,620,700,350,480,\
520,550,600,730,400,530,610,580,\
490,720,510,650,470,380,560,640,\
390,460,590,710,370,540,500,620,\
670,510,680,420,570,630,460,\
690,580,410,530])
mu = np.mean(data) # Mean
sigma=np.std(data,ddof=1) #St dev
# Step 3: Define flood threshold
flood1=700 #Flood threshold (m3/s)
# Step 4: Flood risk probability
z_flood=(flood1-mu)/sigma #z-score
p_flood=1-norm.cdf(z_flood) #P(X>700)
flood_prcnt = p_flood * 100
cdf_at_700=norm.cdf(z_flood) #CDF@700
# Step 5: Display results
```

```
print(f"Mean ={mu:.2f} m3/s")
print(f"St Dev={sigma:.2f} m3/s")
print(f"Flood Risk (P(X>700))=\
{flood_prcnt:.2f}%")
# Define x values for plotting
x=np.linspace(mu-4*sigma,mu+4*sigma,500)
# Compute PDF and CDF
pdf = norm.pdf(x, mu, sigma)
cdf = norm.cdf(x, mu, sigma)
# Subplot 1: Plot PDF
plt.figure(figsize=(12, 5))
plt.subplot(1, 2, 1)
plt.plot(x,pdf,'k-',linewidth=2.5)
plt.fill_between(x,pdf,where=\
(x >=flood1), color='m',
alpha=0.3,label=f'P(X>700)=\
{flood_prcnt:.2f}%')
plt.axvline(flood1,color='b',\
linestyle='--',linewidth=2)
plt.ylim([0, 0.0041])
plt.grid(True, linewidth=1.5)
# Subplot 2: Plot CDF
plt.subplot(1, 2, 2)
plt.plot(x, cdf, 'k-', linewidth=2.5)
plt.axvline(flood1,color='b',\
linestyle='--',linewidth=2)
plt.axhline(cdf_at_700,xmin=0,xmax=flood1,\
color='b',linestyle='--',linewidth=2)
plt.ylim([0, 1.05])
plt.grid(True, linewidth=1.5)
# Show plots
plt.tight_layout()
plt.show()
# End of Python code.
```

Python Output:

```
Mean (mu)=546.75 m3/s
St Dev (sigma)=106.06 m3/s
Flood Risk (P(X>700))=7.42%
This script also plots Figure 8.9.
```

8.6 PROBLEMS

Problem 8.1: In structural engineering, the compressive strength of concrete is a key property. Suppose a structural engineer tests 42 concrete samples and records their strengths (MPa):

24.5 MPa, 27.1, 30.3, 25.8, 28.9, 32.5, 29.7, 31.2, 26.3, 33.1, 30.8, 27.5, 34.2, 25.8, 29.3, 25.1, 28.4, 32.9, 31.7, 29.9, 27.0, 26.5, 30.0, 28.7, 35.4, 29.5, 33.7, 24.9, 31.5, 27.8, 28.2, 26.0, 32.1, 34.7, 30.5, 25.6, 29.1, 33.4, 28.0, 31.9, 26.7, 30.9.

Group the data into the following class intervals: 24.0–26.0 MPa, 26.0–28.0 MPa, 28.0–30.0 MPa, 30.0–32.0 MPa, 32.0–34.0 MPa, and 34.0–36.0 MPa, where the upper limit is excluded from each interval (e.g., 28.0 MPa belongs to 28.0–30.0 MPa). Plot a histogram to visualize the distribution using MATLAB and Python. Identify the most frequent strength range and interpret the results in terms of concrete quality and consistency.

Answer:

Range (MPa)	Count
24.0 - 26.0	5
26.0 - 28.0	9
28.0 - 30.0	10
30.0 - 32.0	9
32.0 - 34.0	6
34.0 - 36.0	3

The most frequent strength range is 28.0 - 30.0 MPa with a count of 10.

Problem 8.2: In hydrology and civil engineering, understanding river flow rates is crucial for designing bridges, flood control systems, and managing water resources. Suppose a hydrologist measures the daily average flow rate (in cubic meters per second, m³/s) of a river over a 45-day period during the rainy season. The recorded flow rates are as follows:

12.3, 18.7, 25.4, 14.9, 22.6, 30.1, 17.5, 27.8, 19.2, 35.6, 15.8, 23.9, 28.4, 13.6, 20.5, 16.4, 31.7, 24.2, 29.5, 21.3, 18.9, 26.7, 33.2, 19.8, 37.4, 22.1, 25.9, 14.2, 30.8, 17.9, 23.5, 27.1, 34.9, 20.8, 15.3, 28.9, 32.6, 21.7, 19.4, 36.2, 16.8, 24.8, 29.3, 18.2, 31.4

Using MATLAB and Python, plot the data as bar charts to visualize the variations in flow rates over the measurement period.

Problem 8.3: Referring to data presented in Problem 8.2, group the data into the following class intervals: 10–15 m³/s, 15–20 m³/s, 20–25 m³/s, 25–30 m³/s, 30–35 m³/s, 35–40 m³/s, where the upper limit is excluded from each interval (e.g., 20 m³/s belongs to 20–25 m³/s). Using both MATLAB and Python, plot a histogram to visualize the distribution of flow rates, calculate the frequency of days in each interval and display the results in a table, and identify the most frequent flow rate range.

Answer:

Range (m^3/s)	Count
10.0 - 15.0	4
15.0 - 20.0	12
20.0 - 25.0	10
25.0 - 30.0	9
30.0 - 35.0	7
35.0 - 40.0	3

Most frequent flow rate range: 15.0 - 20.0 m^3/s with a count of 11.

Problem 8.4: Using MATLAB and Python, calculate the mean (μ) and standard deviation (σ) for the daily average river flow data presented in Problem 8.2. Plot the original data as circles and show the mean and standard deviation in a graph.

Answer: μ = 23.94 m^3/s, and σ = 6.71 m^3/s.

Problem 8.5: Consider the following scenario on a construction site where two machines, a digger and a bulldozer, are used in the construction process. The failure probabilities for each machine are as follows:

- The digger fails 12% of the time: $P(\text{Digger}) = 0.12$.
- The bulldozer fails 18% of the time: $P(\text{Bulldozer}) = 0.18$.

Assume the failures of the two machines are independent of each other.

- **a)** Calculate the probability that both the digger and the bulldozer fail at the same time.
- **b)** Calculate the probability that at least one machine (either the digger or the bulldozer) fails during the construction process.
- **c)** Calculate the probability that both machines are working (i.e., neither the digger nor the bulldozer fails).
- **d)** If the construction project cannot continue unless both the digger and the bulldozer are working, what is the likelihood that the project will proceed without failure during a work shift?

Answer: a): 2.2%, b): 27.8%, c): 72.2%, d): 72.2%.

Problem 8.6: In traffic engineering, road pavement is designed to withstand the wear and tear caused by vehicle loads. Consider a stretch of highway with a pavement that faces two independent risks:

- Excessive load, which causes damage to the pavement 4% of the time: $P(L) = 0.04$.
- Temperature fluctuation, which accelerates pavement wear 7% of the time: $P(T) = 0.07$.

You are tasked with determining the following:

- **(a)** The probability that neither excessive load nor temperature fluctuation causes damage to the pavement in a given year.
- **(b)** The probability that only one of the two risks affects the pavement during a given year.
- **(c)** The probability that both risks affect the pavement simultaneously, causing significant damage.
- **(d)** Over the course of 5 years, what is the probability that the pavement will survive without any major damage from either risk?

Answer: a): 89.3%, b): 10.4%, c): 0.3%, d): 56.7%.

Problem 8.7: You are given the following dataset of concrete compressive strength (MPa) obtained from laboratory tests on 33 concrete samples. The data is assumed to follow a normal distribution:

24.5 MPa, 26.1, 23.8, 25.7, 27.3, 24.9, 26.8, 25.4, 27.1, 24.3, 23.5, 28.0, 26.5, 25.1, 27.6, 23.9, 28.2, 24.7, 26.0, 25.9, 27.9, 23.7, 26.3, 24.8, 27.4, 25.2, 28.1, 23.6, 26.7, 24.6, 27.0, 25.0, 26.2

Calculate the following:

- **a)** Employ MATLAB and Python to plot the Probability Density Function (PDF) and Cumulative Distribution Function (CDF) for the given data, assuming it follows a normal distribution.
- **b)** Calculate the probability that the concrete compressive strength is below 24 MPa.
- **c)** Interpret the results in a short paragraph, explaining what the probability value means in the context of concrete strength and structural safety.

Answer: a): $\mu = 25.8$ MPa, $\sigma = 1.4$ MPa, b): 10.1%.

Problem 8.8: Considering the concrete compressive strength data presented in Problem 8.7, calculate the following:

- **a)** The probability that the concrete compressive strength is between 25 MPa and 27 MPa.
- **b)** The strength value that corresponds to the 90th percentile of the distribution.

Answer:

a) 51.5% (see Figure 8.10 for the solution). Here, we used the equation:

$$P(a < f_c < b) = P(f_c < b) - P(f_c < a) = \text{CDF}(b) - \text{CDF}(a)$$

where CDF is the Cumulative Distribution Function (CDF) for the normal distribution of f_c (see the answer in Figure 8.10).

b) 27.6 MPa.

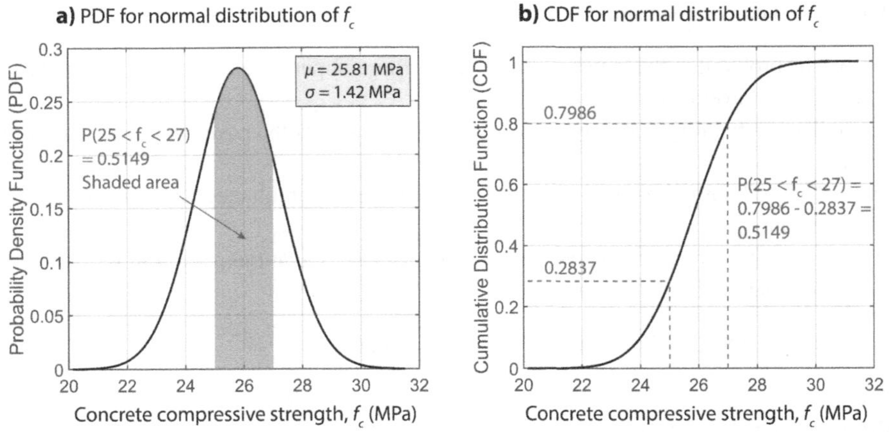

FIGURE 8.10 **a)** Probability Density Function (PDF) for compressive strength of concrete samples (f_c) showing the probability that f_c is between 25 MPa and 27 MPa. **b)** The corresponding Cumulative Distribution Function (CDF) curve showing the same.

Problem 8.9: In traffic engineering, the free-flow speed of vehicles on highways is often modeled as a normal distribution. Suppose the free-flow speed of vehicles on a highway segment follows a normal distribution with a mean (μ) of 110 km/h and a standard deviation (σ) of 10 km/h. The highway design requires that vehicles travel at speeds greater than 90 km/h for safety. Complete the following tasks:

- **(a)** Calculate the probability that a randomly selected vehicle travels below 90 km/h.
- **(b)** Determine the speed corresponding to the 85th percentile of the distribution.

Answer: a): 2.3%, b): 120.4 Km/h.

Problem 8.10: For the compressive strength data of concrete samples (f_c) presented in Problem 8.1, determine whether the requirement that at least 90% of samples exceed 26 MPa is satisfied based on the observed data.

Answer: $P(f_c < 26) = 0.1130\,(11.3\%)$. $P(f_c > 26) = 0.8870\,(88.7\%)$. Requirement is not met: Less than 90% of samples exceed 26 MPa.

REFERENCES

[1] Mohammad Heidarzadeh et al. "Field surveys of tsunami runup and damage following the January 2024 Mw 7.5 Noto (Japan sea) tsunamigenic earthquake." In: *Ocean Engineering* 307 (2024), p. 118140.

[2] Erwin Kreyszig. *Advanced engineering mathematics 9th edition with Wiley plus set*. Vol. 334. John Wiley & Sons US, 2007.

Index

STRATEGIC FORESIGHT

Organisations today are being challenged to make sense of changes in environments that, now more than ever, are described as VUCA (volatile, uncertain, complex, and ambiguous). They are also being driven to understand how the future will evolve and what impact it will have not only on the organisations themselves but also on industries and societies. In recent decades a field has emerged to support organisations in addressing these challenges: strategic foresight.

This book is a comprehensive introduction to strategic foresight. It presents a history of the field and explains the main principles in thinking about the future. The book describes how organisations can apply strategic foresight and explains how it relates to other fields such as strategy, innovation, and leadership, highlighting the relevance of strategic foresight not only for organisations but also for individuals, particularly managers and leaders. Grounded in the theoretical foundations of strategic foresight, the book reflects the latest academic research and explores practical applications in different contexts. It draws on more than two decades of experience that the author has in the field as a researcher and as a consultant in the corporate context.

This is essential reading for managers and leaders of public and private organisations who want to establish strategic foresight practices as well as students of foresight and managers in the fields of innovation, research & development, and marketing.

Jan Oliver Schwarz is a professor of Strategic Foresight and Trend Analysis and head of the Bavarian Foresight-Institute at the Technische Hochschule Ingolstadt. His academic and consulting work focuses on various aspects of strategic foresight.

STRATEGIC FORESIGHT

An Introductory Guide to Practice

Jan Oliver Schwarz

Routledge
Taylor & Francis Group
LONDON AND NEW YORK

Cover image: Getty Images / piranka

First published 2024
by Routledge
4 Park Square, Milton Park, Abingdon, Oxon OX14 4RN

and by Routledge
605 Third Avenue, New York, NY 10158

Routledge is an imprint of the Taylor & Francis Group, an informa business

British Library Cataloguing-in-Publication Data
A catalogue record for this book is available from the British Library

Library of Congress Cataloging-in-Publication Data
Names: Schwarz, Jan Oliver, author.
Title: Strategic foresight : an introductory guide to practice /
Jan Oliver Schwarz.
Description: Abingdon, Oxon ; New York, NY : Routledge, 2023. |
Includes bibliographical references and index. |
Identifiers: LCCN 2022060536 (print) | LCCN 2022060537 (ebook) |
ISBN 9781032299211 (hardback) | ISBN 9781032299235 (paperback) |
ISBN 9781003302735 (ebook)
Subjects: LCSH: Strategic planning. | Organizational change.
Classification: LCC HD30.28 .S34798 2023 (print) | LCC HD30.28 (ebook) |
DDC 658.4/012—dc23/eng/20230411
LC record available at https://lccn.loc.gov/2022060536
LC ebook record available at https://lccn.loc.gov/2022060537

ISBN: 978-1-032-29921-1 (hbk)
ISBN: 978-1-032-29923-5 (pbk)
ISBN: 978-1-003-30273-5 (ebk)

DOI: 10.4324/9781003302735

Typeset in Joanna
by codeMantra

CONTENTS

PREFACE

As a management student about two decades ago I found an advertisement for an internship in strategic foresight in the HR division of Deutsche Bank in Frankfurt, Germany. I applied for and accepted the internship. I cannot recall what initially drew me to this ad, but my fascination with strategic foresight has stayed with me ever since.

After this internship, I had the privilege to write my master's thesis and later my PhD dissertation under the supervision of Franz Liebl who had already done substantial work on strategic foresight, or *Strategische Frühaufklärung* in German. Completing a second master's degree in Futures Studies at the Institute for Futures Research at the University of Stellenbosch, South Africa only deepened my fascination with the topic.

I was fortunate to gain practical insights into applying strategic foresight. For instance, I conducted business wargames at the consulting firms Booz Allen Hamilton and A.T. Kearney and implemented strategic foresight at the insurance company Allianz SE; I also oversaw scenario planning projects at Decision Strategies International and the Institute for Innovation and Change Methodologies. What followed were many more engagements with organisations to help them develop foresight in many ways.

I have managed to keep a balance between these two elements, theory and practice, over the years and have found it extremely enriching to have theory inform practices and vice versa. My most recent appointment as a professor of Strategic Foresight and Trend Analysis at the Technische Hochschule Ingolstadt, Germany, allows me to incorporate this combination of theory and practice even more into my teaching and research.

I have never lost my fascination with foresight. The future is a difficult topic to grasp, because it is not predictable. For this reason, we should not talk about 'the future' but of 'the futures', as there are several plausible ones. I will be using the term 'future' in this book in light of this assessment.

However, just as the future is not predetermined, it can be changed. This is true for us as individuals but of course also for organisations. Further, in every decision we make as individuals or as organisation, we anticipate the future. In other words, dealing with the future is essential for individuals and organisations alike. In the end, we will spend the rest of our lives in the future.

In this book, I do not attempt to cover everything that there is to know about foresight; I present my perspective on this field, reflecting my work on theory and practice. Admittedly, there are also other books (e.g. Bishop and Hines 2012; Hines and Bishop 2015; Lustig 2015; Gidley 2017; Poli 2019; De Toni, Siagri, and Battistella 2020; Slaughter 2020) that are introductions to foresight. However, I hope that this book provides a useful overview on strategic foresight from the perspective of theory and practice. I believe that this is essential because theory allows us to understand the mechanics of strategic foresight and to reflect our practice, but foremost, we can learn from theory how to rise to the challenges of using strategic foresight by learning from others.

For whom is this book? This book is intended for practitioners who want to make a difference by implementing strategic foresight in a sustainable way in their organisation. Further, this book is suited for anyone who is interested in gaining a solid understanding of strategic foresight.

Jan Oliver Schwarz
Munich, Germany, 2022

References

Bishop, P., and A. Hines. 2012. *Teaching about the Future*. London: Palgrave Macmillan.

De Toni, A. F., R. Siagri, and C. Battistella. 2020. London: *Corporate Foresight: Anticipating the Future*. Taylor & Francis.

Gidley, J. 2017. *The Future: A Very Short Introduction*. Oxford: Oxford University Press.

Hines, A., and P. Bishop. 2015. *Thinking about the Future: Guidelines for Strategic Foresight*. Houston: Social Technologies.

Lustig, P. 2015. *Strategic Foresight: Learning from the Future*. Axminster: Triarchy Press.

Poli, R. 2019. *Working with the Future. Ideas and Tools to Govern Uncertainty*. Milano: Bocconi University Press.

Slaughter, R. 2020. *The Knowledge Base of Futures Studies 2020*. Washington: Association of Professional Futurists.

INTRODUCTION

For decades, scholars, managers, and students have discussed the rapidly changing business environment and the associated notion of complexity, ambiguity, or the uncertainty concerning, for instance, the future. Indeed, this description of the business environment is not new. Over half a century ago the argument has been made that the 'age of discontinuity' has already begun (Drucker 1969), and that change is occurring more rapidly than in the past. The situation in which organisations find themselves was later described as 'the mess' (Ackoff 1981), perceived by organisations as the future implied by their own behaviour and that of their environment.

Organisations today face a more complex and dynamic environment than ever, one that is characterised by discontinuities and an uncertain future – a state that is most likely to continue. The major task for managers today is to make decisions, and then formulate and execute strategic management systems in that environment. It is obvious that the imperative of 'predict and prepare', as the foundation of the neoclassical school of management (Gharajedaghi 1999), is no longer appropriate for organisations.

DOI: 10.4324/9781003302735-1

The financial crises in 2008, the COVID-19 pandemic, and most recently the war in Ukraine underline the notion that organisations are faced with a VUCA (volatile, uncertain, complex, and ambiguous) environment. The acronym is now used to frame the challenges that organisations face. Ramirez and Wilkinson (2016) have coined the acronym TUNA (turbulence, uncertainty, novelty, and ambiguity), suggesting that we have already moved beyond VUCA in terms of increasing uncertainty and dynamics in the business environment. Overall, the argument is being made that companies need to be more vigilant and that strategic foresight is part of this effort (Day and Schoemaker 2019).

A way of observing VUCA is to pay attention to the trends and issues in the environment of an organisation. Liebl and Schwarz (2010) introduce what they call the entrepreneurial view on strategy which draws on three generic sources for the formulation of strategic options:

- The customers and their 'worlds', for example, their knowledge, imaginings, perceptions, and experiences
- The resources and competences of the focal company
- The trends and issues in a company's environment (e.g., the future of an industry)

Each of these areas can inspire strategy formulation; however, every option created from one source has to be tested against the other two (Liebl 2004).

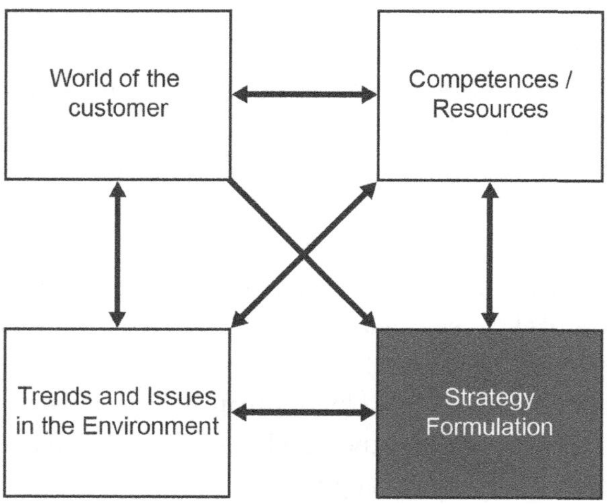

Figure I.1 The entrepreneurial view on strategy.

New offerings based on a firm's competences are meaningful only if they are perceived and valued as beneficial by the customers. Conversely, ideas about how to gain competitive advantage from the viewpoint of a customer must be realistic with respect to a company's resources and competences. Further, trends and issues in the business environment are not important in themselves. They can only be meaningfully interpreted and assessed if their implications for the customers' worlds – and accordingly, for customer behaviour – are considered. Conversely, ideas about how to gain competitive advantage from the viewpoint of a customer must be tested against the trends in the socio-political, socio-cultural, environmental, economic, or technological environments, as these may represent drivers of change in the future. In addition, a firm's resources and competences may need to be re-evaluated due to emerging trends and issues (Liebl 2002).

In light of the COVID-19 pandemic, the relevance of developing foresight has been reemphasised (Scoblic 2020). However, in the past (Hamel and Prahalad 1994; Courtney 2001) the relevance of foresight has been underlined, for instance, in the context of strategy. The main claim in this context is that organisations need to develop foresight to compete for the future and be a viable part of the future.

But what is foresight? The following definition guides this book:

> Corporate foresight is identifying, observing and interpreting factors that induce change, determining possible organization-specific implications, and triggering appropriate organizational responses. Corporate foresight involves multiple stakeholders and creates value through providing access to critical resources ahead of competition, preparing the organization for change, and permitting the organization to steer proactively towards a desired future.
>
> *(Rohrbeck, Battistella, and Huizingh 2015: 2)*

This definition has the advantage of focusing not only on perceiving changes in the business environment, in the form of weak signals and trends, but that it also emphasises the relevance of deciphering how these changes might develop into the future (prospecting) and how organisations need to change to address the insights derived from perceiving and prospecting. Later it allows us to discuss foresight in the context of change management (Kotter 2012, 2014).

In discussing the relationship between foresight and change, turning to the case of Kodak is insightful. Often the downfall of iconic brand Kodak is associated with the inability of the company to have recognised the rising trend of digital photography. Actually, the opposite is true. Kodak was one of the first firms to file patents for digital photography. However, the organisation failed to transform itself from a high-margin film-producing company to a low-margin digital equipment company (Krupp and Schoemaker 2014; Agarwal and Satish 2021). Similar perspective can be taken on the decline of the Encyclopaedia Britannica (Greenstein 2017). What this implies is that foresight is not enough to have an impact, foresight needs to be also perceived in the context of transforming an organisation or as part of a change journey.

Further, I perceive foresight as an activity which is predominantly geared towards the long term. This allows us to conceptualise foresight as an activity which can be perceived as strategic. I hereby emphasise the following definition of strategy: 'Strategy is the direction and scope of an organization over the long term, which achieves advantage in a changing environment through its configuration of resources and competences with the aim of fulfilling stakeholder expectations' (Johnson, Scholes, and Whittington 2006: 3).

Therefore, the argument can be made that the terms 'foresight' or 'strategic foresight' can be used interchangeably. I will continue to use the term 'strategic foresight' to emphasise the strategic nature of foresight. While the above definition centres on the application of foresight in a corporate setting, this definition can also be perceived as being relevant to other forms of organisations. However, while also mentioning the relevance of foresight for different types of organisations, this book will focus on the application of foresight in the corporate context, also emphasising the relevance of corporate foresight for strategy and management (Fergnani 2022).

In this book, I discuss three dimensions of foresight:

1. Individual mindset to develop foresight
2. Tools to develop strategic foresight
3. Processes to develop strategic foresight which may be a combination of tools

In considering the tools used in strategic foresight, I will be combining tools from different fields, also acknowledging that these might have developed from other fields and/or before foresight as a field has emerged, or

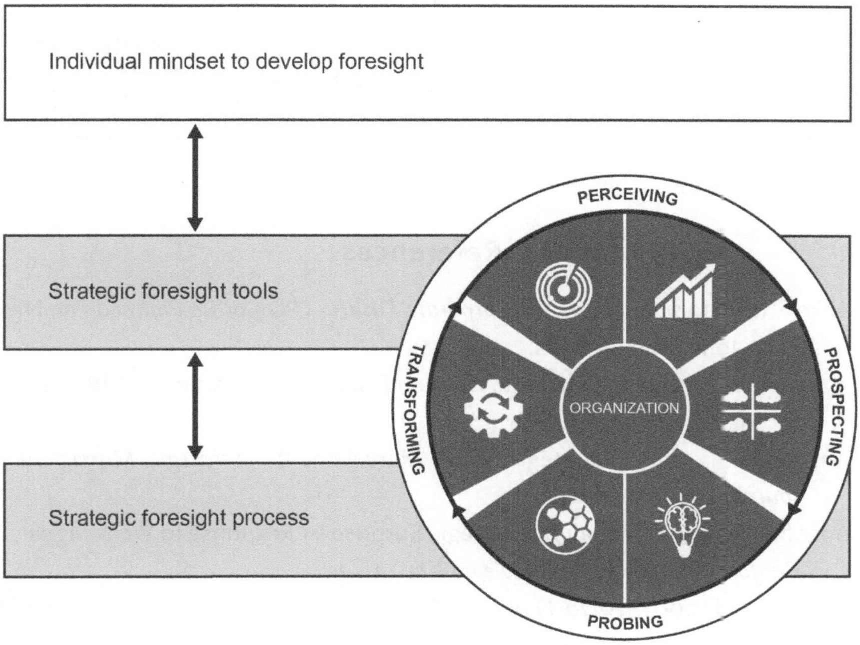

Figure I.2 Three dimensions of strategic foresight.

in parallel. In this respect to tools one would refer, for instance, to scenario planning (Schoemaker 1995; Wilkinson and Kupers 2013), trends and weak signals (Ansoff 1975; Liebl and Schwarz 2010), the Delphi method (Linstone and Turoff 1975; von der Gracht 2012), or business wargaming (Schwarz 2009; Oriesek and Schwarz 2020). But other fields such as strategic issue management (Ansoff 1980) or competitive intelligence (Gilad 2004; Fuld 2006) have also contributed to our understanding of strategic foresight.

Following the three dimensions of strategic foresight, the book has this structure (see Figure I.3): Part I presents a historical overview of the development of strategic foresight and reflects the cognitive aspects of thinking about the future. Part II describes the process of strategic foresight in more detail. This discussion both explains how foresight is connected to other fields and how strategic foresight can be connected to generate valuable input. Part III concludes the book with a discussion of the practical challenges of applying strategic foresight.

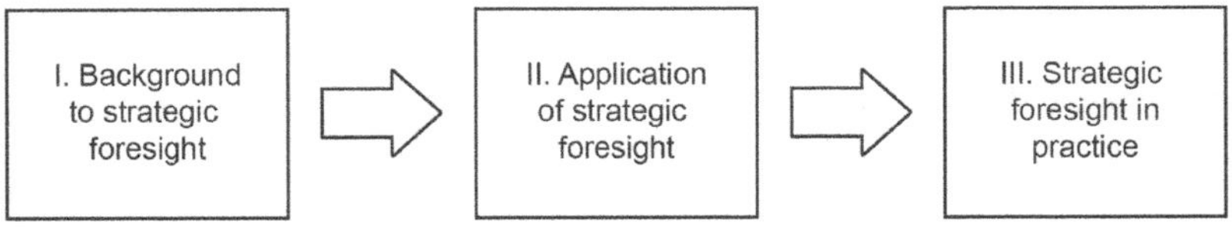

Figure I.3 Structure of the book.

References

Ackoff, R. L. 1981. *Creating the Corporate Future – Plan or Be Planned For.* New York: John Wiley & Sons.

Agarwal, M., and D. Satish. 2012. *Rise and Fall of Kodak*. Hyderabad: IBS Center for Management Research.

Ansoff, H. I. 1980. "Strategic Issue Management." *Strategic Management Journal* 1 (2): 131–148.

Ansoff, I. H. 1975. "Managing Strategic Surprise by Response to Weak Signals." *California Management Review* 18 (2): 21–33.

Courtney, H. 2001. *20/20 Foresight: Crafting Strategy in an Uncertain World.* Boston: Harvard Business School Publishing.

Day, G. S., and P. J. H. Schoemaker. 2019. *See Sooner, Act Faster: How Vigilant Leaders Thrive in an Era of Digital Turbulence.* Management on the Cutting Edge. Cambridge: MIT Press.

Drucker, P. F. 1969. *The Age of Discontinuity: Guidelines to Our Changing Society.* London: Heinemann.

Fergnani, A. 2022. "Corporate Foresight: A New Frontier for Strategy and Management." *Academy of Management Perspectives* 36 (2): 820–844.

Fuld, L. 2006. *The Secret Language of Competitive Intelligence.* New York: Crown Business.

Gharajedaghi, J. 1999. *Systems Thinking.* Boston: Butterworth Heinemann.

Gilad, B. 2004. *Early Warning: Using Competitive Intelleignce to Anticipate Market Shifts, Control Risk, and Create Powerful Strategies.* New York: AMACOM.

Greenstein, S. 2017. "The Reference Wars: Encyclopædia Britannica's Decline and Encarta's Emergence." *Strategic Management Journal* 38 (5): 995–1017. https://doi.org/10.1002/smj.2552.

Hamel, G., and C. K. Prahalad. 1994. *Competing for the Future.* Boston: Harvard Business School Press.

Johnson, G., K. Scholes, and R. Whittington. 2006. *Exploring Corporate Strategy.* Harlow: FT Prentice Hall.

Kotter, J. P. 2012. *Leading Change*. Boston: Harvard Business Review Press.

———. 2014. *Accelerate: Building Strategic Agility for a Faster-Moving World*. Boston: Harvard Business Review Press.

Krupp, S., and P. J. H. Schoemaker. 2014. *Winning the Long Game: How Strategic Leaders Shape the Future*. New York: Public Affairs.

Liebl, F. 2002. "The Anatomy of Complex Societal Problems and Its Implications for OR." *Journal of the Operational Research Society* 53 (2): 161–184.

———. 2004. "Knowledge Management for Strategic Marketing." In *Developments in Marketing Science XXVII, Proceedings of the Annual Conference of the Academy of Marketing Science*, edited by H. E. Spotts, 48–57. Coral Gables: Springer.

Liebl, F., and J. O. Schwarz. 2010. "Normality of the Future: Trend Diagnosis for Strategic Foresight." *Futures* 42 (4): 313–327.

Linstone, H. A., and M. Turoff. 1975. "Evaluation: Introduction." In *The Delphi Method: Techniques and Applications*, edited by H. A. Linstone and M. Turoff, 229–235. Reading: Addison-Wesley Publishing.

Oriesek, D. F., and J. O. Schwarz. 2020. *Winning the Uncertainty Game: Turning Strategic Intent into Results with Wargaming*. London: Taylor & Francis.

Ramirez, R., and A. Wilkinson. 2016. *Strategic Reframing: The Oxford Scenario Planning Approach*. Oxford: Oxford Publishing.

Rohrbeck, R., C. Battistella, and E. Huizingh. 2015. "Corporate Foresight: An Emerging Field with a Rich Tradition." *Technological Forecasting and Social Change* 101 (December): 1–9. https://doi.org/10.1016/j.techfore.2015.11.002.

Schoemaker, P. 1995. "Scenario Planning : A Tool for Strategic Thinking." *Sloan Management Review* 36 (2): 25–40. https://doi.org/10.1016/0024-6301(95)91604-0.

Schwarz, J. O. 2009. "Business Wargaming: Developing Foresight within a Strategic Simulation." *Technological Analysis and Strategic Management* 21 (3): 291–305.

Scoblic, J. P. 2020. "Learning from the Future." *Harvard Business Review* (July–August), https://hbr.org/2020/07/learning-from-the-future.

von der Gracht, H. A. 2012. "Consensus Measurement in Delphi Studies. Review and Implications for Future Quality Assurance." *Technological Forecasting and Social Change* 79 (8): 1525–1536. https://doi.org/10.1016/j.techfore.2012.04.013.

Wilkinson, A., and R. Kupers. 2013. "Living in the Futures." *Harvard Business Review* (May), https://hbr.org/2013/05/living-in-the-futures.

Part I

BACKGROUND TO STRATEGIC FORESIGHT

1

HISTORICAL DEVELOPMENT OF STRATEGIC FORESIGHT

Theresa Schropp[1]

Dealing with the future is probably one of the greatest challenges to individuals and corporations alike, as it is easier to understand the future when it has arrived in the present (Schwarz 2009). However, managers and employees have to make future-related decisions before they know what the future holds.

For over five decades, and in view of a VUCA-environment, managers and researchers alike are considering the question of how corporates can manage for today and still prepare for tomorrow (Liebl and Schwarz 2010). This is shown by the tremendous interest of researchers, elaborating on ideas and measurements to act upon an unpredictable future. In this regard and even though the term 'strategic foresight' has not been coined until the late 1990s (e.g., Slaughter 1999), scientific literature has already discussed foresight-related management tools and concepts like 'environmental scanning' and 'early warning' much earlier (e.g., Aguilar 1967; Ansoff 1975).

DOI: 10.4324/9781003302735-3

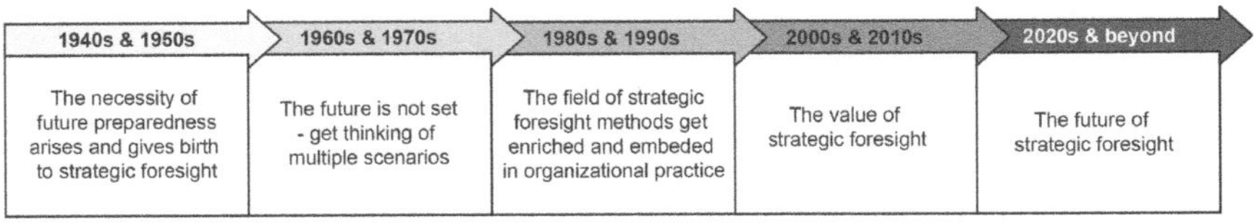

Figure 1.1 History of strategic foresight.

Based on studies from different researchers (e.g., Rohrbeck, Battistella, and Huizingh 2015; Fergnani 2019, 2020; Gordon, Ramic, Rohrbeck, and Spaniol 2020), it is possible to define and assign research themes to five phases.

The following section offers a more detailed review of each of these phases, highlighting their most striking developments, the practical application, and the research of strategic foresight.

1.1 The 1940s and 1950s: The necessity of future preparedness arises and gives birth to strategic foresight

The starting point of the foresight discipline is unknown, as humanity has always thought about the future. Nevertheless, researchers agree on the emergence of foresight in the context of World War II and the U.S. military's forecasting practices (Hines 2020). These activities, however, focused on the military and should thus not be equated with strategic foresight in a corporate environment. Project RAND, and its offspring the foresight school, as well as the French prospective school can be considered as the two roots of strategic foresight, instead (Rohrbeck, Battistella, and Huizingh 2015). The two schools, however, differ in their fundamental philosophy.

Triggered by the impressions of World War II and the assumption that R&D activities, collaboration of military, government agencies, industries, and universities will matter in terms of future wars and conflicts, Project RAND was initiated immediately after the war. Project RAND, a kind of an external think tank, brought together expert opinions, refined the Delphi technique, and set the foundation for the institutionalisation of foresight, as with the establishment of the foresight school based on the work of Herman Kahn, a member of Project RAND (Rohrbeck, Battistella, and Huizingh 2015; Hines 2020). The foresight school proposed sophisticated methods, with a strong focus on the Delphi technique as well as on the engagement of experts (Rohrbeck, Battistella, and Huizingh 2015).

In addition to Project RAND, the French prospective school, founded by Gaston Berger, is the second root of strategic foresight. Berger stressed inclusion of and collaboration among diverse actors (Rohrbeck, Battistella, and Huizingh 2015). In contrast to Project RAND, the French prospective school does not focus on R&D activities and the military, but on corporate decision-making (Rohrbeck, Battistella, and Huizingh 2015). According to Rohrbeck, Battistella, and Huizingh (2015), Berger's methods are designed as workshops where systems-thinking is collaboratively performed. They further facilitate joint reflection, decision-making, and "future-oriented sensemaking" (Rohrbeck, Battistella, and Huizingh 2015:3). Additionally, Berger has introduced a premise of the foresight discipline in his research: the idea of desired ends that facilitate planning, leading, and steering towards desired futures (Rohrbeck, Battistella, and Huizingh 2015). This assumption is related to the scenario planning approach developed in the next phase.

1.2 The 1960s and 1970s: The future is not set – get thinking of multiple scenarios

With the 1960s and 1970s, foresight practices became embedded in the operational practice of corporations and the discipline of corporate foresight emerged. Practitioners, however, applied linear models since the environment was perceived as stable (Rohrbeck, Battistella, and Huizingh 2015). At the same time, they recognised that their forecasts lacked accuracy and thus were no longer sufficient to perform planning in a more complex and competitive marketplace (Rohrbeck, Battistella, and Huizingh 2015).

This changed perception initiated the development of more extensive and holistic foresight methods that consider not only technological factors but also socio-political, economic, and other external factors (Gordon, Ramic, Rohrbeck, and Spaniol 2020). And as corporations discerned that their traditional forecasting methods, often based on the assumption that the future is simply a replication of the past, no longer put forward meaningful and sufficient results, they became receptive to new methods (Gordon, Ramic, Rohrbeck, and Spaniol 2020). The scenario planning approach thus entered the corporate foresight field (Rohrbeck, Battistella, and Huizingh 2015).

One of the most prominent examples of scenario planning in a corporate setting is that of Royal Dutch/Shell. This programme brought forward multiple plausible scenarios, serving as a foundation for internal discussion on the future (Rohrbeck, Battistella, and Huizingh 2015). As Shell's application of scenario planning was a proof of concept, other firms implemented this technique in addition to their traditional business and corporate planning systems (Rohrbeck, Battistella, and Huizingh 2015). Since then, scenario planning has become one of the most popular and most powerful strategic foresight methods, frequently applied and discussed until today.

Research work conducted in the 1960s and 1970s mirrors the changing perception on forecasting methods. Consequently, researchers discussed the failure to integrate strategic issues into traditional, technology-focused forecasting (e.g., Swager 1972) and how these strings could be woven even more closely together. The process and the result of technology forecasting were thus transferred into questions on organisational, sales, and venture planning and intertwined with decision-making (Gordon, Ramic, Rohrbeck, and Spaniol 2020). With the greater variety of application areas and environmental complexity, researchers discussed the need to include other external aspects, besides technology trends, in strategic planning (Gordon, Ramic, Rohrbeck, and Spaniol 2020).

1.3 The 1980s and 1990s: The field of strategic foresight methods get enriched and embedded in organisational practice

Driven by the increase of uncertainty, dynamics, complexity, and ambiguity and by the challenges of globalised markets, the strategic importance of foresight practice became acknowledged in the 1980s and 1990s (Rohrbeck, Battistella, and Huizingh 2015). To stay competitive, firms integrated foresight into their operational practice and the application area of foresight procedures was extended. In other words, strategic foresight was no longer restricted to support decision-making but also of increased importance to innovation management (Rohrbeck, Battistella, and Huizingh 2015). Think tanks, supposed to inform the strategy department or drive innovation, were established in large corporations such as Daimler or Deutsche Bank (Rohrbeck, Battistella, and Huizingh 2015).

Furthermore, and in light of uncertain, dynamic, and more complex markets, continuous foresight practices, such as environmental scanning, were implemented (Rohrbeck, Battistella, and Huizingh 2015). These developments mark a turning point of the strategic foresight discipline, as foresight projects were performed on a large scale and with long pauses of several months to years before (Rohrbeck, Battistella, and Huizingh 2015).

The 1980s and 1990s were characterised by an extension of foresight application areas and by an enrichment of foresight methods (Rohrbeck, Battistella, and Huizingh 2015). Consequently, other techniques, such as technology roadmapping, were integrated into the foresight practice, a development that has continued.

Scientific literature of the 1980s and 1990s is concerned with the characteristics of the VUCA-world, supporting the findings made in the previous phase that traditional, linear forecasting methods are inappropriate for planning in view of uncertainty (e.g., Mahajan and Wind 1989). In this regard, researchers discussed how a firm's innovative strength and competitiveness could be improved by integrating foresight methods (e.g., signal scanning) (Gordon, Ramic, Rohrbeck, and Spaniol 2020).

1.4 The 2000s and 2010s: The value of strategic foresight

Since 2000, the integration and the interdisciplinarity of foresight practices in corporations have become embedded into organisational routines (Rohrbeck, Battistella, and Huizingh 2015). At the same time, Rohrbeck's study (2010) on the foresight maturity of companies reveals that companies still struggle in translating their future insights into strategic measures. According to Rohrbeck, Battistella, and Huizingh (2015), this situation can be explained by two phenomena. On the one hand, foresight processes generate a vast amount of information that must be interpreted by knowledgeable staff, ideally by top management. On the other hand managers often lack the time to analyse these insights.

Both foresight practitioners and researchers have raised the question of how foresight practices and insights could be integrated into corporations (Rohrbeck, Battistella, and Huizingh 2015). As early as the 1990s, the foresight research was still concerned with the integration of technology roadmapping in strategic and business-related procedures (Gordon, Ramic, Rohrbeck, and Spaniol 2020). The scientific literature discusses how scanning, communication, and knowledge management could be structurally integrated into strategic planning (Gordon, Ramic, Rohrbeck, and Spaniol 2020).

Rohrbeck, Battistella, and Huizingh (2015) suggest assigning an orchestrator role to leverage the sensing, interpretation, and planning capabilities performed by and allocated to different corporate functions (e.g., R&D, innovation management, strategy, or risk management).

Furthermore, literature reviews indicate that research still lacks an empirical and quantitative prove of the actual value contribution (Iden, Methlie, and Christensen 2017; Marinković, Al-Tabbaa, Khan, and Wu 2022). This statement, however, only partially holds true, as researchers (e.g., Rohrbeck and Schwarz 2013; Jissink, Huizingh, and Rohrbeck 2014, 2015; Paliokaite and Pačesa 2015; Rohrbeck and Kum 2018) have demonstrated the positive effects of foresight in a corporate setting. Nonetheless, more theoretical underpinnings of strategic foresight's positive influence are required not only to improve the standing of foresight in science but also to underscore its importance for corporate application.

This leads to the last phase that should not be omitted, as foresight is concerned with the future – *the future of strategic foresight*.

1.5 The 2020s and beyond: The future of strategic foresight

Strategic foresight has become an individual, separate research stream and is, according to Fergnani (2019), the biggest research stream of the foresight discipline. As strategic foresight is gaining more attention the question is, where the strategic foresight stream might develop, which research themes are still untapped, and how strategic foresight will be embedded in the organisational setting.

It can therefore be said that strategic foresight has gained momentum. The challenges that individuals, society, governments, and corporates face have underscored the importance of foresight and its application. This awareness may result in an implementation wave, where many corporates establish strategic foresight routines. At the same time, many firms, and especially SMEs, experience difficulties in performing foresight successfully (Iden, Methlie, and Christensen 2017). This observation raises questions on how SMEs that often lack financial and human resources can be supported in conducting foresight, how the corporate culture and leadership should ideally look to foster a sufficient 'foresight-climate', and what foresight tools offer the greatest benefit.

In addition, technological advancements and digitalisation drive the generation of big data that serve as a data basis for artificial intelligence (AI)-based foresight tools. AI, thereby, might help organisations by detecting weak signals sooner or by improving corporate sensemaking capacities.

Based on this application-related considerations on the future of strategic foresight, several research trajectories can be identified. Future foresight research might grapple with the question of which factors drive the implementation and application of foresight in a corporate setting (Gordon, Ramic, Rohrbeck, and Spaniol 2020). Moreover, and as a logical consequence of technological advancements and the digitalisation, future foresight research probably also deals with the question on how new technologies like AI will affect foresight methods and processes. Fergnani and Chermack (2021) state that there is more demand for theoretical underpinning of the strategic foresight discipline that was rather application-oriented in the past. In addition, strategy-related research increasingly returns to the concept of dynamic capabilities (see Teece, Pisano, and Shuen 1997), instead of to the traditional resource-based theory. With this change, there is more research to be conducted on strategic foresight as a dynamic capability (Fergnani 2020; Marinković, Al-Tabbaa, Khan, and Wu 2022).

Another future research trajectory is concerned with the cognition of individuals and organisations, and its connection to future thinking. In this regard, there is a need for aligning the strategic foresight community in its perception of what 'future skills' a good foresight practitioner must possess.

Note

1 Theresa Schropp has a master's degree in International Business Development from the ESB Business School, Reutlingen. In 2022, she started her current position as research fellow at the Bavarian Foresight-Institute of Technische Hochschule Ingolstadt where she performs research on corporate foresight.

References

Aguilar, F. 1967. *Scanning the Business Environment. Studies of the Modern Corporation.* New York: Macmillan.

Ansoff, H. I. 1975. "Managing Strategic Surprise by Response to Weak Signals." *California Management Review* 18 (2): 21–33. https://doi.org/10.2307/41164635.

Fergnani, A. 2019. "Mapping Futures Studies Scholarship from 1968 to Present: A Bibliometric Review of Thematic Clusters, Research Trends, and Research Gaps." *Futures* 105: 104–123. https://doi.org/10.1016/j.futures.2018.09.007.

———. 2020. "Corporate Foresight: A New Frontier for Strategy and Management." *Academy of Management Perspectives* 36 (2): 820–844. https://doi.org/10.5465/amp.2018.0178.

Fergnani, A., and T. J. Chermack. 2021. "The Resistance to Scientific Theory in Futures and Foresight, and What to Do about It." *Futures & Foresight Science* 3 (3–4): e61. https://doi.org/10.1002/ffo2.61.

Gordon, A. V., M. Ramic, R. Rohrbeck, and M. J. Spaniol. 2020. "50 Years of Corporate and Organizational Foresight: Looking Back and Going Forward." *Technological Forecasting and Social Change* 154. https://doi.org/10.1016/j.techfore.2020.119966.

Hines, A. 2020. "When Did It Start? Origin of the Foresight Field." *World Futures Review* 12 (1): 4–11. https://doi.org/10.1177/1946756719889053.

Iden, J., L. B. Methlie, and G. E. Christensen. 2017. "The Nature of Strategic Foresight Research: A Systematic Literature Review." *Technological Forecasting and Social Change* 116: 87–97. https://doi.org/10.1016/j.techfore.2016.11.002.

Jissink, T., E. K. Huizingh, and R. Rohrbeck. 2014. *Corporate Foresight: Antecedents and Contributions to Innovation Performance.* https://pure.au.dk/portal/en/publications/corporate-foresight-antecedents-and-contributions-to-innovation-performance(ae3cf14d-a749-4d3a-8684-66ab9313d016).html.

———. 2015. "Corporate Foresight and Performance: A Chain-of-Effects Model." *Corporate Foresight and Performance: A Chain-of-Effects Model.* https://pure.au.dk/portal/en/publications/corporate-foresight-and-performance-a-chainofeffects-model(739ff02d-ba51-4cf6-acdb-52efa842d6ac).html.

Liebl, F., and J. O. Schwarz. 2010. "Normality of the Future: Trend Diagnosis for Strategic Foresight." *Futures* 42 (4): 313–327.

Mahajan, V., and J. Wind. 1989. "Market discontinuities and strategic planning: a research agenda." *Technological Forecastasting and Social Change* 36 (1–2): 185–199. https://doi.org/10.1016/0040-1625(89)90023-1.

Marinković, M., O. Al-Tabbaa, Z. Khan, and J. Wu. 2022. "Corporate Foresight: A Systematic Literature Review and Future Research Trajectories." *Journal of Business Research* 144: 289–311. https://doi.org/10.1016/j.jbusres.2022.01.097.

Paliokaite, A., and N. Pačesa. 2015. “The Relationship between Organisational Foresight and Organisational Ambidexterity.” *Technological Forecasting and Social Change* 101: 165–181. https://doi.org/10.1016/j.techfore.2014.03.004.

Rohrbeck, R. 2010. *Corporate Foresight: Towards a Maturity Model for the Future Orientation of a Firm*. Physica-Verlag.

Rohrbeck, R., C. Battistella, and E. Huizingh. 2015. “Corporate Foresight: An Emerging Field with a Rich Tradition.” *Technological Forecasting and Social Change* 101: 1–9. https://doi.org/10.1016/j.techfore.2015.11.002.

Rohrbeck, R., and J. O. Schwarz. 2013. “The Value Contribution of Strategic Foresight: Insights from an Empirical Study of Large European Companies.” *Technological Forecasting and Social Change* 80(8): 1593–1606. https://doi.org/10.1016/j.techfore.2013.01.004.

Rohrbeck, R., and M. E. Kum. 2018. “Corporate Foresight and Its Impact on Firm Performance: A Longitudinal Analysis.” *Technological Forecasting and Social Change* 129: 105–116. https://doi.org/10.1016/j.techfore.2017.12.013.

Schwarz, J. O. 2009. “The Symbolism of Foresight Processes in Organizations.” In *Handbook of Research on Strategy and Foresight*, edited by L. A. Costanzo and R. B. MacKay, 82–89. Edward Elgar Publishing Limited.

Slaughter, R. A. 1999. “A New Framework for Environmental Scanning.” *Foresight* 1 (5): 441–451. https://doi.org/10.1108/14636689910802331.

Swager, W. L. 1972. “Strategic Planning I: The Roles of Technological Forecasting.” *Technological Forecasting and Social Change* 4(1): 85–99. https://doi.org/10.1016/0040-1625(72)90049-2.

Teece, D. J., G. Pisano, and A. Shuen. 1997. “Dynamic Capabilities and Strategic Management.” *Strategic Management Journal* 18 (7): 509–533. https://doi.org/10.1002/(SICI)1097-0266(199708)18:7<509::AID-SMJ882>3.0.CO;2-Z.

2

THINKING ABOUT THE FUTURE

Here, I discuss not only the challenges for individuals in thinking about the future but also the relevance of thinking about the future for organisations. It has been argued that the attitude towards the future of an individual determines how the present and the past are defined (Rappaport 1991). The human capacity to imagine future outcomes and to devise the means to handle them are a basic human capacity (Reading 2004). Cognitive activities can be perceived as based upon concepts of the future (Ingvar 1985). Overall, foresight has been perceived as an ongoing cognitive activity (de Jouvenel 1967).

In the context of foresight, one finds the assumption that the future cannot be predicted. Part of this argument is that there is no such thing as a single future. Rather, one would need to refer to futures, as we need to admit that the future is not yet predetermined. However, one can argue that the predictability of the future is related to what one is trying to predict, how far into the future, and under what circumstances (Tetlock and Gardner 2015). This might imply that a temporary prediction in a narrow

DOI: 10.4324/9781003302735-4

setting might be possible, but not for longer time periods in complex settings.

Turning to the individual level also allows a better understanding of the possible challenges to organisations in developing foresight To reemphasise this, the connection between the individual capacity to think about the future and the linkage to an organisation, I refer to the perception that psychiatric disorders can be used to describe organisational disorders (Kets de Vries and Miller 1984, 1986). Kets de Vries (2004: 184) states that when looking at organisations, the field of psychology can be very helpful: 'Rejecting a psychoanalytical informed approach to studying human issues is a mistake, plain and simple. After all, it is individuals that make up organisations and create the units that contribute to social process'. Before discussing time and the inner future, from the perspective of psychiatry and psychotherapy, I will discuss cognitive barriers to developing foresight.

2.1 Cognitive barriers to developing foresight

Several authors have pointed to cognitive barriers as the cause of failures in developing foresight and one of the great challenges in developing foresight (e.g., de Geus 1997; Bazerman and Watkins 2004; Day and Schoemaker 2004; MacKay and McKiernan 2004; Seidl and van Aaken 2004). Organisations often fail to perceive weak signals of change or trends because they do not fit the mental models of the individual members of these organisations and are, therefore, rejected. According to Senge (1990: 8), mental models can be defined as 'deeply ingrained assumptions, generalisations, or even pictures or images that influence how we understand the world and how we take action'.

Day and Schoemaker (2006: 37) describe obstacles to developing foresight as the 'powerful tendency to ignore warning signals and pretend that all is well. The more intelligent we are, the better we also are at rationalizing away important signals of impending doom'. The relevance of cognitive barriers in foresight is also underscored by the notion that cognitive activities are based upon concepts of the future (Ingvar 1985) and that humans can only imagine the future in ways which are related to how the past has been experienced and understood (Reading 2004).

Understanding how cognitive factors can interfere with developing foresight is crucial. Day and Schoemaker (2004: 138) identify the following cognitive barriers:

- '*Mental filters:* Research shows that people tend to force the world into their own frames. Weak signals that do not fit are distorted or ignored. Humans see what they expect to see, not necessarily what is there.
- *Overconfidence:* A demonstrated tendency to be too certain makes people tend to believe that the view they hold is correct.
- *Penchant for confirming as opposed to disconfirming evidence:* It is more difficult to detect disconfirming evidence than confirming evidence, so the mind is more likely to accept than to reject an idea.
- *Dislike for ambiguity:* People dislike ambiguity, particularly in organisations in which managers are expected to have answers to questions.
- *Groupthink:* Members of organisations take comfort from belonging to the majority and seeing the world in the same way, so there is a tendency to go along with what others say, rather than to use an individual mind to find flaws in the group's thinking'.

2.2 Cognitive aspects of thinking about the future

Thinking about and dealing with the future is an important activity for organisations and for humans (Lombardo 2006). As Melges (1982: 34) points out: 'Many forms of mental illness are characterized by a bleak, foreshortened or fragmented future time perspective'. He argues that the importance of time distortions in psychiatric illness is that they can cloud the future of an individual and distort their view of that future, thereby disrupting goal-directed behaviour. While emphasising the importance of considering the future when attempting to understand human behaviour, Melges (1982: XX) states:

> I believe that the effect of the personal future on the present has to be added to the traditional psychiatric approaches of the effect of the past on the present. In particular, it appears that how a person conducts his outlook on his personal future determines, in large measure, the time patterns of his life and the ways in which he derives meaning from life.

Melges (1982) conceptualised the inner future of an individual as follows:

1. Humans are goal-directed organisms.
2. A person attempts to gain control over his own future through an interplay of future images, plans of action, and emotions
3. Distortions of psychological time disrupt a person's sense of control over the future and can lead to psychological vicious cycles.
4. Correction of time distortions and the harmonisation of future images, plans of action, and emotions restore a sense of control over the personal future.

Melges (1982) refers to spirals when the future becomes clouded, inducing a vicious cycle, since the individual's goal-directed behaviour becomes impaired. This situation may lead to a progressive mismatch between future images, plans of action, and emotion and to a lack of control over the future. It is possible to argue that argue the future is dynamically active and enables individuals to be proactive, rather than being reactive (de Jouvenel 1967; Melges 1982; Ingvar 1985).

2.3 Enabling futures thinking

On the one hand, the discussion allows us to understand the relevance of assessing how individuals think about the future; it also points out to the danger when organisational leaders have psychiatric disorders such as schizophrenia, paranoia, and depression (Schwarz 2007) and what the effects on the organisation could be. On the other hand, this discussion allows the identification of approaches to enable futures thinking on an individual level which of course also has organisational implications.

Melges (1982: XXI) suggests that

> the restoration of control over the personal future can be accomplished through the correction of time distortions and the harmonisation of future images, plans of action, and emotions. The restoration of control over the future is the key to interrupting psychopathological spirals.

Melges's (1982) assumption is that only the future can be controlled, since the past has already happened, the present quickly becomes the past, and control over action has to be directed at the future. To control the future, Melges (1982) suggests bringing the future into the psychological present by visualising future possibilities. This process is exemplified in terms of envisioning the future, picturing what might happen, or sketching an actual scenario.

The argument can be made that this approach is like the tools used in foresight that are discussed later in this book. For example, in the context of futures literacy laboratories the approach is described as making assumptions about the future explicit, reframing these assumptions and then rethinking or imaging different futures. As I will discuss later, the tools used in strategic foresight such as scenario planning have been associated with challenging, testing, and changing the assumptions decision-makers make about their present and future business environment (Wilson 1998). Scenario planning has been described as a fundamental way of thinking about the future (Schoemaker 1992).

Ingvar (1985) has established the concept of 'memory of the future', that the human brain is able to store different pictures of the future. The more memories of the future that are stored, the more receptive an individual can be to signals from the outside world. Reading (2004) remarks that humans can only imagine the future in ways that relate to how their past has been experienced and understood. This thought implies, making a case for the 'memory of the future', that humans are unable to conceive ideas that do not fit preconceived models of the world. The 'memory of the future' is also relevant in establishing the notion that foresight, especially scenario planning, is not about making predictions but about challenging assumptions and mental models about the present and the future and about creating memories of the future which allow leaders to take action sooner (de Geus 1997; van der Heijden, Bradfield, Burt, Crains, and Wright 2002). Similar thoughts are articulated in the discussions on agile leadership and management.

Schoemaker, Krupp, and Howland (2013) have developed a concept for strategic thinking in leadership which is of particular interest in strategic foresight. They mention six actions that entrepreneurial leaders should take in VUCA environments: anticipate, challenge, interpret, decide, align, and learn. The capability of anticipation is linked to Schoemaker's work (Day and Schoemaker 2005, 2006; Schoemaker and Day 2009) on developing peripheral vision to detect weak signals or trends in the business environment.

Schoemaker (2018) has emphasised that individual leaders need to direct attention to the periphery and ensure sufficient mental slack. Meissner and Wulf (2013) pointed out that Schoemaker (1993) and Bradfield (2008) were the first to analyse the influence of the scenario method on cognitive biases, thereby reducing overconfidence bias and confirmation bias (Spaniol and Rowland 2018). This also explains how strategic foresight should help individuals overcome cognitive limitations in perceiving changes in the business environment and prospecting the future.

This not only implies that applying foresight tools such as scenario planning is helpful to overcome the cognitive limitations of members of an organisation. Individuals need to be trained to think about the future. An empirical investigation by Schwarz, Rohrbeck, and Wach (2020) has found that corporate foresight training of managers has a positive impact on the dynamic capabilities of a firm. Implying that leaders need to be trained in sensing change or developing peripheral vision, making decisions under uncertainty, being confronted with an uncertain future, and how to reconfigure strategic resources accordingly.

Besides the organisational aspects of strategic foresight that I will discuss in this book, the aspects in this chapter highlight the relevance of considering the ways in which foresight can be developed not only through foresight process and tools but also through systematic training for the members of an organisation.

2.4 Futures literacy

The idea of futures literacy, promoted by Riel Miller and his work at UNESCO, has gained significant attention. The goal of the UNESCO in advocating futures literacy is to contribute to human resilience, map how people around the world think about the future, and to help people to become more futures literate (Miller 2018c). The concept of futures literacy focusses on the challenges in thinking about the future.

The concept of futures literacy draws on the research on anticipation.

> Hunting for different kinds of futures runs directly into the obstacle that by definition the future cannot exist in the present, since if it did it would no longer be the future but the present. And yet, as everyone knows, the

> future plays a role in the present. How can something that does not exist have an impact? One answer to this problem is the idea of anticipation. The future does not exist in the present but anticipation does. The form the future takes in the present is anticipation.
>
> *(Miller 2018b: 2)*

It can also be argued that all attempts to think about the future can be described as forms of anticipation (Miller, Poli, and Rossel 2018).

Miller (2018b: 15) defines futures literacy as

> a capability. A futures literate person has acquired the skills needed to decide why and how to use their imagination to introduce the non-existent future into the present. These anticipatory activities play an important role in what people see and do. Developing a detailed description of this capability to "use-the-future" calls for an analytical framework that can clarify the nature of different anticipatory systems and guide both research into [futures literacy] and its acquisition as a skill. Such a framework is presented in this chapter, focusing on the sub-set of anticipatory systems and processes that humans use when they consciously imagine the future.

The application of the futures literacy concepts, in UNESCO, can be observed in futures literacy laboratories, which will be discussed in Chapter 10. Miller (2018a) argues that the first step in any futures literacy process is to find a way to make anticipatory assumptions about the future explicit and observable. One can argue that this is also supposed to happen in scenario planning projects (van der Heijden 1996; van der Heijden, Bradfield, Burt, Crains, and Wright 2002; Rowland and Spaniol 2021). The basic structure of futures literacy laboratories is (1) to revel the assumption about the future of the participants; (2) reframe those assumptions; and (3) allow participants to rethink the future (Bergheim 2018).

References

Bazerman, M. H., and M. D. Watkins. 2004. *Predictable Surprises.* Boston: Harvard Business School Press.

Bergheim, S. 2018. "The Futures Literacy Laboratory- Novelty (FLL-N) Case Studies." In *Transforming the Future: Anticipation in the 21st Century,* edited by R. Miller, 110–229. London: Taylor & Francis.

Bradfield, R. M. 2008. "Cognitive Barriers in the Scenario Development Process." *Advances in Developing Human Resources* 10 (2): 198–215. https://doi.org/10.1177/1523422307313320.

Day, G. S., and P. J. H. Schoemaker. 2004. "Driving Through the Fog: Managing at the Edge." *Long Range Planning* 37 (2): 127–142.

———. 2005. "Scanning the Periphery Scanning the Periphery." *Harvard Business Review* 83 (11): 135–148. https://doi.org/10.1225/R0511H.

———. 2006. *Peripheral Vision: Detecting the Weak Signals That Will Make or Break Your Company*. Boston: Harvard Business School Press.

de Geus, A. 1997. *The Living Company*. Boston: Harvard Business School Press.

de Jouvenel, B. 1967. *The Art of Conjecture*. London: Basic Books.

Ingvar, D. H. 1985. "Memory of the Future: An Essay on the Temporal Organization of Conscious Awareness." *Human Neurobiology* 4 (3): 127–136.

Kets de Vries, M. 2004. "Organizations on the Couch: A Clinical Perspective on Organizational Dynamics." *European Management Journal* 22 (2): 183–200. https://doi.org/10.1016/j.emj.2004.01.008.

Kets de Vries, M. F. R., and D. Miller. 1984. *The Neurotic Organization*. New York: Harper Business.

———. 1986. "Personality, Culture, and Organization." *The Academy of Management Review* 11 (2): 266–279. https://doi.org/10.2307/258459.

Lombardo, T. 2006. "Thinking Ahead: The Values of Future Consciousness." *The Futurist* 40 (1): 45–50.

MacKay, R. B., and P. McKiernan. 2004. "The Role of Hindsight in Foresight: Refining Strategic Reasoning." *Futures* 36 (2): 161–179.

Meissner, P., and T. Wulf. 2013. "Cognitive Benefits of Scenario Planning: Its Impact on Biases and Decision Quality." *Technological Forecasting and Social Change* 80 (4): 801–814. https://doi.org/10.1016/j.techfore.2012.09.011.

Melges, F. T. 1982. *Time and the Inner Future*. New York: John Wiley & Sons.

Miller, R. 2018a. "Futures Literacy Laboratories (FLL) in Practice: An Overview of Key Design and Implementation Issues." In *Transforming the Future: Anticipation in the 21st Century*, edited by R. Miller, 95–109. London: Taylor & Francis.

———. 2018b. "Introduction, Futures Literacy: Transforming the Future." In *Transforming the Future: Anticipation in the 21st Century*, edited by R. Miller, 1–12. London: Taylor & Francis.

———. 2018c. "Sensing and Making-Sense of Futures Literacy: Towards a Futures Literacy Framework (FLF)." In *Transforming the Future: Anticipation in the 21st Century*, edited by R. Miller, 15–50. London: Taylor & Francis.

Miller, R., R. Poli, and P. Rossel. 2018. "The Discipline of Anticipation: Foundations for Futures Literacy." In *Transforming the Future: Anticipation in the 21st Century*, edited by R. Miller, 51–65. London: Taylor & Francis.

Rappaport, H. 1991. "Measuring Defensiveness against Future Anxiety: Telepression." *Current Psychology* 10 (1): 65–77. https://doi.org/10.1007/BF02686781.

Reading, A. 2004. *Hope & Despair: How Perceptions of the Future Shape Human Behavior.* Baltimore: Johns Hopkins University Press.

Rowland, N., and M. Spaniol. 2021. "The Strategic Conversation, 25 Years Later: A Retrospective Review of Kees van Der Heijden's Scenarios: The Art of Strategic Conversation." *Futures & Foresight Science* 4: e102. https://doi.org/10.1002/ffo2.102.

Schoemaker, P. J. H. 1992. "How to Link Strategic Visions to Core Capabilities." *MIT Sloan Management Review* 34 (1): 67–81.

———. 1993. "Multiple Scenario Development: Its Conceptual and Behavioral Foundation." *Strategic Management Journal* 14 (3): 193–213. https://doi.org/10.1002/smj.4250140304.

———. 2018. "Attention and Foresight in Organizations." *Futures & Foresight Science* 1: e5. https://doi.org/10.1002/ffo2.5.

Schoemaker, P. J. H., and G. S. Day. 2009. "How to Make Sense of Weak Signals." *MIT Sloan Management Review* 50 (3): 81–89.

Schoemaker, P. J. H., S. Krupp, and S. Howland. 2013. "Strategic Leadership: The Essential Skills." *Harvard Business Review* (January–February), https://hbr.org/2013/01/strategic-leadership-the-esssential-skills.

Schwarz, J. O. 2007. "Assessing Future Disorders in Organizations: Implications for Diagnosing and Treating Schizophrenic, Depressed or Paranoid Organizations." *Foresight* 9 (2): 15–26.

Schwarz, J. O., R. Rohrbeck, and B. Wach. 2020. "Corporate Foresight as a Microfoundation of Dynamic Capabilities." *Futures & Foresight Science* 2 (2): e28. https://doi.org/10.1002/ffo2.28.

Seidl, D., and D. van Aaken. 2004. "Anticipating Critique and Occasional Reason: Modes of Reasoning in Face of Radically Open Future." *2nd International Conference on Organizational Foresight*. Strathclyde Business School, University of Strathclyde, Glasgow.

Senge, P. M. 1990. *The Fith Discipline*. London: Century Business.

Spaniol, M. J., and N. J. Rowland. 2019. "Defining Scenario." *Futures & Foresight Science* 1: e1. https://doi.org/10.1002/ffo2.3.

Tetlock, P., and D. Gardner. 2015. *Superforecasting: The Art and Science of Prediction*. New York: Random House.

van der Heijden, K. 1996. *Scenarios: The Art of Strategic Conversation*. Chichester: John Wiley & Sons.

van der Heijden, K., R. M. Bradfield, G. Burt, G. Crains, and G. Wright. 2002. Chichester: *The Sixth Sense*. John Wiley & Sons.

Wilson, I. 1998. "Mental Maps of the Future: An Intuitive Logics Approach to Scenarios." In *Learning from the Future: Competitive Foresight Scenarios*, edited by L. Fahey and R. M. Randall, 81–108. San Francisco: John Wiley and Sons.

Part II

APPLICATIONS OF STRATEGIC FORESIGHT

3

CONCEPTUALISING STRATEGIC FORESIGHT

Following the idea of this book that strategic foresight can be developed on an organisational and individual level, I will now turn to the organisational level. The organisational level has gained the most attention in the research on strategic foresight. However, the individual level is of no less importance (Schwarz, Rohrbeck, and Wach 2020). I will structure the discussion on the application of strategic foresight in organisations along the process of strategic foresight and the tools used in developing foresight.

As I will discuss in Section 3.2, many tools have been created to develop foresight in an organisation. This book does not describe all of the available tools. Instead, I will focus on a selection of tools that I find relevant. In some cases, these tools will be described by experts in the field.

I will describe these tools as part of a process of strategic foresight. This allows us not only to assess in which phase of the strategic foresight process it is feasible to apply a certain tool; it also allows discussing strategic foresight as a comprehensive activity that should be understood as an ongoing process, instead of an one-time exercise.

DOI: 10.4324/9781003302735-6

Before detailing the process of strategic foresight, it is necessary to discuss the focus of such an activity. The focus of a strategic foresight process should arguably be as broad as possible. The rationale is that given that the future is uncertain, we cannot know from which area new trends are likely to emerge or how they might influence other trends. Therefore, a broad perspective is a prerequisite. The political, economic, social, and technological (PEST) framework is helpful (Burt, Wright, Bradfield, Crains, and van der Heijden 2006). A PEST framework can also serve as a reminder to consider all areas when developing foresight.

Further, organisations might engage in customer, technology, or competitor foresight (Rohrbeck and Thom 2008). While technology foresight is applied in many organisations, customer foresight might be more of an emerging field (Schweitzer, Hofmann, and Meinheit 2019; Eller, Hofmann, and Schwarz 2020). The objective of customer foresight is to understand customers' future circumstances and wishes for tomorrow's products and services by combining two relevant, yet mostly separate fields: the investigation of customer needs based on their everyday usage and the dynamics of change and projections.

The linkage between business wargaming and foresight (Oriesek and Schwarz 2008; Schwarz 2009; Oriesek and Schwarz 2020) and more specific scenario planning (Schwarz, Ram, and Rohrbeck 2018) has stressed the competitive dimension or perspective in foresight which is also reflected in the process of strategic foresight.

3.1 Process of strategic foresight

The process of strategic foresight has been conceptualised and labelled differently in the past. The research on strategic foresight has been labelled in the German context as 'Strategische Frühaufklärung' (Krystek and Müller-Stewens 1993; Liebl 1996, 2000), and I have also used the term 'strategic early warning' (Schwarz 2005). In the article 'Organizations as Interpretation Systems', Daft and Weick (1984) have described a basic process that is of relevance for our discussion: organisations perceive their environment (step 1: 'scanning – data collection'), translate what they find into organisational implications (step 2: 'interpretation – data giving meaning'), and develop responses based on their insights into their environment (step 3: 'learning – action taken').

The guiding definition in this book on strategic foresight indicates the phases of strategic foresight: perceiving, prospecting, and probing. I build on the process model, which distinguishes three phases (Højland and Rohrbeck 2018; Rohrbeck and Kum 2018), while adding a fourth (transforming), as shown in Figure 3.1.

The *perceiving* phase describes practices that firms use to identify the factors that drive environmental change. These firms aim to identify trends and weak signals ahead of competition to gain a lead-time advantage. The underlying assumptions in this phase of the strategic foresight process is that discontinuities do not emerge without warning. These warning signs can be described as 'weak signals'. The concept of 'weak signals' (Ansoff 1975) aims at early detection of those signals which could lead to strategic surprises and to an event which has the potential, for instance to jeopardise an organisation's strategy. The nature and importance of 'weak signals' are best described as follows: 'As any adult knows, a magician cannot produce a rabbit unless it is already in (or very near to) his hat. In the same way, surprises in the business environment almost never emerge without a warning' (Wack 1985: 148). We will later take a closer look at the concept

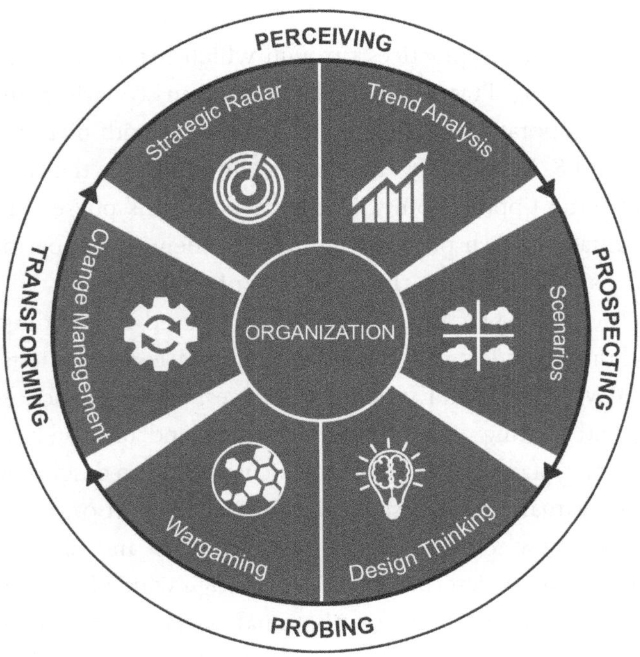

Figure 3.1 Systematic approach to strategic foresight.

of weak signals and their relationship to trends in Section 4.1, 'Conceptualising weak signals and trends'.

'Weak signals' are detected by scanning the organisational environment. Environmental scanning (Aguilar 1967; Daft, Sormunen, and Parks 1988; Boyd and Fulk 1996; Slaughter 1999) is a process whereby the environment in which an organisation operates is systematically scanned for weak signals and trends. The purpose is to identify early indications of possible and ongoing environmental changes (Lester and Waters 1989).

There are two methods of environmental scanning. The outside-in method uses a 360-degree focus to avoid blind spots. This approach acknowledges that finding the individual and organisational blind spots is essential for detecting 'weak signals'. Unfortunately, this approach is hindered by the problem of information overload.

The second and more widely used method is 'inside-out' scanning. Here, top management determines in advance the fields to be scanned. Inside-out scanning limits the amount of information gathered but risks limiting the focus of the organisation and increasing the number and/or gravity of blind spots. If an environmental scan has detected weak signals, it is important to monitor them. Environmental scanning is the first phase of strategic foresight.

The *prospecting* phase describes practices through which firms engage in sensemaking and strategising. Practices include working in scenario planning or futures literacy laboratories. The book elaborates on both methods later, in Chapters 7 and 8. In addition, these firms aim to foresee the right time to act by identifying tipping points. The purpose of this phase is to gain an insight advantage, which permits the firms to identify a superior course of action that is different from the status quo of the industry.

The *probing* phase describes practices that are often enacted in dedicated accelerator units and may include prototyping, R&D projects, consumer tests, internal and external venturing, and strategic initiatives. Probing practices aim at legitimising and starting a new course of action and ultimately at gaining a competitive advantage. In this book, the focus will be on describing tools such as business wargaming or design thinking that support organisations in identifying these new courses of action, see Chapters 9 and 10.

Finally, the *transforming* phase describes practices through which firms enhance, combine, and reconfigure their organisational assets – its resources and capabilities, eventually changing or even transforming an organisation.

3.2 Tools used in the process

Before discussing the process of strategic foresight, I want to provide an overview of tools used in strategic foresight. The description of the process will also highlight some tools. Other tools will also be discussed later.

Based on a review of 886 foresight studies, Popper (2008) offers not only an overview of foresight tools but also classifies them in terms of nature and capability. Nature refers to qualitative, quantitative, or semi-quantitative tools and capabilities, meaning the ability to gather or process information based on evidence, expertise, interaction, or creativity.

Qualitative methods provide meaning to events and perceptions. Such interpretations tend to be based on subjectivity or creativity that is often difficult to validate, for example, opinions, judgements, beliefs, and attitudes. Quantitative methods generally measure variables and apply statistical analyses, using or generating – at least in theory – reliable and valid data, such as socio-economic indicators. Semi-quantitative methods are those that apply mathematical principles to quantify subjectivity, rational judgements, and viewpoints of experts and commentators (i.e. weighting opinions and probabilities) (Popper 2008).

Popper (2008) defines the capabilities of foresight tools as follows:

- **Creativity** refers to the mixture of the original and the imaginative. These methods rely heavily on the inventiveness and ingenuity of very skilled individuals, such as science fiction writers or the inspiration that emerges from groups of participants in brainstorming sessions.
- **Expertise** refers to the skills and knowledge of individuals in a particular area or subject. It is frequently used to support top-down decisions, provide advice, and make recommendations.
- **Interaction** recognises that expertise often benefits considerably from being brought together and challenged to articulate with other expertise.
- **Evidence** recognises that it is important to attempt to explain and/or forecast a particular phenomenon with the support of reliable documentation and means of analysis of, for example, statistics and various types of measurement indicators.

These attributes are then the building blocks to what Popper (2008) calls the foresight diamond, depicted in Figure 3.2.

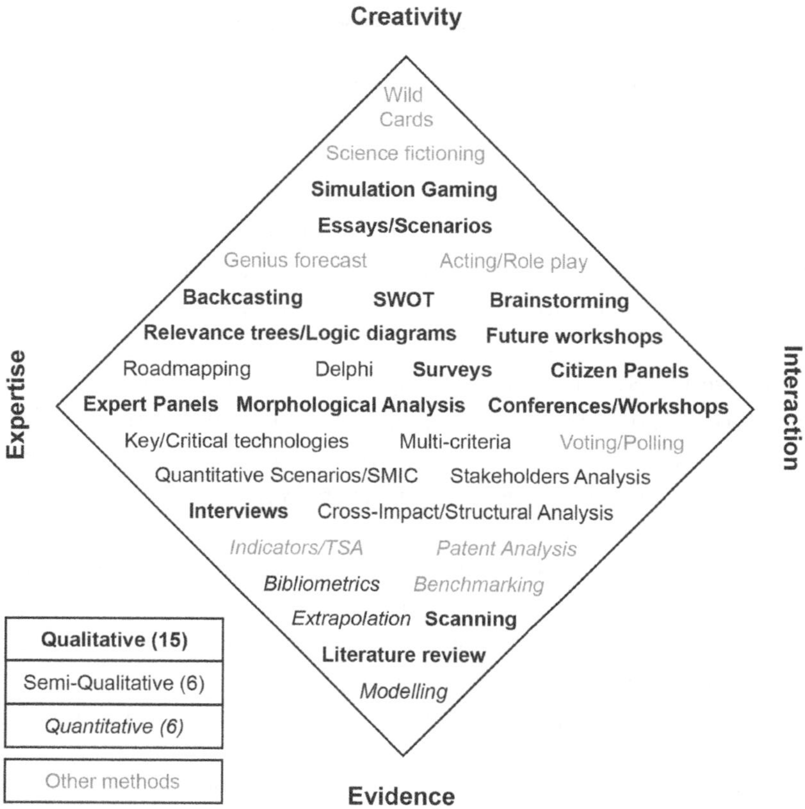

Figure 3.2 The foresight diamond (Popper 2008).

Based on a literature review, Schwarz (2005) comes to the conclusion that the tools in foresight such as trend research, scenario technique, Delphi technique, and simulation as well as gaming are mentioned frequently. The longer in use tools such as the Delphi method, scenario technique, and, in particular, business wargaming have their roots in defence research (Linstone and Turoff 1975; Pohl 1996; Oriesek and Schwarz 2020).

As described in Figure 3.3, I will focus on tools that I have either used in developing strategic foresight or that I find relevant. While describing the process, the tools will be discussed.

Weak signals and trends are the basis for any foresight activity and therefore are discussed here in the phase of *perceiving*. Along with this discussion, I will highlight the relevance of applying science fiction to

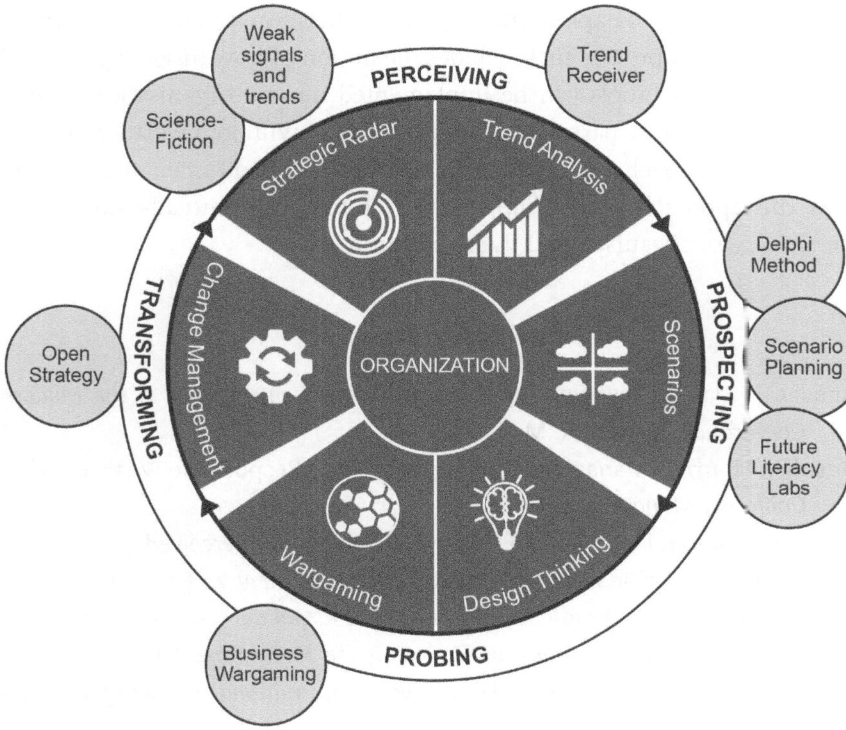

Figure 3.3 Foresight tools discussed in more detail.

strategic foresight. The concept of a trend receiver adds to the perception of changes in the business environment and builds on weak signals and trends.

The Delphi method and scenario planning, while very different, help to *prospect*, in terms of understanding how the future could evolve. In addition, futures literacy laboratories will be discussed as a tool that emerged from the futures literacy school of thought. Business wargaming and design thinking can be perceived as tools of *probing*, better understanding what a certain set of actions could imply for the future of an organisation. Design thinking, for instance, is of course a tool that we know from the context of design or innovation, but here I will also discuss the connection between foresight and design thinking.

Finally, in the phase of *transforming*, change management plays a role. As mentioned in the beginning of this book, the aim is not only to understand foresight as a way to explore futures, but furthermore to understand strategic foresight as a process that includes the notion of how insights generated through foresight tools can be implemented into an organisation. Transforming is therefore part of the process. While I will not be elaborating on the huge field of change management, I will focus on open strategy and how this approach can be supported by foresight, eventually supporting the process of organisational transformation.

References

Aguilar, F. J. 1967. *Scanning the Business Environment. Studies of the Modern Corporation*. New York: Macmillan.

Ansoff, I. H. 1975. "Managing Strategic Surprise by Response to Weak Signals." *California Management Review* 18 (2): 21–33.

Boyd, B. K., and J. Fulk. 1996. "Executive Scanning and Perceived Uncertainty: A Multidimensional Model." *Journal of Management* 22 (1): 1–21.

Burt, G., G. Wright, R. Bradfield, G. Crains, and K. van der Heijden. 2006. "The Role of Scenario Planning in Exploring the Environment in View of the Limitations of PEST and Its Derivatives." *International Studies of Management & Organization* 36 (3): 50–76.

Daft, R. L., J. Sormunen, and D. Parks. 1988. "Chief Executive Scanning, Environmental Characteristics, and Company Performance: An Empirical Study." *Strategic Management Journal* 9 (2): 123–139.

Daft, R. L., and K. E. Weick. 1984. "Toward a Model of Organizations as Interpretation Systems." *Academy of Management Review* 9 (2): 284–296.

Eller, E., R. Hofmann, and J. O. Schwarz. 2020. "The Customer Foresight Territory." *Marketing Review St Gallen* 27 (3): 888–895.

Højland, J., and R. Rohrbeck. 2018. "The Role of Corporate Foresight in Exploring New Markets – Evidence from 3 Case Studies in the BOP Markets." *Technology Analysis & Strategic Management* 30 (6): 734–746. https://doi.org/10.1080/09537325.2017.1337887.

Krystek, U., and G. Müller-Stewens. 1993. *Frühaufklärung Für Unternehmen: Identifikation Und Habung Zukünftiger Chancen Und Bedrohungen*. Stuttgart: Schäfer-Poeschel.

Lester, R., and J. Waters. 1989. *Environmental Scanning and Business Strategy*. London: British Library Research and Development Department.

Liebl, F. 1996. *Strategische Frühaufklärung: Trends - Issues - Stakeholders*. München: Oldenbourg.

———. 2000. *Der Schock Des Neuen: Entstehung Und Management von Issues Und Trends*. München: Gerling Akademie Verlag.

Linstone, H. A., and M. Turoff. 1975. "Evaluation: Introduction." In *The Delphi Method: Techniques and Applications*, edited by H. A. Linstone and M. Turoff, 229–235. Reading: Addison-Wesley Publishing.

Oriesek, D. F., and J. O. Schwarz. 2008. *Business Wargaming: Securing Corporate Value*. Aldershot: Gower.

———. 2020. *Winning the Uncertainty Game: Turning Strategic Intent into Results with Wargaming*. London: Taylor & Francis.

Pohl, F. 1996. "Thinking about the Future." *The Futurist* 30 (5): 8–13.

Popper, R. 2008. "'How Are Foresight Methods Selected?' Edited by Maurits Butter, Felix Brandes, Michael Keenan, and Rafael Popper." *Foresight* 10 (6): 62–89. https://doi.org/10.1108/14636680810918586.

Rohrbeck, R., and M. E. Kum. 2018. "Corporate Foresight and Its Impact on Firm Performance: A Longitudinal Analysis." *Technological Forecasting and Social Change* in press (January): 1–30. https://doi.org/10.1016/j.techfore.2017.12.013.

Rohrbeck, R., and N. Thom. 2008. "Strategic Foresight at Deutsche Telekom AG." *Proceedings of the 4th IEEE International Conference on Management of Innovation and Technology, ICMIT*, September. https://doi.org/10.1109/ICMIT.2008.4654329.

Schwarz, J. O. 2005. "Pitfalls in Implementing a Strategic Early Warning System." *Foresight* 7 (4): 22–30.

———. 2009. "Business Wargaming: Developing Foresight within a Strategic Simulation." *Technological Analysis and Strategic Management* 21 (3): 291–305.

Schwarz, J. O., C. Ram, and R. Rohrbeck. 2018. "Combining Scenario Planning and Business Wargaming to Better Anticipate Future Competitive Dynamics." *Futures* 105: 133–142. https://doi.org/10.1016/j.futures.2018.10.001.

Schwarz, J. O., R. Rohrbeck, and B. Wach. 2020. "Corporate Foresight as a Microfoundation of Dynamic Capabilities." *Futures & Foresight Science* 2 (2): e28. https://doi.org/10.1002/ffo2.28.

Schweitzer, N., R. Hofmann, and A. Meinheit. 2019. "Strategic Customer Foresight: From Research to Strategic Decision-Making Using the Example of

Highly Automated Vehicles." *Technological Forecasting and Social Change* 144 (April): 49–65. https://doi.org/10.1016/j.techfore.2019.04.004.

Slaughter, R. A. 1999. "A New Frame Work for Environmental Scanning." *Foresight* 1 (5): 441–451.

Wack, P. 1985. "Scenarios: Shooting the Rapids." *Harvard Business Review* 63 (6): 139–150.

4

WEAK SIGNALS AND TRENDS

As we learned in Chapter 3, trends are fundamental to the process of strategic foresight. The identification of trends stands at the beginning of a strategic foresight process. Moreover, trends are relevant input for many other foresight approaches such as scenario planning (Chapter 7) or business wargaming (Chapter 9).

Ansoff (1975) introduced the concept of weak signals and thus laid the basis for trend management (Liebl 2000). Ansoff (1975: 12) described weak signals as '… warnings (external or internal), events and developments which are still too incomplete to permit an accurate estimation of their impact and/or to determine their full-fledged responses'. Ansoff's (1975) concept aimed at the early detection of indicators of environmental changes that could lead to strategic surprises and events and that have the potential to jeopardise an organisation's strategy. He described weak signals as external or internal signs, occurrences, and developments that are too immature to have their impact or complete reactions assessed (Ansoff 1982). A weak signal can also be perceived as an upcoming trend.

DOI: 10.4324/9781003302735-7

Buck, Herrmann, and Lubkowitz (1998) highlight two definitions: The classic definition describes a trend as an ongoing and fundamental societal change over an extended period. In contrast, the modern definition treats a trend as a short-term phenomenon, such as in fashion or music. Liebl (2002) offers a third definition of a trend as a phenomenon that is always complex and that cannot be perceived simply as a fashion; neither can its lifespan be measured accurately. A trend is a connection, an association that is defined by its crossing of contextual borders.

Few terms have caused as much confusion in the context of foresight, strategy, and management as 'trend' has. It has been used widely in the practitioners' literature in the fields of marketing and general management, for instance, in 'megatrends', fashion trends, and youth culture. 'Pop futurists' (Slaughter 1999), for instance, Naisbitt (1984) or Popcorn (1991), sell trend labels like 'cocooning' to both a larger audience and the business community without a deeper scientific background. Here the term 'trend' is used to address the intuitive feeling that something is of importance.

Therefore, clarifying what trends are and more importantly how they can be conceptualised is essential for our discussion on strategic foresight.

4.1 Conceptualising weak signals and trends

Following Liebl and Schwarz (2010), trends shall be conceptualised as (socio-cultural) innovations. Trend analysis is therefore understood as innovation research or the science of the new. If the nature of innovation is taken seriously, then strategic trend diagnosis must deal with two different aspects, invention and diffusion. Both aspects can be approached by asking questions:

- 'How can the new be identified? And what constitutes this new? This refers to the aspect of invention.
- 'Will the new become widespread? This refers to the aspect of diffusion.

Liebl and Schwarz (2010) argue that to detect strategic potentials, a trend should not be conceived of as a quantifiable phenomenon, as something that has already has been clearly operationalised – such an activity must be regarded as a strategic late warning system. Rather, a trend can be thought of as a vague silhouette, whose (new) quality and outline must be identified and evaluated. This outline has much to do with the line between normality and abnormality. Therefore, it is important to focus, first, on the detection of abnormal phenomena and, second, on the forces that drive the subsequent normalisation process – which may lead to a social convention.

4.1.1 The new

To better understand what constitutes the new, Liebl and Schwarz (2010) refer to the work of the art theorist Boris Groys (2008) who researched the anatomy of the new in a context of cultural economy. According to Groys (2008) the fundamental characteristic of the new can be seen in the fact that objects or concepts are transferred into another context. The invention therefore is best described by transgressing the boundaries of contexts, by connecting formerly separated contexts. Groys (2008) discusses several examples: Duchamp exhibited everyday objects in galleries and museums, a ready-made strategy which transformed these objects into valuable pieces of art. Variations of this strategy have been applied by other famous artists like Andy Warhol or Jeff Koons. Groys (2008) also uses examples from other fields: Freud connected the nearly forgotten myth of Oedipus to psychoanalysis, thereby reviving its popularity. This logic is by no means limited to art or cultural history, as the following examples illustrate:

- By establishing the brand Y-3, Yohji Yamamoto and Adidas merged sportswear with avant-garde prêt-à-porter-fashion and created a three-stripe-couture that became a category killer in the fashion industry.
- Quadrophonic sound, a music reproduction technique with a very small software range, was a market disaster, until its revival as a device for surround-sound in the booming multimedia home-entertainment and DVD market.
- Finally, innovative fields in humanities and science including cognitive science, bionics, or biological chemistry are based on the transgression of disciplinary boundaries.

These examples underline that the new is neither about something which has never been in the world before, nor that there is nothing new under the sun. What becomes clear is the power of the context: contexts influence human perception, structure expectations, and evoke new forms of interpretation and new ways of using things. Trends therefore are not a merely one-dimensional development in each direction, but rather they represent a hitherto non-existing combination of contexts.

4.1.2 Diffusion

To elaborate on their conceptualisation of trends, Liebl and Schwarz (2010) discuss the aspect of diffusion in more detail by introducing Mathews and Wacker's (2002) lifecycle of trends. This lifecycle of trends both helps to enrich the definition of trends and has practical implications for trend research.

Figure 4.1 shows the first half of the life cycle. Mathews and Wacker (2002) point out that significant trends – those that change mainstream behaviour – are born on the 'fringes' of society in the minds of deviant outsiders and then move to the 'edge'. This implies that trend research wants to outrun competition and gain a competitive advantage, and therefore

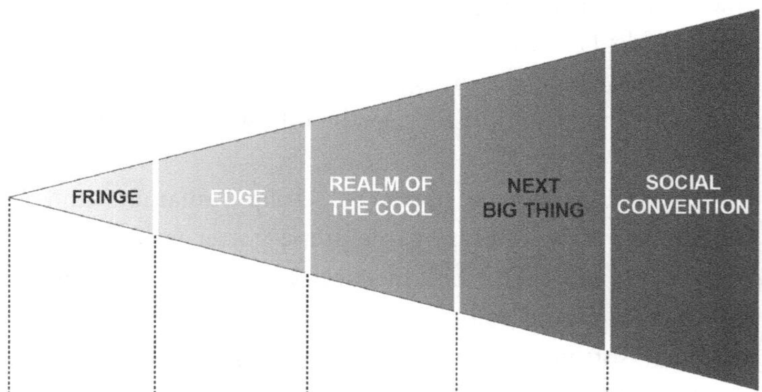

Audience Size	One	Limited	Identifiable cohort	Multiple cohorts	Mass market
Communication Vehicle	Originating deviant	Word of mouth	Events, feature stories	Mainstream news coverage	Advertising and market
Commercial Potential	0%	1-5%	10-25%	30-40%	70-100%

Figure 4.1 Life cycle of trends, part I (Liebl and Schwarz 2010).

needs to focus on these stages. Others have labelled this focus as 'abnormalities', arguing that these weak signals might not be purely speculative, not named and not widely known (Reeves, Goodson, and Whitaker 2021).

The 'realm of the cool' follows the 'fringe' and requires that these trends have been adopted by groups who see themselves as a marginalised avant-garde (the 'edge'). Trend researchers or 'cool hunters' attempt to forecast the next stage (i.e., the 'next big thing') for their clients; this implies that these consultants are late in picking up trends, and the resulting lead times are often too short to make the strategic moves for capitalising on the identification of trends (Harris and Zeisler 2002; Waters 2005, 2006).

When it is becoming obvious that the next phase of 'social convention' is almost certain to be reached, then the intense competition among the players will no longer allow pioneer advantages by claiming 'street credibility'. Stated differently, the bizarre phenomena of today are the starting point for tomorrow's mainstream. In the final analysis, Mathews and Wacker (2002) do not so much address the quantity of adherents or attention in the various lifecycle stages but rather who is involved and what the implications are for marketers and strategists.

Usually, trend research tends to concentrate on the rising branch of the lifecycle. This is for two reasons: (1) only 'emerging trends' are perceived as of interest and of news value in mass media; and (2) the 'in' easily tips into a 'mega-out'. However, Mathews and Wacker (2002) do not stop there, but go on to ask what happens with developments which already have become 'social convention'. They also explain how the second half of the life cycle of a trend comes into play, a phase of disposal in the form of post-normalisation (see Figure 4.2). From a strategic point of view this phase seems to be no less interesting, since it may offer considerable opportunities for recycling and endgames. The way in which these opportunities are used is essential for the further development of a mainstream phenomenon: Will it freeze in a cliché? Will it develop into a cultural icon? Will it end up as a rare example of an archetype? Or will it simply fall into oblivion? An important strategic implication is that different dispositions in the customer segments come significantly into play. An old trend may have at the same time several incarnations; depending on the context these may be cliché, icon, or archetype. It is thus not less relevant than an emerging trend.

To discuss the process of normalisation in the first half of Mathews and Wacker's (2002) lifecycle of trends, Liebl and Schwarz (2010) cite the

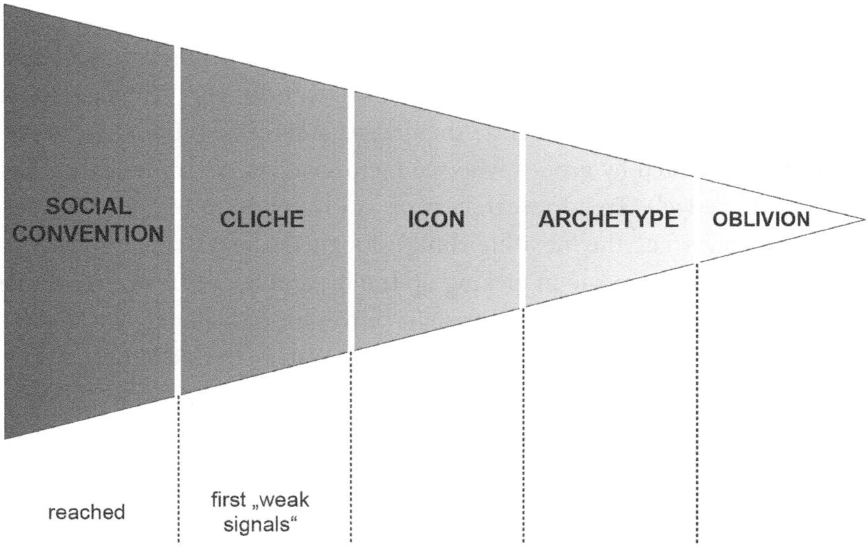

Figure 4.2 Life cycle of trends, part II (Liebl and Schwarz 2010).

research on the phenomena of normality and the emergence and drift of societal standards (Link 1999) and describe this part of the lifecycle of a trend as a threefold process of normalisation:

- 'As a new socio-cultural practice diffuses into society, it undergoes a reassessment: what starts as an abnormal behaviour ('perversity') of a (stigmatised) group becomes the behaviour ('preference') of a broader audience and even part of the establishment.
- 'The normalisation of a socio-cultural practice can be regarded as change in general value schemes: even individuals who do not adopt a certain practice tend to accept it as normal.
- 'While such a practice is spreading, it tends to lose its extreme forms and to assume more mass-compatible flavours – which allows even larger groups to relate to this practice. Therefore, the trend becomes a moving target.

The transformation of a psychological category ('perversity') into an economic one ('preference') points to the fact that new markets are being created here. In other words: where taboo-zones are shrinking, new markets begin to emerge as a consequence of normalisation. Therefore, when we monitor a trend, we have to monitor the forces that are negotiating and shaping the boundary between normality and abnormality – with respect to all three aspects of normalisation.

To give an example of how a trend can move through the trend life cycle, Liebl and Schwarz (2010) provide two cases on hip-hop and skateboarding (see Figure 4.3).

The hip-hop culture consists of four central elements: disc jockeying (DJing), break dancing, graffiti art, and rapping. We see from this list that the contexts involved are manifold: new practices of presenting music, new forms of dancing, new styles of art, and new ways to present lyrics and communicate with an audience.

According to legend, hip-hop started in the last week of August 1973 at a house party in New York City's Bronx neighbourhood (Chang 2005). DJ Kool Herc, known as the father of hip-hop, used two turntables and a microphone to isolate and combine the rhythm parts of two songs and laid

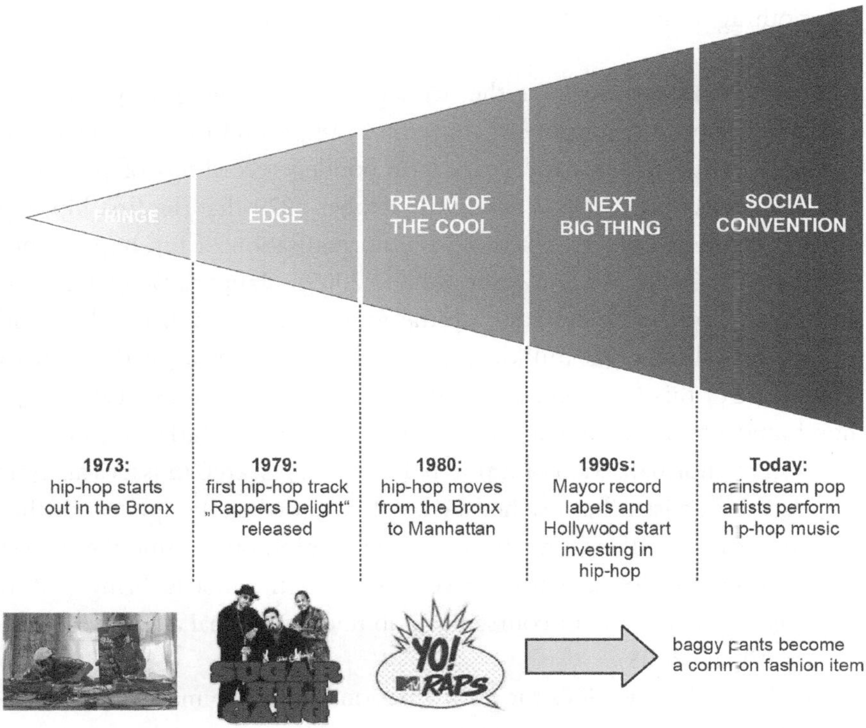

Figure 4.3 Life cycle of trends, part I, case study hip-hop (Liebl and Schwarz 2010).

his voice over both, creating a new musical style by using technical equipment in an unintended way. According to the threefold process of normalisation, the following applies to the normalisation of hip-hop:

- 'Hip-hop had emerged from being stigmatised and perceived as an abnormality to being the preference of a broader audience. Obviously, this transformation has, from an economic point of view, not only created a significant market for the music industry, but also for the fashion industry (McCracken 2006).
- 'While not universally accepted and liked, hip-hop has emerged, of course due to its many different forms and styles, to a music genre described as 'normal'.
- 'Hip-hop has changed over the years and lost its original forms from the 1970s; styles of music, outfit, and attitude have diversified. Major record labels today promote a broad range of hip-hop artists, including mainstream hip-hop performers who lack street credibility and have nothing to do with the original performers in the Bronx.

The importance of media as the driving force for emerging youth subcultures has been emphasised (Thornton 2018), a finding also relevant to hip-hop. While in the early years only bootleg recordings of parties in the Bronx circulated, it was not until October 1979 that the first hip-hop single, 'Rapper's Delight', advanced the normalisation of hip hop, changing the rules of the hip-hop game (Chang 2005). 'Rapper's Delight' spread from New York to Black Radio, up the American Top 40 list and around the world, selling eight million copies. While it can be argued that with the release of this hip-hop track, and of course because of its success, hip-hop found a broader audience, the release of the single marks only an early point in the normalisation of hip-hop. Empirical research in issues analysis suggests that an issue has reached a take-off point when it migrates to other communicative milieus. This happened when hip-hop became the subject of a movie: the 1982 docudrama *Wild Style* was the first to bring graffiti art, break dancing, and of course hip-hop music to global attention on the screen (Chang 2005).

Finally, in the late 1980s the normalisation of hip-hop made another significant leap, when hip-hop was featured on MTV, the cable television music network. In 1988 the show *MTV Raps* was launched, making hip-hop music available to millions of youths, especially white youth. Approximately a year later, hip-hop permeated American media because of parents' complaints about the violent and sexually explicit hip-hop lyrics. This campaign peaked in the early 1990s, forcing the major record labels to reconsider moving ahead with hip-hop artists, in defiance of public pressure. To the public, hip-hop had become a commodity – and subject to all problems and consequences of product liability.

A final important aspect in the normalisation of hip-hop is the paradoxical form of its globalisation: it is a perfect example of glocalisation. On the one hand, superstar-based American hip-hop has been spreading all over the globe; but on the other hand, it has been adapted and reconfigured, creating specific local varieties. This includes both the appropriation of local music traditions and lyrics in languages which significantly differ from English in their rhythms and flows (e.g., German, French or Turkish); and it has brought the emergence of local superheroes who in their home countries may sell even more records than their American blueprints.

While the evolution of hip-hop seems to fit quite nicely into the lifecycle of trends (Mathews and Wacker 2002), one also needs to mention, that a trend not automatically moves from the fringe to social convention, or from a weak to a strong signal. Liebl and Schwarz (2010) mention skateboarding as an example for a trend that represents a lifecycle deviation, characterised by backlashes and revivals, and combining processes of normalisation and post-normalisation.

4.1.3 Further aspects

Here I want to highlight further aspects which help us to conceptualise trends. Liebl and Schwarz (2010) add another criterion for the conceptualisation of trends: the paradoxical form of a trend. The paradoxical form underlines the novelty, but also the potential diffusion of a trend. The history of successful products supports this line of thought. It is evident that innovations are most likely to be successful when they incorporate contradictory features. Examples are Post-It notes, which are both adhesive and removable, and Nivea cream, a moisturiser that is not oily.

The paradox is the driver of innovation and therefore the driver of trends and their diffusion. Liebl and Schwarz (2010) cite *The Savage Girl* by Alex Shakar (2001), a novel about trend researchers, for 'paradessence', the paradoxical nature of a product or brand. Paradessence is the combination of mutually exclusive phenomena or features, which is precisely the description of a trend. This is why trends must be conceptualised as more than simply a development. Rather, the new and surprising qualities of trends are important – and reconcile hitherto contradictory elements.

This idea of paradoxical phenomena is also manifested in countertrends (Liebl 2000; Waters 2006; Weiner and Arnold 2006). Once a trend has been detected and formulated, it is most likely to have at least one simultaneously operating countertrend. Conversely, when no countertrend to a given trend has been formulated, then the complexity of a trend has probably not been adequately addressed. 'Countertrends don't happen despite trends; they happen because of them' (Weiner and Arnold 2006: 23). And a supposedly identified and well-understood trend may take on an entirely new meaning when extended by its countertrend. In this case we are not dealing with two developments that are separate or evolving in opposite directions, but rather with a new – hybrid and more complex – phenomenon. Below are a few examples of this phenomenon:

- 'Glocalisation (i.e., local specialisation in a globalising economy)
- 'Bobos (Bourgeois Bohemians), who combine a subcultural attitude with a bourgeois lifestyle
- 'Mainstream of the marginalised (i.e., groups labelling themselves marginalised subcultures but showing no significant differences from mainstream groups)

Each of these examples can be characterised as a paradoxical and complex development, which displays characteristics formerly perceived as contradictory or mutually exclusive; instead, they have to be regarded as forms of pluralisation, individualisation, and hybridisation.

Liebl and Schwarz (2010) summarise their approach to trend research and diagnoses as a means of identifying possible 'future normalities':

- 'Conceptualisation of trends as socio-cultural innovations
- 'The new' as the first aspect of innovation, described by a transgression of contextual boundaries
- 'Diffusion as the second aspect of innovation, operationalised by a threefold process of normalisation

- "Paradessence' with its linkage function between the new and diffusion
- 'A two-part life cycle of trends, consisting of consecutive stages of normalisation and different – but not mutually exclusive – stages of post-normalisation (including revivals)

Further, the question 'in which direction will a trend evolve?' can be replaced by the following four sets of questions for exploring future normalities (Liebl and Schwarz 2010):

- '*Whither?* Which contexts are involved in a trend? What can be regarded as new in this constellation? What is the current stage in the life cycle of this trend – normalisation or post-normalisation?
- '*What's missing?* What is the paradoxical element of the trend? What is the corresponding countertrend? Are there any contexts which would be needed to allow the trend to advance in its process of normalisation?
- '*Why not?* Are there any contexts which could prevent the trend from moving on in its normalisation process? How could the trend relate to other trends? In other words, does the trend share some contexts with other trends?
- '*So what?* What aspects of the trend match our expectations or seem familiar? What is unexpected or represents an anomaly? What aspects of the trend challenge the premises of the current strategy? In what contexts will the trend cause the maximum damage? In what contexts will the trend provide the greatest strategic potential?

4.1.4 Megatrends and a constructivist perspective

Von Groddeck and Schwarz (2013) add a perspective to the conceptualisation of trends and weak signals by stressing the issue of semantics in trend management but also by focusing on the concept of megatrends (Vejlgaard 2008). A term coined by U.S. trend researcher John Naisbitt (1984), distinguishing relevant trends from very relevant or mega-relevant ones. Vejlgaard (2008: 21) explains that 'some people use the word "megatrend" when discussing cultural, economic, political, or technological shifts that are just about to happen, with the implication that these megatrends will affect all or almost all of society'. Von Groddeck and Schwarz (2013) argue for a more critical approach to these definitions in respect to constructivism.

After examining the literature on trends and weak signals, Rossel (2009) argues that it is striking how many authors do not recognise that the concept of weak signals is only a metaphor. Rossel (2009: 314) points out that weak signals do not appear in simple forms such as 'words, sentences, images or events, but rather as bundles of concepts and processes, systemic transitions, hybrid pattern of shifting influences, cultural change dynamics, etc.'. Seidl (2004) introduces a constructivist perspective to dealing with trends. He criticises the epistemological assumptions about weak signals as naïve in the sense that cognitions are conceptualised as direct representations of the external word. He states: 'Weak signals accordingly would have to be conceptualised as experiences with one's current constructions, which in some way or other point at future problems' (Seidl 2004: 151).

Von Groddeck and Schwarz (2013) argue that Seidl (2004: 156) makes an argument for a deliberate discourse-theoretical approach to trend management:

> Weak signals do not exist as such in the outside environment waiting to be picked up by the (members of the) organisation. Rather they have to be conceptualised cognitively, i.e. as cognitive phenomena, determined by the structures of the cognitive system.

He adds, from a constructivist angle, 'Cognitive systems interact with their environments, but it is the cognitive system – and not the environment – that determines how and in what way it interacts' (Seidl 2004: 157).

By drawing on Ernesto Laclau and Chantal Mouffe's (1985) discourse theory and its elaborations (Laclau 1996), von Groddeck and Schwarz (2013) assess the implications of labelling a trend as a megatrend. In this context the notion of empty signifier is especially relevant to conceptualise megatrends. The notion of an empty signifier is used in reference to a semantic element that tries to name and limit a discourse but always fails to do so. Trends and megatrends can be perceived as empty signifiers, enabling us to understand why operative trend management has dysfunctional effects. The empty signifier is a signifier without a signified meaning, making it applicable to a semantic form without a specific meaning. The empty signifier is thus 'meaningless as such' (Zizek 1993: 178).

Von Groddeck and Schwarz (2013) go on to argue that megatrends can be perceived as empty signifiers because they hold together a number of heterogeneous discursive elements as equal contributions to a certain discourse. The price of empty signifiers is that they are so overloaded with meaning that they cannot transport clear information, which is precisely the analogy we see between the notions of empty signifiers and megatrends. Megatrends, in parallel to empty signifiers, are a semantic form that works as an abstract bracket for heterogeneous societal changes, which leads to an overdetermination of meaning. Consequently, working with megatrends may limit the capacity of organisations to develop foresight.

Von Groddeck and Schwarz (2013) suggest when dealing with megatrends, such as urbanisation or demographic change, one should try to reconstruct the ways in which these signifiers are sought, given value, and fixated in different discourses. This provides an organisation with material or a horizon to detect locations that have the character of nodal points (i.e., where new rules are established and where societal transformation is accompanied by the production of new meaning). Furthermore, discourse analysis can be used to detect those locations in society where the new emerges.

If an empty signifier is only imagined to be likely to limit, end, or freeze a discourse, the strategic disadvantage of working with megatrends becomes even more apparent. The danger is that managers undervalue the fact that megatrends, as empty signifiers, only semantically limit a discourse, but beneath this abstract address, the discourse is permanently changing. Megatrends increase the probability that an organisation will take them to be a solid, inevitably emerging future.

The description of the life cycle (Mathews and Wacker 2002) demonstrated that a trend constantly moves through society and that a trend itself is subject to constant change, as are all objects related to that trend. Outside the perspective of an organisation, therefore, a megatrend does not increase strategic foresight; rather, it increases blind spots, especially when the associated discourse is frozen.

4.2 Narratives and science fiction

In the past the relevance of cultural products for trend research has occasionally been pointed out (Schwarz 2011, 2015; Schwarz and Liebl 2011; Liebl and Schwarz 2012). Liebl (2003) argued that numerous trends originated in literary works and were disseminated by trend research institutes. One example is Douglas Coupland's novel *Generation X*, which coined terms such as Downnesting, Decade Blending, Lessness, and McJobs. Other disciplines, such as design, already seem to be more advanced, using cultural products in the sense of science fiction prototyping (Johnson 2011; Graham, Greenhill, and Callaghan 2013, 2014; Schwarz and Liebl 2013; Schwarz, Kroehl, and von der Gracht 2014), or using science fiction as an inspiration for the design of interfaces (Shedroff and Noessel 2012). For example, science fiction films and series such as *Star Trek* have arguably been an inspiration and role model for generations of engineers and product developers.

4.2.1 The role of narratives in the construction of reality

The social construction of reality is based largely on cultural products such as literature and movies. Czarniawska (2006: 249) refers to 'the constructive role of popular culture'. According to Crotty (2003: 58),

> ... social constructionism emphasises the hold our culture has on us: it shapes the way in which we see things (even the way in which we feel things!) and gives us a quite definite view of the world. This shaping of our minds by culture is to be welcomed as what makes us human and endows us with the freedom we enjoy.

In the science fiction genre it has been pointed out that these narratives can be used as prototypes (Johnson 2011; Schwarz and Liebl 2013; Schwarz, Kroehl, and von der Gracht 2014). Kirby (2010: 66) has described how film makers and science consultants have created cinematic representations of technological possibilities with the effect of stimulating the desire for these technologies among the audience, but also with an eye on generating funding opportunities for these technologies. 'Fiction's lack of constraints and film makers' creative assistance provides an open "free" space to put forward speculative conceptualisations; it also amends these speculations within narrative that treats these ideas as already actualized within social context'.

The relevance of cultural products in constructing reality becomes more evident if we focus on a single cultural product: literature. While there is scant evidence that literature has been applied in developing foresight, some evidence exists in respect to science fiction literature. Literature has attracted the notice of scholars in consumer research and organisational studies (Barry and Elmes 1997; de Monthoux and Czarniawska-Joerges 2006).

On the one hand, literature can be attributed a role model function: Literature influences its readers and is at the same time an expression of the reader or society. This means that literature both influence and reflects a society. Images, fictions, fantasies, and dreams shape our reality and just as often these images feed on what we have read. Of course, literary texts can be read as a scenario, stimulating and guiding the readers' imagination. Not only are different futures formulated in scenarios; so are inventions in the near or distant future. Literature can be understood as a cultural reservoir of topics. If literary texts are read as scenarios, new spheres of possibility can be described and alternative images of the future or scenarios evoked, even provoked. A novel can therefore be assigned the task and function of inventing new possibilities and thus creating new hypothetical truths. These possibilities, utopias, and scenarios contribute to the social construction of a cultural reservoir of themes. The function of literature as a gateway to the inner world is also intriguing: literary texts not only describe externalities, actions, or contexts, but also the inner worlds of individual protagonists. Consequently, unlike in films, there is access, for example, to the feelings or inner worlds of the characters.

Bauer (2005: 181) describes how a literary text can influence its readers or change their perceptions:

> The reader first tries to classify the text within his or her familiar framework of understanding. If the schematized views elude this order, he must, as it were, modify the frames of understanding, he must correct his own schemata. Since this will not remain completely without consequences for his further perception of himself and the world, he emerges changed from the interaction with the text.

It becomes clear that with the use of cultural products in foresight, the supposed separation between fact and fiction, between science and fantasy, which exists above all in companies, is also weakened. With regard to reading literary texts, Eco (1996: 103) refers to the contract in fiction that a reader tacitly concludes with the author: '...the author simply pretends to be telling the truth, and we accept the fiction contract and pretend that what the author tells has actually happened'. At the same time, one cannot be sure that this contract will be kept. A reader can therefore save the information gained from a novel as facts. The role of fiction thus becomes the focus of interest. According to the American sociologist Richard Sennett (Belk 1986: 24): 'One can learn more about the complexity of motives and mutual perceptions from a reasonably good novel than from a "solid" piece of social-science research'.

Schwarz (2015) has argued for a 'narrative turn' in developing foresight and that organisations should focus less on storytelling and more on storylistening: listening to stories in the organisational environment. This notion becomes even more apparent in the genre of science fiction.

4.2.2 Science fiction in foresight and science fiction prototyping

For several years, science fiction had become increasingly relevant to management. In addition to the references to the use of science fiction at Intel (Johnson 2011), Peper's (2017) article 'Why Business Leaders Need to Read More Science Fiction' in the *Harvard Business Review* argues that science fiction not only inspires innovation, but above all encourages managers to question their assumptions about the present and the future: 'Science fiction isn't useful because it's predictive. It's useful because it reframes our perspective on the world' (Peper 2017). In 2019, Charles-Edouard Bouée, CEO of Roland Berger Consulting, advocated the use of science fiction in strategy development in a LinkedIn post, 'The Power of Science Fiction: Shaping the Present with Visions of the Future'.

Recent examples of the use of science fiction can be found at car manufacturer Audi (Schwarz and Hofmann 2019) or software company SAP. The book *Science Fiction: A Starship for Enterprise Innovation* (Rosenberg 2019) describes why science fiction has played an important role in innovation in the past, and why 'science fiction thinking' will be important in developing innovation. Deutsche Telekom reports on a project carried out in collaboration with the Fantastic Library in Wetzlar (von Reventlow, Thesen, Le Blanc, and Haas 2019). In this project, science fiction was evaluated to find innovative ideas. Scenarios were then developed from these ideas with the aim of using them in design thinking workshops.

Cultural products, and especially science fiction, provide a different kind of input with an independent value for trend research. Through literary texts and films, knowledge can be made explicit, and terms assigned to images. In this way, it is possible to develop a feeling for the imaginary worlds or soundboards that exist outside an organisation. Therefore, the use of literature allows the pursuit of an essential goal of trend research: to perceive collective resonance spaces or addressability and thus changes in the socio-cultural environment of an organisation.

Schwarz, Kroehl, and von der Gracht (2014) report on the multiple usage of text excerpts taken from the science-fiction novel *Super Sad True Love Story* (Shteyngart 2011) and summarise the functions of narratives science-fiction for foresight: As science fiction novels provide a new frame of reference for managers, these narratives can serve as vehicles for researching weak signals and trends and for identifying corresponding business opportunities as well as for ensuring the link to customers. Further, they point out to a central characteristic of literature: polyvalence. The meaning of a text is not exclusively derived from the text itself; it is constructed through reading (Groeben and Vorderer 1986), reflecting the individual context of the recipient. According to the notion of polyvalence in literature, a text can be interpreted in many ways.

This means that when a science-fiction narrative in a workshop is used, multiple perspectives can be developed. When confronted with complex and ambiguous issues, organisations need to find means to increase their perspective, for instance, by conducting cross-industry projects (Seidl and Werle 2017). And this multitude of perspectives can then be achieved by referring narratives, for instance, from science fiction.

References

Ansoff, I. 1982. "Strategic Response in Turbulent Environments." *Working Paper No. 82-35*. European Institute for Advanced Studies in Management.

Ansoff, I. H. 1975. "Managing Strategic Surprise by Response to Weak Signals." *California Management Review* 18 (2): 21–33.

Barry, D., and M. Elmes. 1997. "Strategy Retold: Toward a Narrative View of Strategic Discourse." *The Academy of Management Review* 22 (2): 429–452.

Bauer, M. 2005. *Romantheorie Und Erzählforschung: Eine Einführung*. Stuttgart: J. B. Metzler.

Belk, R. W. 1986. "Art Versus Science as Ways of Generating Knowledge About Materialism." In *Perspectives on Methodology in Consumer Research*, edited by D. Bringberg and R. J. Lutz, 3–36. New York: Springer.

Buck, A., C. Herrmann, and D. Lubkowitz. 1998. *Handbuch Trendmanagement*. Frankfurt: Frankfurter Allgemeine Zeitung.

Chang, J. 2005. *Can't Stop Won't Stop*. London: Ebury Press.

Crotty, M. 2003. *The Foundations of Social Research: Meaning and Perspective in the Research Process*. London: Sage Publications.

Czarniawska, B. 2006. "Doing Gender unto the Other: Fiction as a Mode of Studying Gender Discrimination in Orgnization." *Gender, Work and Organization* 13 (3): 234–253.

de Monthoux, P. G., and B. Czarniawska-Joerges. 2006. "Introduction: Management Beyond Case and Cliché." In *Good Novels, Better Management: Reading Organizational Realities*, edited by P. G. de Monthoux and B. Czarniawska-Joerges, 1–16. London: Routledge.

Eco, U. 1996. *Im Wald Der Fiktionen: Sechs Streifzüge Druch Die Literatur*. München: Deutscher Taschenbuch Verlag.

Graham, G., A. Greenhill, and V. Callaghan. 2013. "Exploring Business Visions Using Creative Fictional Prototypes." *Futures* 50: 1–4.

———. 2014. "Technological Forecasting and Social Change Special Section: Creative Prototyping." *Technological Forecasting and Social Change* 84 (0): 1–4. https://doi.org/10.1016/j.techfore.2013.11.007.

Groeben, N., and P. Vorderer. 1986. "Empirische Literaturpsychologie." In *Psychologie Der Literaur*, edited by R. Langer, 105–143. Weinheim: Psychologie-Verlags-Union.

Groys, B. 2008. *Art Power*. Cambridge: The MIT Press.

Harris, S. D., and S. Zeisler. 2002. "Weak Signals: Detecting the Next Big Thing." *The Futurist* 36 (6): 21–28.

Johnson, B. D. 2011. *Science Fiction for Prototyping: Designing the Future with Science Fiction*. San Rafael: Morgan & Claypool.

Kirby, D. 2010. "The Future Is Now: Diegetic Prototypes and the Role of Popular Films in Generating Real-World Technological Development." *Social Studies of Science* 40 (1): 41–70.

Laclau, E. 1996. *Emancipation(S)*. London: Verso (Phronesis).

Laclau, E., and C. Mouffe. 1985. *Hegemony and Socialist Strategy: Towards a Radical Democratic Politics*. London: Verso.

Liebl, F. 2000. *Der Schock Des Neuen: Entstehung Und Management von Issues Und Trends*. München: Gerling Akademie Verlag.

———. 2002. "The Anatomy of Complex Societal Problems and Its Implications for OR." *Journal of the Operational Research Society* 53 (2): 161–184.

———. 2003. "Woher Kommt Der Trend?" *Brandeins*. https://www.brandeins.de/magazine/brand-eins-wirtschaftsmagazin/2003/2004-und-weiter/kolumne-woher-kommt-der-trend.

Liebl, F., and J. O. Schwarz. 2010. "Normality of the Future: Trend Diagnosis for Strategic Foresight." *Futures* 42 (4): 313–327.

———. 2012. "'Art Facts': Zur Nutzung Kultureller Originalitätsproduktion Für Die Strategische Frühaufklärung." In *FOCUS-Jahrbuch 2012: Prognose, Trend- Und Zukunftsforschung*, edited by W. J. Koschnick, 277–301. München: FOCUS Magazin Verlag.

Link, J. 1999. *Versuch Über Den Normalismus: Wie Normalität Produziert Wird*. Opladen: Westdeutscher Verlag.

Mathews, R., and W. Wacker. 2002. *The Deviant's Advantage*. London: Crown Publishing Group.

McCracken, G. 2006. *Flock and Flow: Predicting and Managing Change in a Dynamic Marketplace*. Bloomington: Indiana University Press.

Naisbitt, J. 1984. *Megatrends: Ten New Directions Transforming Our Lives*. New York: Warner Books.

Peper, E. 2017. "Why Business Leaders Need to Read More Science Fiction." *Harvard Business Review*, 2017.

Popcorn, F. 1991. *The Popcorn Report*. London: Random House.

Reeves, M., B. Goodson, and K. Whitaker. 2021. "The Power of Anomaly: To Achieve Strategic Advantage, Scan the Market for Surprises." *Harvard Business Review* (July-August): 94–101.

Rosenberg, A. 2019. *Science Fiction: A Starship for Enterprise Innovation*. Copenhagen: Startup Guide World.

Rossel, P. 2009. "Weak Signals as a Flexible Framing Space for Enhanced Management and Decision-Making." *Technology Analysis & Strategic Management* 21 (3): 307–320.

Schwarz, J. O. 2011. *Quellcode Der Zukunft: Literatur in Der Strategischen Frühaufklärung*. Berlin: Logos.

———. 2015. "The 'Narrative Turn' in Developing Foresight: Assessing How Cultural Products Can Assist Organisations in Detecting Trends." *Technological Forecasting and Social Change* 90 (Part B): 510–513. https://doi.org/10.1016/j.techfore.2014.02.024.

Schwarz, J. O., and F. Liebl. 2011. "Quellcode Der Zukunft." *GDI Impuls* 4: 82–87.

———. 2013. "Cultural Products and Their Implications for Business Models: Why Science Fiction Needs Socio-Cultural Fiction." *Futures* 50: 66–73.

Schwarz, J. O., and R. Hofmann. 2019. *Is Today's Science Fiction Tomorrow's Science Fact? On the Relevance of Science-Fiction for Trend Research: Insights from the Audi Brand Fiction Space Project*. Munich: Foresight Academy, http://www.motivesandfiction.com/cms-media/media-3952720.pdf.

Schwarz, J. O., R. Kroehl, and H. A. von der Gracht. 2014. "Novels and Novelty in Trend Research - Using Novels to Perceive Weak Signals and Transfer Frames of Reference." *Technological Forecasting and Social Change* 84 (May): 66–73. https://doi.org/10.1016/j.techfore.2013.09.007.

Seidl, D. 2004. "The Concept of 'Weak Signals' Revisited: A Re-Description from a Constructivist Perspective." In *Managing the Future: Developing Strategic Foresight in the Knowledge Economy*, edited by H. Tsoukas and J. Shepherd, 153–170. Oxford: Blackwell.

Seidl, D., and F. Werle. 2017. "Inter-Organizational Sensemaking in the Face of Strategic Meta-Problems: Requisite Variety and Dynamics of Participation." *Strategic Management Journal* 39 (3): 830–858. https://doi.org/10.1002/smj.2723.

Shakar, A. 2001. *The Savage Girl*. New York: HarperCollins.

Shedroff, N., and C. Noessel. 2012. *Make It So: Interaction Design Lessons from Science Fiction*. New York: Rosenfeld Media.

Shteyngart, G. 2011. *Super Sad True Love Story*. New York: Random House.

Slaughter, R. A. 1999. "A New Frame Work for Environmental Scanning." *Foresight* 1 (5): 441–451.

Thornton, S. 2018. *Club Cultures: Music, Media and Subcultural Capital*. Oxford: Wiley.

Vejlgaard, H. 2008. *Anatomy of a Trend*. New York: McGraw-Hill.

von Groddeck, V., and J. O. Schwarz. 2013. "Perceiving Megatrends as Empty Signifiers: A Discourse-Theoretical Interpretation of Trend Management." *Futures* 47: 28–37.

von Reventlow, C., P. Thesen, T. Le Blanc, and M. Haas. 2019. *Wenn Die Zukunft Zum Heute Wird*. Berlin: Deutsche Telekom Design.

Waters, R. 2005. *The Trendmaster's Guide*. New York: Portfolio.

———. 2006. *The Hummer and the Mini*. New York: Portfolio.

Weiner, E., and B. Arnold. 2006. *Future Think*. Upper Saddle River: Prentice Hall.

Zizek, S. 1993. *Grimassen Des Realen: Jacques Lacan Oder Die Monstrosität Des Aktes*. Köln: Kiepenheuer & Witsch.

5

TREND RECEIVER

Rupert Hofmann[1]

The future arises from the interplay of continuity and change, sometimes disruptive, but never in a vacuum. This is just as true for substantial changes in our everyday lives as it is for groundbreaking new product projects or in the course of the realignment of a traditional company. Exploring future needs and wishes, it seems obvious to survey knowledge and opinions from different target and population groups within a study. Selection of interviewees plays a decisive role both in qualitative and quantitative research projects. But who could those people be? Who can provide credible trend information on the topics of strategy, brand, products, service, and communication? Who can help to make future customer expectations more tangible? How can we include as many perspectives and the needs of our fellow world members as possible? How can the resulting information be translated into action? These questions – and answers! – are the focus of customer foresight research.

DOI: 10.4324/9781003302735-8

5.1 The trend receiver method

In customer foresight practice, it is fascinating to research people who elicit change – because they combine things that had previously been separated, thus bringing something new into the world or spreading an invention. Also relevant are people aware of change dynamics in an early phase and who are able to assess the potential of the 'new'. Following concepts in trend research, a typology of 'agents of the new' (see Figure 5.1) was developed. Among these agents are lead users, trend receivers, experts, early adopters, and influencers.

Trend receivers are especially relevant for future topics. They perceive changes and potentials of the new in a specific domain in a highly sensitive and differentiated way (Hofmann 2015). Trend receivers have special antennae for changes and needs. Their visionary competence enables them to develop notions and projections, ideas and scenarios from signals and observations. They are very well connected, have profound opinions about what motivates individuals, and observe social changes in a very reflective way.

The trend receiver method focuses on ways of identifying these persons, who bring everyday needs and experience together with a sensitive perception of changes and potentials of the new. It enables practitioners to find suitable dialogue partners by tailored search profiles based on topic and

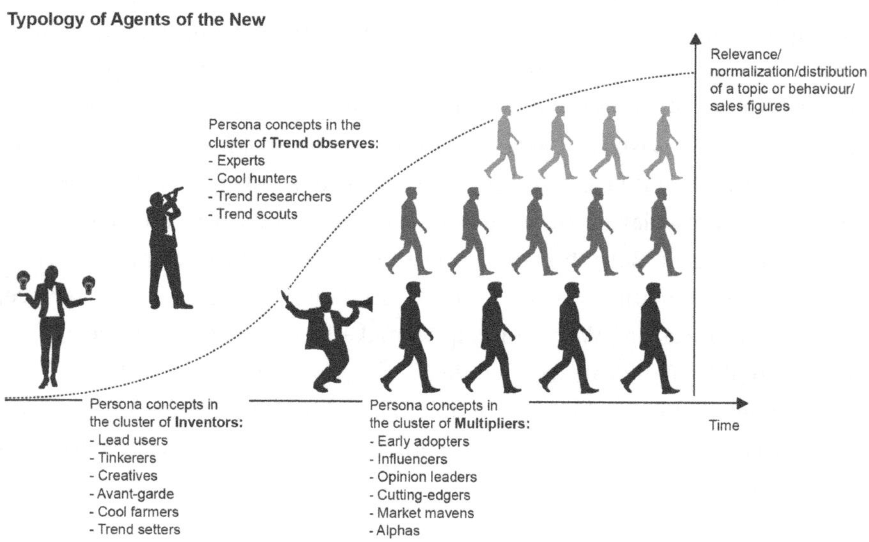

Figure 5.1 Typology of 'Agents of the New' (Hofmann 2015).

context. It also provides a guideline on how to conduct the dialogue with these conversation partners. Originally developed by the exemplary practice needs of Audi, the method has since proven useful in various entrepreneurial, economic, and socio-political studies. The method is especially recommendable if the research topic involves mid- and long-term market developments and if market conditions are likely to change substantially. The trend receiver concept has a high degree of transferability because it provides for a tailored determination that is equally appropriate to the respective context, the specific objectives, and the complexity of the issue. At the same time, it demands a certain level of qualitative research skills and curiosity among the responsible persons of the given research projects in order to implement such a 'made by measure study design' which is epistemologically based on Kant's reflective judgement (Hofmann 2011).

5.2 Identification and implementation of trend receivers

Four characteristics proved typical of trend receivers across topics and industries: own user and context experience, self-abstraction ability and pattern recognition, curiosity and heterogeneity, and selection competence (Hofmann 2015). Furthermore, biographical discontinuities and communication competence are often helpful characteristics (Hofmann 2020). These characteristics enable trend receivers to identify new trends early and provide a coherent description of a possible or desirable everyday life in the future. Yet, it is important to note that no one can oversee all areas of life. The identification of trend receivers for the respective context is therefore crucial. Catalogues of criteria like in publications can serve as the basis for the formulation of a search profile, customised to the research questions. Following the creation of such a tailored description, suitable intermediaries come into play to check their network. The search process is based on 'pyramiding' (Prügl 2006; Schweitzer, Hofmann, and Meinheit 2019).

The subsequent dialogue requires as much attention. Discussions must be prepared in a targeted manner, both in content and communication. The first decisive step is how to present the initial situation and questions, and how to offer one's own hypotheses and ideas. Then we should listen well and enter discussions where necessary. In the process, observations are presented, arguments and counterarguments are weighed, and together we think further into the future. The conversations should be understood as a journey in which each person and the group itself continue to enrich and clarify the emerging projections and visions.

The trend receiver method, as already mentioned earlier, has been implemented in different entrepreneurial, economic, and socio-political contexts. The projects allowed the integration of visionary customers into vision finding and strategy as well as brand, product and service development processes. The following two case studies will explain the application of the method.

5.3 'Easy Rider' study – autonomous driving from the market and customers' perspective

Few topics hold so much disruptive power as that of autonomous driving. From 2013, Audi's market and trend research sought to understand potentials and concerns from the customer's point of view – not an easy task, since everyday mobility life with semi-autonomous or fully autonomous cars would differ significantly from today's (even more so from the perspective of 2013.). And while many quantitative broad-based surveys revealed great fears and rejections, reflected ideas of future driving and mobility situations were sought. This was intended to answer the question of what an attractive car and system would look like under expanded technical possibilities.

Therefore, 29 trend receivers in the United States, Germany, and China were identified through a customised search profile and various intermediaries. Initially, individual in-depth interviews were conducted; between these individual interviews, the content and structure of the interview guideline were continuously developed. Parallel to wide-ranging desk research, the iterative study design (Figure 5.2) consisted of a behavioural economic analysis, a cultural and innovation history analysis, and a scenario analysis. Additional expert interviews were conducted with city planners and pilots.

Impulses and hypotheses on user behaviour and descriptions of desirable future mobility situations or vehicle characteristics arising in discussions with trend receivers were quantified in the second part of the study in cooperation with the research institute of a partner university (Hahn et al., 2016). The condensed interim results were converted into 12 use cases and

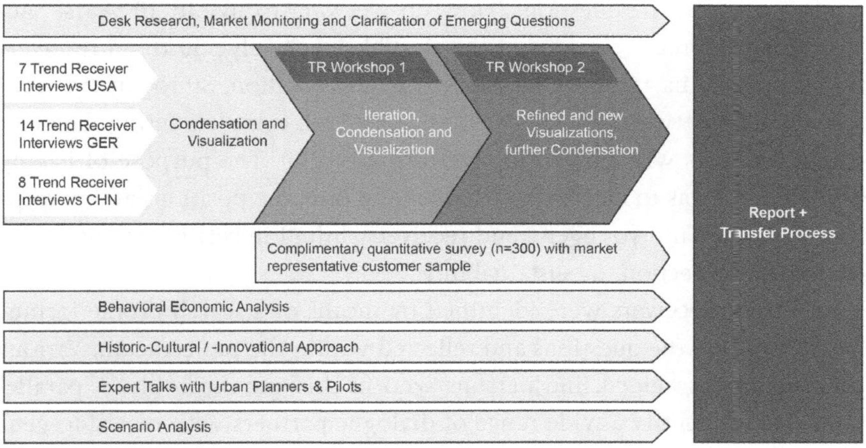

Figure 5.2 Iterative study design of the easy rider project (Hahn et al. 2016).

Figure 5.3 'Easy Rider' workshop situation and concept model Audi AICON (gravity GmbH, Audi).

visualised. In workshops with individual and group formats (see Figure 5.3) these representations were then questioned, discarded, enriched, and refined. The results were used by specialised departments on the Board of Management and by various areas such as Audi design, pre-development, brand strategy, and product marketing. It influenced, amongst others, the concept car AICON (Figure 5.3) and further examination of the future topic of autonomous driving.

5.4 Cross-industry project: 'How do we want to live in 10 years?'

In 2017/2018, the companies BSH, Hornbach, Ikea, and Audi carried out a major international life worlds study, which sought cross-industry access and centred on the question: 'How do we want to live in 10 years?' According to a broad 'corridor of the possible and likely', 30 trend receivers in Europe, China and the USA participated. In addition, 30 rookie trend receivers aged between 20 and 30 were consulted, complementing the older trend receivers who ranged in age from 30 to 60. The purpose of having two cohorts was to determine whether and how the positions and perceptions of older and younger trend receivers shifted in relation to digitalisation, data protection, or sustainability.

All trend receivers were identified by means of a search profile formulated based on the questions and reflected within the project team. Various intermediaries helped find persons who fit this search profile. A parallel aim was to identify a wide range of dialogue partners with regard to gender, age, background, family situation, interests, experience, and career. The iterative process became particularly intensive in this project, not least because the responsible representatives of the participating companies took part in the discussions and trips and, together with the lead agency, continuously reflected on the hypotheses and implications.

The results and statements on future life worlds were brought together and visualised in six chapters (see Figure 5.4), short films and a central presentation. The project has been a continuous and sustainable source of

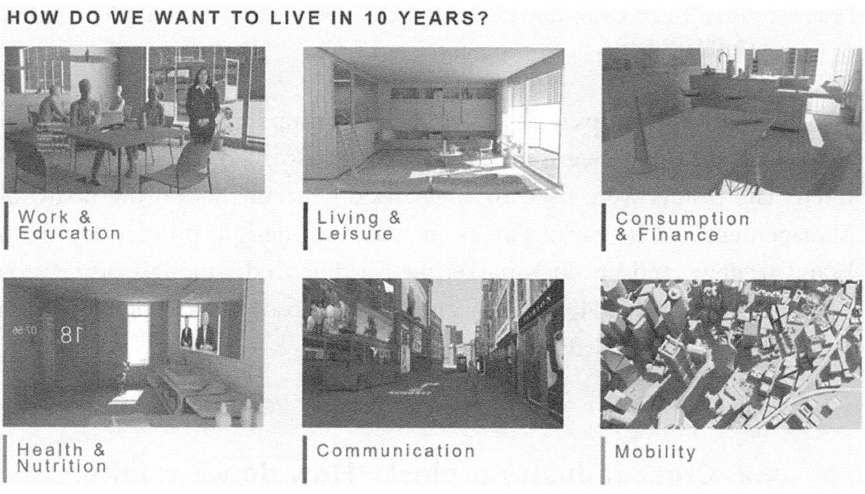

Figure 5.4 Areas of research and projected life worlds (cross-industry study 'How do we want to live in 10 years?', graphics by gravity GmbH).

inspiration and information for the development of brands, products, services, trade networks, business models, and organisation. Moreover, it led to the launch of the Foresight Academy – a cross-industry platform for future-oriented work and customer foresight expertise in 2019. Thirteen leading companies from a variety of sectors were involved in Foresight Academy 2019–2020 and in 2021 a further round started with 16 major brands (www.foresightacademy.com).

5.5 Conclusion

These two exemplary projects illustrate that it is worthwhile to use the visionary competence and foresight abilities of trend receivers for important future topics. In every context, persons can be found who are particularly well attuned to changes and needs. Dialogues with them help us to create visions and provide guidance for development processes and decisions.

And especially as we experience enormously growing influence and potentials of algorithms and data, on our way into the future we should combine these new opportunities with the 'human factors', from which the trend receiver method draws. Let's use both detailed fact knowledge and skillful data modelling as well as the power of intuition, ideas, associations, empathy, and vision!

Note

1 Dr. Rupert Hofmann works for Audi Business Innovation and is an observational draftsman. He studied at Munich Academy of Fine Arts, Passau University, Columbia University in New York City and at Federal University of Juiz de Fora, Brazil. In his PhD he developed the Trend Receiver concept, a novel qualitative method that combines insight and foresight research. Rupert Hofmann is the founder of the cross-industry platform Foresight Academy and has been lecturing and writing on different topics related to trend research, branding and strategy. Rupert's interest in people and their environment, behaviour, and motives also drives his observational drawings of people, modern everyday life, city perspectives, stores, beaches, and bars (www.ruperthofmann.com).

References

Hahn, A., R. Hofmann, V. Bilgram, J. O. Schwarz, A. Meinheit, and F. Johann. 2016. "Easy Rider: Die Erschließung der Welt des autonomen Fahrens aus Sicht von Markt und Kunde." In *Die frühe Phase des Innovationsprozesses*, edited by T. Abele, 75–98. Wiesbaden: Springer (FOM Edition).

Hofmann, R. 2011. *Trend Receiver – qualifizierte Visionskraft. Kriterien und Vorgehensweisen der Befragtenauswahl und Dialoggestaltung bei Studien zu zukünftigen Konzepten am Beispiel der AUDI AG*. Göttingen: Cuvillier.

———. 2015. "Visionary Competence for Long-term Development of Brands, Products, and Services: The Trend Receiver Concept and Its First Applications at Audi." *Technological Forecasting and Social Change* 101: 83–98.

———. 2020. "Customer Foresight Practice. How to Access Future Markets Through Extraordinary People." *Marketing Review St. Gallen* 3: 36–45.

Prügl, R. 2006. Die Identifikation von Personen mit besonderen Merkmalen: eine empirische Analyse zur Effizienz der Suchmethode Pyramiding (Dissertation). Wien: Wirtschaftsuniversität.

Schweitzer, N., R. Hofmann, and A. Meinheit. 2019. "Strategic Customer Foresight: From Research to Strategic Decision-making Using the Example of Highly Automated Vehicles." *Technological Forecasting and Social Change* 144: 49–65.

6

THE DELPHI METHOD

HOW EXPERTS SEE THE FUTURE

Heiko von der Gracht[1]

The climate is in crisis: Will Barcelona sink into the sea? Or Hamburg? Considering the rising sea levels, in which coastal regions will people be able to live at all in ten years?

As the saying goes, predictions are difficult, especially when they concern the future. Nevertheless, very little has interested people over the past 40,000 years as much as their future. That could explain the use of practices like fortune telling and divination… didn't anyone tell them to ask the experts? One of the best-known expert questioning techniques is the Delphi method, based on the location of the Greek oracle.

6.1 Historical development of the method

The Delphi method was developed in the 1950s by researchers at the American RAND (Research ANd Development) Corporation. On October 1, 1945, Henry Arnold, Edward L. Bowles, Donald Douglas, Arthur C. Raymond,

DOI: 10.4324/9781003302735-9

and Frank Collbohm met to form Project RAND under a special contract with the Douglas Aircraft Company. The organisation was established immediately after World War II to coordinate military planning with decisions in research and development. In May 1946, Project RAND published its first report, 'Preliminary Design of an Experimental World-Circling Spaceship' (RAND Corporation 1946), which discussed the design, performance, and application of satellites. On May 14, 1948, Project RAND broke away from the Douglas Aircraft Company and became an independent, non-profit organisation (RAND Corporation 2022).

The first experiment in which the Delphi method was used at the RAND Corporation, was to determine potential targets for attack in the USA and the number of atomic bombs needed to destroy them (Dalkey and Helmer 1963). The first civilian Delphi study, 'Report on a Long-Range Forecasting Study' followed several years later in 1964 (Gordon and Helmer 1964). The panel evaluated more than 100 future occurrences and developments with a time horizon up to the year 2020 in several written discussion rounds, making it a very long-range study. This highly regarded study launched the international dissemination of the method, which continues to this day. The Delphi method is still used by scientists and foresight managers around the world.

Of course, much has changed since the 1950s. In the past, experts on the panel submitted written assessments, so-called 'paper and pencil studies'. Today, this can be done faster and more directly online, saving an enormous amount of time during the evaluation. Apart from that, the development of the method proceeds.

6.2 Design aspects of the method

What is the Delphi method? 'Delphi may be characterised as a method for structuring a group communication process so that the process is effective in allowing a group of individuals, as a whole, to deal with a complex problem' (Linstone and Turoff 1975: 3).

According to some sources, the goal of Delphi studies is for a panel of experts to reach consensus. But this is a misconception. In 2011, the two pioneers of the method, Harold A. Linstone and Murray Turoff, issued a clarification:

> Over the years, as a reviewer and editor-in-chief of *TFSC* (Technological Forecasting & Social Chance) it has often been necessary to correct the mistaken impression that the aim of Delphi is consensus. Our 1975 book clearly states that Delphi is 'a method for structuring a group communication process', not a method aimed to produce consensus'.
>
> *(Linstone and Turoff 2011: S. 1714)*

In other words, a Delphi panel discussion can lead to consensus, but it does not have to.

A Delphi research design follows the method's four core elements: anonymity, iteration, controlled feedback, and group response (Beiderbeck, Frevel, von der Gracht, Schmidt, and Schweitzer 2021b).

6.2.1 Anonymity

Delphi is an anonymous written survey. The participating experts are, of course, known by name to the moderator/study director but not to the other experts on the panel. This is because when the experts in the Delphi survey rounds discuss each other's assessments, a renowned name, for example, should not influence the discussion. This avoids the negative psychological effects that are known from almost all group discussions, for example, the bandwagon effect ('follow the crowd'): As soon as a prominent expert gives their assessment in the discussion, everyone or nearly everyone agrees.

6.2.2 Iteration

Delphi studies are conducted in several rounds. In the classic method, according to RAND, the first round is a brainstorming session with the experts on the question: What do the panel members believe are the relevant topics? In the second round, the same experts comment on and rate these topics. The important thing here is that the participants not only evaluate, but also defend their assessments. In other words, an expert not only explains what they think the future holds, but also presents arguments for and against it. In the third round, all panel experts see each other's assessments and the statistical group opinion. In light of the panel colleagues' assessments, each expert can then modify and/or support their own assessment it with further arguments, if they wish. According to the textbook, these 'revision rounds' should be repeated until the results are stable, meaning that all of the experts are done with making changes to their assessments (Dajani, Sincoff, and Talley 1979). This makes obvious sense: No one wants unreliable results.

In the past, however, modified Delphi studies deviated from the classical approach by conducting either more or fewer iterations, sometimes predetermined or based on predefined termination criteria. The number of iterations depended on many variables, such as the composition of the panel, the nature of the problem to be solved, and the type of feedback given after each round (Erffmeyer, Erffmeyer, and Lane 1986). In practice, there are usually one to two revision rounds.

6.2.3 Controlled feedback

We have already touched on this characteristic element of every Delphi study. After each round, the study director forwards the evaluated assessments of the individual experts to all the others in a 'controlled' manner, with the statistical group assessment (Belton et al. 2021).

6.2.4 Group response

The group of experts is given an 'answer' with each round, usually as boxplot diagrams, which also reflect statistical outlier assessments and statistical measures of group assessment. It also includes the degree of consensus and dissent in the panel (von der Gracht 2012).

6.2.5 Different scales in Delphi studies

Researchers traditionally ask about the probability of an event or development, usually using a rating scale of 0–100 per cent. In addition, they usually ask participants about the impact of the event with the specified probability on an industry, a company, their company, or the national economy. It is also customary to inquire about the desirability of the event, often along a 5-point Likert scale (1 = 'I do not wish this to occur at all' to 5 = 'absolutely desirable'). Other variables are feasibility (especially for technological developments), urgency for action, or confidence (the expert's certainty about their assessment).

6.2.6 *Types of Delphi questions*

Most Delphi questionnaires include future-oriented theses, also called Delphi projections. In other words, not questions, but statements such as: 'Projection 9: The oil price will rise to $400 per barrel in 2030'. Each projection is assessed by the experts along the evaluation scales with arguments for and/or against this development.

How the projections are formulated is central to the validity of subsequent results (Markmann, Spickermann, von der Gracht, and Brem 2021). For example, it is important when formulating the projections that they are unambiguous, and therefore understood by all experts and, above all, understood in the same sense. The point is to avoid ambiguity. If projections are formulated in a way that is too long or complicated, the result can be increased misunderstanding on the part of the experts and partially or completely unreliable assessments. There is an underlying dilemma: If a projection is too short, it is usually too superficial and so are the experts' answers. There are actually studies that determine the optimal number of words for each Delphi statement (e.g., Salancik, Wenger, and Helfer 1971).

6.3 Selecting the experts

The initial question in any Delphi study is: Who is an expert in this field? There are no official directories for this. It always requires detective work in conference programme brochures, databases, publications on relevant topics, awards ceremonies, and the like. Another way is through recommendations: One expert recommends another, because experts know each other.

6.3.1 *Significance of the selection*

The importance of a correct selection of experts is obvious: Delphi results are always only as good as their experts (Mauksch, von der Gracht, and Gordon 2020). The selection is not about representativity, as in many classical empirical studies. Rather, it is about the greatest possible expertise that should be represented in the panel. Unlike representative studies, in which hundreds if not thousands of respondents are interviewed, there are usually 'only' 20–100 experts in a Delphi study. Furthermore, these experts come from a range of fields.

This is crucial, because diversity is key to the success of the Delphi and to the quality of the statements (Spickermann, Zimmermann, and von der Gracht 2014). The more diverse the panel, the better for most research purposes. If the experts were only from one area or related areas, the statements about the future would certainly show a tendency and blind spots: The more eyes there are, the more they will see. And different sets of eyes will see even more into the future.

That is why good panels often include experts from business, politics, science, and associations, with corresponding sub-groups. Another important question is: Do you want to make comparisons between the groups later, on the basis of the results? In that case, there should be at least 15–20 experts in each group so that the results are statistically reliable.

6.3.2 Special form: Real-time Delphi

In this special form, the expert sits in front of a computer, completes the assessment, and then, as soon as they click 'Next', sees what everyone else has said about this projection (Gordon and Pease 2006; Aengenheyster et al., 2017). This projection is automatically evaluated and passed on by the software in real time. Targeted studies have shown that this special form significantly streamlines the Delphi process while maintaining the quality of the data (Gnatzy, Warth, von der Gracht, and Darkow 2011). Another advantage is that there is no limit to the number of discussion rounds. Each expert can log into the projections as many times as they desire within a specified time limit.

6.4 Application example: European Generics and Biosimilars Industry

In 2021, the Delphi study 'The Future of the European Generics and Biosimilars Industry 2030plus' was published (von der Gracht, Kisgen, Lange, and Jalufka 2021). It is a good example of how a structured survey of experts can give the commissioning decision-makers of an entire industry a strategic perspective over nine years and beyond. In turbulent times like these, such a perspective is crucial for success.

6.4.1 The study

For the study, more than 60 experts were interviewed over several months. They were not only from the industry, but also from pharmaceutical wholesalers, start-ups, pharmacies, hospitals, universities, and ministries. The survey included 12 projections, for example: 'Medicines are produced specifically for each patient on the 3D printer at decentralised supply points (e.g., pharmacy, doctor's practice, retailer, at the patient's home, etc.)'.

6.4.2 The questions

For each projection, the probability of occurrence was inquired. This is a special feature of this study for four years: 2025, 2030, 2035, and 2040. The study benefited from a lively discussion, which brought to light more than 800 contributions.

Three arguments in favour of the above proposition were: '3D printing could become The Next Big Thing'. 'Individual printlets [pills from the 3D printer] are already marketable today'. 'The emerging mega-trend of personalised medication is driving technology development'.

Three arguments against were: '3D printing is more expensive than volume production'. 'Printing drugs with 3D in the living room?' 'Biosimilars [copycat bio-pharmaceutical drugs, like insulin] cannot be printed!'

Ultimately, the panel was unanimous that by 2030 the probability of the projection occurring is only 16 per cent. By 2040, the panel set the probability at 38 per cent. Scepticism obviously predominated with this projection. If one looks at the individual expert groups separately, it becomes clear that the panel members from science were more optimistic that the occurrence would come to pass than the panellists from business and politics were. Viewed across the entire panel, the development outlined by the projection is definitely desirable.

6.4.3 Other special characteristics of the study

The special thing about the design of the study was that the experts were also asked how sure they were about their assessment of an individual projection. Inquiries about the behavioural aspects of each expert were also made, since personal predisposition naturally influences assessment tendency (see also Kovaleva 2012). Do experts believe that they can influence the state of the world (internal locus of control)? Or is 'fate' unchangeable (external locus of control)? Here, too, it is obvious that an expert with a strong internal locus of control assesses future events differently from someone who takes the course of events at face value.

Delphi studies should therefore include the personal psychological predispositions of experts and their world views. These parameters influence an expert's assessment of and should therefore not be neglected in the interpretation (Loye 1980; Spickermann, Zimmermann, and von der Gracht 2014; Beiderbeck, Frevel, von der Gracht, Schmidt, and Schweitzer 2021b).

6.5 Challenges of the method

A search for 'Delphi method' in Google Scholar returns more than 7,000 hits in 2022); in 2021, there were more than 17,000, and in 2018, over 20,000. Even compared to other scientific methods, these figures indicate its popularity. On average, there are between 1,000 and 2,000 publications on the method every month! That is impressive.

6.5.1 Technical shortages: Not enough experts

It is less impressive that Delphi studies still have technical deficiencies. One reason is that many researchers shy away from interviewing a large number of experts. Then they discover that there is a study format in which only 20 experts were interviewed, which is probably enough for our purposes. Some also argue that in the first RAND Delphi study, only seven experts were interviewed, and the result was historic.

The design of this historical study cannot be generalised. In other words, if a researcher is lucky enough to be able to interview ten Nobel laureates, there is no need for ten more. That is the point: each study needs only as many experts as is necessary to validate the findings. What does 'exactly as many' mean?

Practical experience indicates usually 40–80 experts (von der Gracht, Gnatzy, Darkow, Gordon, and Glenn 2011). With participation of around 40 experts, for example, the variety of arguments no longer increases, but instead the same arguments are repeated.

6.5.2 *Additional technical shortcomings*

It goes without saying that a professional study design and competently moderated panel rounds can avoid the following deficiencies:

- **Limited panel review**: Although they do participate in the Delphi and give their assessments, hardly any discussion takes place afterwards. The majority of the panel remains silent.
- **This results in an artificial consensus**: Due to the lack of discussion, it seems as if the experts agree when they do not. They are merely silent.
- **Ambiguous projections**: Each expert interprets them differently.
- The application of only simple descriptive statistics is used for the evaluation of the expert assessments; **the evaluation remains superficial**. Professional Delphi studies therefore use, for example, statistical group comparisons, outlier and cluster analyses, and other procedures (Warth, von der Gracht, and Darkow 2013).
- **Contestable expert criteria**: The reason an expert is regarded as such and has been selected is not transparent.

6.6 Further development of the method

What do excellent Delphi studies do differently or better, beyond avoiding these shortcomings? Where is the development headed? There are eight key points:

1. **Sentiment analyses**: The personal psychological and ideological predispositions of the experts are made transparent and considered in the interpretation of the results.
2. **Qualitative data analysis**: The textual arguments of the experts are analysed qualitatively and systematically by means of coding and categories. The framework of Förster and von der Gracht (2014: tables 8 and 9) has proven useful for Delphi analyses.
3. **Design layout** under consideration of the dichotomy of end state vs. time intervals. For example, one could ask the experts: 'What is the probability of occurrence for 2030 (end state)?' Or alternatively, 'When do you think the projection will occur?' In addition, metric scales allow for more meaningful analyses than ordinal scales.

'In how many years will the projection occur?' The scale starts at 0 (never) and goes up to 100 years. From this, it is possible to calculate mean values and a distribution curve. This is more precise than the ordinal scaling along periods of time which, averaged, do not allow any conclusion to be drawn about a specific year.

4. **Biases**: Consideration of various cognitive biases among experts, such as the desirability bias (e.g., Ecken, Gnatzy, and von der Gracht 2011; Winkler and Moser 2016; Bonaccorsi, Apreda, and Fantoni 2020). In this case, one expert estimates that 80 per cent of all medicines will come from 3D printers in five years. That many? Yes, because he is an expert on printer manufacturers and is subject to desirability bias. He tends to overrate the probability of occurrence of what he hopes to see happen.
5. **Clustering of Delphi projections**: Not only is each projection evaluated individually, but also in combination with other, related projections and/or in cross-impact, for example, in interaction with other hypotheses (e.g., Beiderbeck, Frevel, von der Gracht, Schmidt, and Schweitzer 2021a).
6. **Confidence or self-rated expertise**: One might ask the experts: 'On a 5-point Likert scale, how do you rate your expertise in this field?' One could also ask about their confidence in making an assessment. This can be used to weight the results of the individual experts as well as at the projection level in the analysis.
7. **The well-founded derivation of hypotheses** along established theories and frameworks: The origin of a hypothesis must be recognisable and comprehensible (for a bibliometric analysis of the use of theories in foresight, see Münch and von der Gracht 2021).
8. **Language:** The formulation of projections can have a lasting influence on an expert's answers (Markmann, Spickermann, von der Gracht, and Brem 2021). This can lead to undesirable bias and tangential discussions. For example, the statement 'The price of oil will rise to $400 per barrel by 2030' can lead to experts arguing over it is actually $400 dollars, $360 or perhaps $420, instead of discussing whether a huge oil price shock is to be expected.

6.7 Adaptation and outlook

Each method has advantages and disadvantages. Therefore, when designing a survey, it is all the more important to answer the question: Does a Delphi survey fit into my research design and the research question? Researchers who are familiar with the Delphi method will find the right response.

Note

1 Dr. Heiko A. von der Gracht is Professor of Futures Research and Foresight at Steinbeis University, School of International Business and Entrepreneurship (SIBE), Germany. He has published more than 50 foresight studies as well as 250 academic and practice-oriented papers and book chapters. Prof. von der Gracht is the founder and editor of several scientific journals. As an associate editor, he heads the international Bureau for Foresight of the leading journal Technological Forecasting & Social Change. He has more than a decade of consulting experience working with multinational corporations.

References

Aengenheyster, S., K. Cuhls, L. Gerhold, M. Heiskanen-Schüttler, J. Huck, and M. Muszynska. 2017. "Real-Time Delphi in Practice—A Comparative Analysis of Existing Software-based Tools." *Technological Forecasting and Social Change* 118: 15–27. https://doi.org/10.1016/j.techfore.2017.01.023.

Beiderbeck, D., N. Frevel, H. A. von der Gracht, S. L. Schmidt, and V. M. Schweitzer. 2021a. "The Impact of COVID-19 on the European Football Ecosystem – A Delphi-based Scenario Analysis." *Technological Forecasting and Social Change* 165: 120577. https://doi.org/10.1016/j.techfore.2021.120577.

———. 2021b. "Preparing, Conducting, and Analyzing Delphi Surveys: Cross-disciplinary Practices, New Directions, and Advancements." *MethodsX* 8: 101401. https://doi.org/10.1016/j.mex.2021.101401.

Belton, I., G. Wright, A. Sissons, F. Bolger, M. M. Crawford, I. Hamlin, ... A. Vasilichi. 2021. "Delphi with Feedback of Rationales: How Large can a Delphi Group Be Such that Participants Are Not Overloaded, De-motivated, or Disengaged?" *Technological Forecasting and Social Change* 170: 120897. https://doi.org/10.1016/j.techfore.2021.120897.

Bonaccorsi, A., R. Apreda, and G. Fantoni. 2020. "Expert Biases in Technology Foresight. Why They Are a Problem and How to Mitigate Them." *Technological Forecasting and Social Change* 151: 119855. https://doi.org/10.1016/j.techfore.2019.119855.

Dajani, J. S., M. Z. Sincoff, and W. K. Talley. 1979. "Stability and Agreement Criteria for the Termination of Delphi Studies." *Technological Forecasting and Social Change* 13 (1): 83–90. https://doi.org/10.1016/0040-1625(79)90007-6.

Dalkey, N., and O. Helmer. 1963. "An Experimental Application of the Delphi Method to the Use of Experts." *Management Science* 9 (3): 458–467. https://doi.org/10.1287/mnsc.9.3.458.

Ecken, P., T. Gnatzy, and H. A. von der Gracht. 2011. "Desirability Bias in Foresight: Consequences for Decision Quality based on Delphi Results." *Technological Forecasting and Social Change* 78 (9): 1654–1670. https://doi.org/10.1016/j.techfore.2011.05.006.

Erffmeyer, R. C., E. S. Erffmeyer, and I. M. Lane. 1986. "The Delphi Technique: An Empirical Evaluation of the Optimal Number of Rounds." *Group & organization studies* 11 (1–2): 120–128. https://doi.org/10.1177/105960118601100110.

Förster, B., and H. A. von der Gracht. 2014. "Assessing Delphi Panel Composition for Strategic Foresight — A Comparison of Panels based on Company-internal and External Participants." *Technological Forecasting and Social Change* 84: 215–229. https://doi.org/10.1016/j.techfore.2013.07.012.

Gnatzy, T., J. Warth, H. A. von der Gracht, and I.-L. Darkow. 2011. "Validating an Innovative Real-time Delphi Approach – A Methodological Comparison between Real-time and Conventional Delphi Studies." *Technological Forecasting and Social Change* 78 (9): 1681–1694. https://doi.org/10.1016/j.techfore.2011.04.006.

Gordon, T. J., and A. Pease. 2006. "RT Delphi: An Efficient, "round-less" Almost Real Time Delphi Method." *Technological Forecasting and Social Change* 73 (4): 321–333. https://doi.org/10.1016/j.techfore.2005.09.005.

Gordon, T. J., and O. Helmer. 1964. *Report on a Long-Range Forecasting Study*. Santa Monica: The RAND Corporation.

Kovaleva, A. 2012. *The IE-4: Construction and Validation of a Short Scale for the Assessment of Locus of Contro.* (Dr. rer. soc.), University of Mannheim, Germany, Mannheim. www.ssoar.info/ssoar/bitstream/handle/document/37119/ssoar-2012-kovaleva-The_IE-4_Construction_and_Validation.pdf?sequence=1&isAllowed=y&lnkname=ssoar-2012-kovaleva-The_IE-4_Construction_and_Validation.pdf.

Linstone, H. A., and M. Turoff. 1975. "Introduction." In *The Delphi Method: Techniques and Applications*, edited by H. A. Linstone and M. Turoff, 3–12. London: Addison-Wesley.

———. 2011. "Delphi: A Brief Look Backward and Forward." *Technological Forecasting and Social Change* 78 (9): 1712–1719. https://doi.org/10.1016/j.techfore.2010.09.011.

Loye, D. 1980. "Personality and Prediction." *Technological Forecasting and Social Change* 16 (2): 93–104. https://doi.org/10.1016/0040-1625(80)90001-3.

Markmann, C., A. Spickermann, H. A. von der Gracht, and A. Brem. 2021. "Improving the Question Formulation in Delphi-Like Surveys: Analysis of the Effects of Abstract Language and Amount of Information on Response Behavior." *Futures & Foresight Science* 3 (1): e56. https://doi.org/10.1002/ff02.56.

Mauksch, S., H. A. von der Gracht, and T. J. Gordon. 2020. "Who Is an Expert for Foresight? A Review of Identification Methods." *Technological Forecasting and Social Change* 154: 119982. https://doi.org/10.1016/j.techfore.2020.119982.

Münch, C., and H. A. von der Gracht. 2021. "A Bibliometric Review of Scientific Theory in Futures and Foresight: A Commentary on Fergnani and Chermack 2021." *Futures & Foresight Science* 3 (3–4): e88. https://doi.org/10.1002/ff02.88.

RAND Corporation (Ed.). 1946. *Preliminary Design of an Experimental World-Circling Spaceship* (Document Number: SM-11827 ed.). Santa Monica: RAND Corporation.

———. 2022. "A Brief History of RAND." www.rand.org/about/history/a-brief-history-of-rand.html.

Salancik, J. R., W. Wenger, and E. Helfer. 1971. "The Construction of Delphi Event Statements." *Technological Forecasting and Social Change* 3 (0): 65–73. https://doi.org/10.1016/s0040-1625(71)80004-5.

Spickermann, A., M. Zimmermann, and H. A. von der Gracht. 2014. "Surface- and Deep-Level Diversity in Panel Selection — Exploring Diversity Effects on Response Behaviour in Foresight." *Technological Forecasting and Social Change* 85: 105–120. https://doi.org/10.1016/j.techfore.2013.04.009.

von der Gracht, H. A. 2012. "Consensus Measurement in Delphi Studies: Review and Implications for Future Quality Assurance." *Technological Forecasting and Social Change* 79 (8): 1525–1536. http://doi.org/10.1016/j.techfore.2012.04.013.

von der Gracht, H. A., S. Kisgen, N. Lange, and J. Jalufka. 2021. *Die Zukunft der europäischen Generika- & Biosimilarsindustrie 2030plus*. Stuttgart: Steinbeis-Edition.

von der Gracht, H. A., T. Gnatzy, I.-L. Darkow, T. J. Gordon, and J. Glenn. 2011. "New Frontiers in Delphi Research - Experiences with Real Time Delphi in Foresight." In *Proceedings of World Future Conference 2011*, edited by World Future Society, 129–160. www.researchgate.net/publication/292524949_New_Frontiers_in_Delphi_Research-Experiences_with_Real_Time_Delphi_in_Foresight.

Warth, J., H. A. von der Gracht, and I.-L. Darkow. 2013. "A Dissent-Based Approach for Multi-Stakeholder Scenario Development — The Future of Electric Drive Vehicles." *Technological Forecasting and Social Change* 80 (4): 566–583. https://doi.org/10.1016/j.techfore.2012.04.005.

Winkler, J., and R. Moser. 2016. "Biases in Future-oriented Delphi Studies: A Cognitive Perspective." *Technological Forecasting and Social Change* 105: 63–76. https://doi.org/10.1016/j.techfore.2016.01.021.

7

SCENARIO PLANNING

Scenario planning is a central tool in strategic foresight and has generated interest beyond the field of foresight, and has been recognised as a central tool for strategy development (Augier, Dew, Knudsen, and Stieglitz 2018). Moreover, scenario planning has been used for over 50 years (McKiernan 2017).

There is a large body of practical and academic literature on scenario planning (e.g., Schoemaker 1993; van der Heijden, Bradfield, Burt, Crains, and Wright 2002; Bradfield, Wright, Burt, Cairns, and van der Heijden 2005; Burt 2010; Chermack 2011; Cairns and Wright 2017). For me the key aspects of scenario planning are that it is predominantly participatory and that it challenges the assumptions of those involved in the process and in creating memories of the future.

> In essence, the scenario process is about enabling managers to visit and experience the future ahead of time, thereby creating what is called 'memories' of the future. These visits to anticipated futures are then

DOI: 10.4324/9781003302735-10

> remembered, creating a matrix in the mind of managers and serving as subconscious guides to make sense of incoming environmental signals and to act on them.
>
> *(van der Heijden, Bradfield, Burt, Crains, and Wright 2002: 177)*

Kees van der Heijden, a former head of the scenario planning team at Royal Dutch Shell, describes scenario planning as the 'art of strategic conversations', emphasising the relevance of managers having conversations on trends and weak signals developing in their business environment and asking questions about the implications of these changes for their organisation.

> It is my experience that scenarios are the best available language for the strategic conversation, as it allows both differentiation in views, but also brings people together towards a shared understanding of the situation, making decision making possible when the time has arrived to take action.
>
> *(van der Heijden 1996: IX)*

By developing scenarios, preparing an organisation for an uncertain future by asking 'what if' (Schwarz 2021). I will later revisit this aspect, in the context of transformation and open strategy in Chapter 13. What this implies is that scenario planning can be understood as an organisational intervention enabling the transformation of an organisation.

7.1 Evolution of scenario planning

The scenario technique dates to the 1950s when Herman Kahn of the RAND Corporation developed the first scenarios. Kahn and his colleague, Anthony J. Wiener, published *The Year 2000: A Framework for Speculations on the Next Thirty-Three Years* (1967). This signalled the inauguration of the scenario technique. Kahn and Wiener (1967: 6) describe scenarios as 'hypothetical sequences of events constructed for the purpose of focusing attention on casual processes and decision-points'.

In the 1970s, Royal Dutch/Shell created what is now known as *scenario planning* by linking the scenario technique with strategic planning. Pierre Wack, who introduced scenario planning to Royal Dutch Shell (Chermack 2017) published two articles in the *Harvard Business Review* (Wack 1985a,b) which are still highly regarded. In contrast to scenario writing, scenario planning is more a qualitative method, which relies on a process conducted by the management of an organisation, not scenario experts. The purpose of scenarios is to identify trends and key uncertainties and combine them into pictures of the future, not covering all eventualities but discovering the boundaries of future outcomes (Schoemaker and van der Heijden 1992). Most importantly, scenarios should cover generically different futures rather than variations of single ones. In other words, scenarios enable managers to think about the unthinkable.

In addition to the important task of providing alternative pictures of the future, scenarios –especially scenario planning – can perform several other functions. Scenario planning restores complexity to the strategic planning process. This is in contrast to methods like forecasting which simplify the planning process (Eden and Ackermann 1998). Another interesting and important effect is that scenario planning could challenge the mental models of participating managers.

Over the years, several schools of scenario planning have developed, such as the intuitive logic school, the probabilistic modified trends school, and the La prospective school (Bradfield, Wright, Burt, Cairns, and van der Heijden 2005; MacKay and McKiernan 2018). While the intuitive logic school is more qualitative and the probabilistic modified trends school more quantitative, the La prospective is a mix of the two. This book is concerned with the intuitive logic methodology, also known as the 'Shell approach'. This school has received the most attention in the literature and is dominant in scenario development in the United States and other countries (Bradfield, Wright, Burt, Cairns, and van der Heijden 2005).

It is based on the premise that business decisions are taken in an environment of complex relationships between dimensions of the PEST framework. It is a qualitative and subjective approach. This school is 'intuitive' because it allows analysis of objective data and 'gut feelings' of experts involved as well as familiarity with the issue of concern of the scenario planning project.

7.2 The process of scenario planning

Based on the rationale of the intuitive logic school, there is a multitude of scenario planning processes (e.g., Schoemaker 1995; van der Heijden 1996;

van der Heijden, Bradfield, Burt, Crains, and Wright 2002; Chermack 2011; Schwartz 2012; Schwenker, Wulf, and Krys 2013). Figure 7.1 builds on elements of these process but is associated with van der Heijden's (van der Heijden 1996; van der Heijden, Bradfield, Burt, Crains, and Wright 2002) approach.

7.2.1 *Scope definition*

The scope of the scenario project must be defined. This step includes the formulation of the question a scenario planning project is supposed to answer, such as 'what is the future of my industry?'

7.2.2 *Key drivers*

The second step is concerned with trend research. Besides desktop research, interviews are a relevant source in this phase. Semi-structured interviews with a diverse set of industry experts and people working in the industry and organisation need to be conducted to identify trends. Interviewing experts within and outside of an organisation also allows to assess what assumptions in an organisation concern the future and how these assumptions might diverge from those of other industry experts. Trends are then allocated to the PEST framework, to ensure that a holistic perspective is taken on the future of industry.

Part of this phase is to differentiate between drivers and key-drivers among the identified trends. Key drivers are those trends that are both impactful and uncertain. The approach used for this step is the relevance and uncertainty matrix: uncertainty along the x-axis and relevance along the y-axis. The upper left-hand section, drivers, contains trends

	1. Scoping the question	2. Key-Drivers	3. Building scenarios	4. Implications
Objectives	Clarify research question/issue to be investigated	• Identify relevant trends • Differentiate between key-drivers (critical uncertainties) and drivers (pre-determined elements)	• Agree on most relevant key drivers • Create scenarios	Asses implications of the scenarios
Deliverables	Scope of the Scenario Planning exercise	Key drivers for scenario workshop	Scenario quadrant / multiple quadrants	Recommendations

Figure 7.1 Process of scenario planning project.

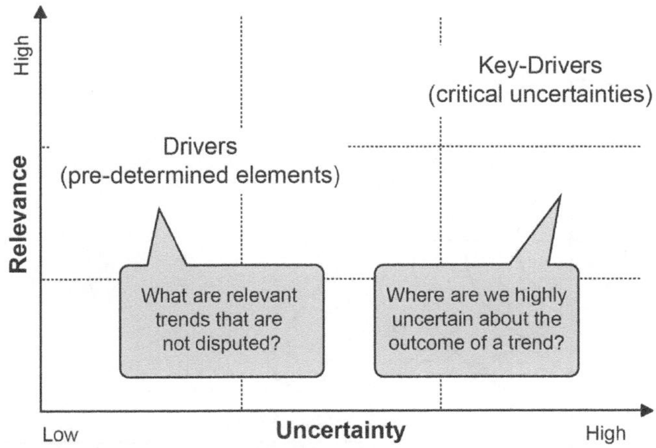

Figure 7.2 Relevance-uncertainty matrix.

with the most relevance for the industry and the least uncertainty. The upper right-hand section, key drivers, contains the most important trends for the scenario development. The goal of the matrix is to identify key drivers, laying the foundation for the four scenario dimensions.

The key drivers are analysed by describing the possible direction in which they could develop, their polar outcomes. Trends could develop in two possible directions. Two independent key drivers with the most relevance to the research questions are then chosen.

7.2.3 Building scenarios

The two key uncertainties with their polar outcomes form the basis for the scenario matrix. As can be seen in Figure 7.3, four distinct scenarios are formed using the scenario matrix. The boundaries are set by two key trends, each with two polar outcomes defined for each scenario dimension.

The scenarios are described by allocating the other trends identified during the process. Each scenario is given a concise name and description, symbolising the developments within it.

One way to describe the scenarios is the scenario blueprint. A scenario blueprint is an auxiliary tool used to list and rate the relevance (1 low – 5 high) of the trends positioned in the two main sections of the relevance and uncertainty matrix. The relevance of each trend is rated for each scenario

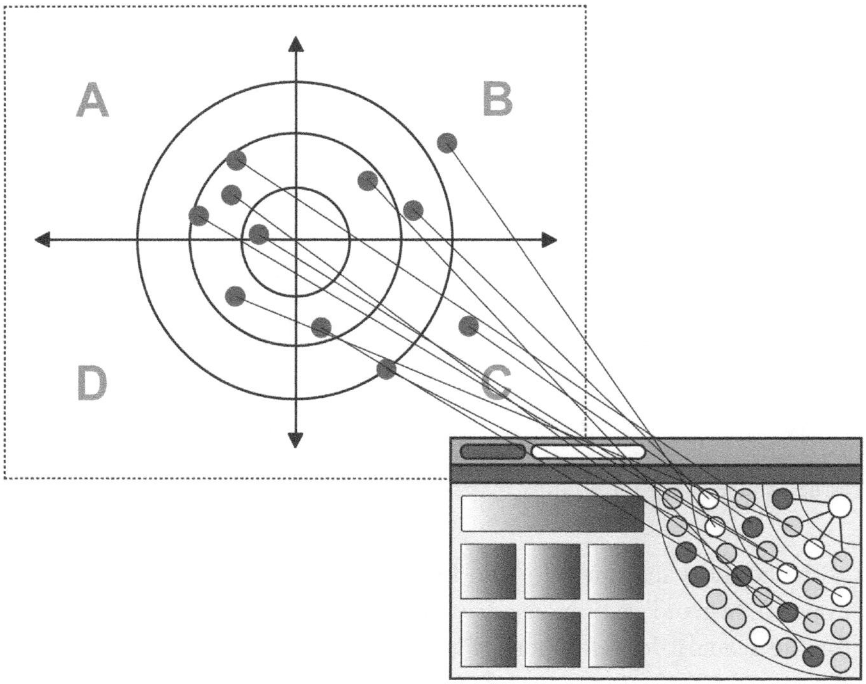

Figure 7.3 Scenario matrix.

and briefly explained. This can also be the basis for creating scenario narratives and alternative pictures of the future. The goal is to present the reader with a picture of the world in that scenario, its dynamics, and the reasons for its development into that state.

In the following, scenarios can be visualised (see example), developed into stories, or translated into short movies explaining the scenarios. The key is to make the scenarios accessible for those who were not involved in the process.

7.2.4 Implications

The purpose of this phase of the scenario planning project is to derive implications from the scenarios for the organisation. These types of conversations can start with questions such as 'what is our preferred scenario?', 'for what scenario are we prepared the best?', or 'which scenario would be for us the most challenging one?'

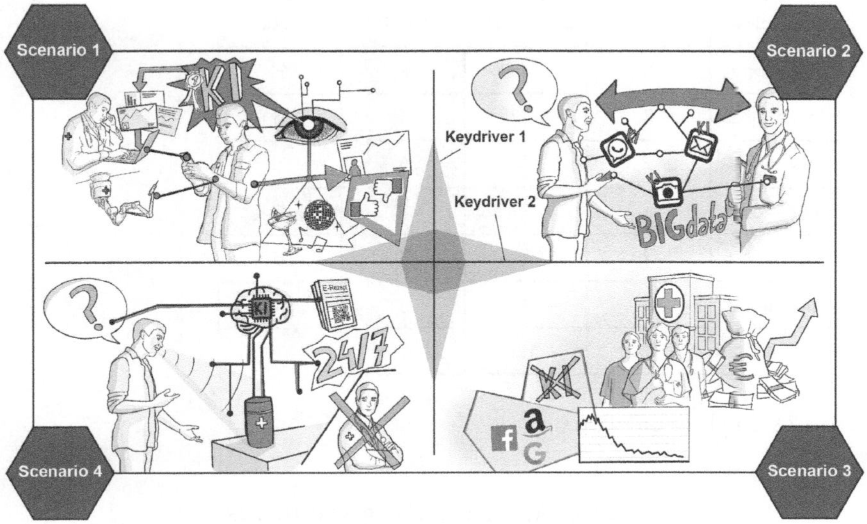

Figure 7.4 Example of the visualisation of scenarios (IICM Institute for Innovation and Change Methodologies).

A more structured approach to working with the scenarios is the implication matrix (Figure 7.5). According to the scenario, several implications are derived. What implications would this scenario have for our organisation? These implications are then briefly described and listed on the implication matrix. After this matrix is created in the 'wind-tunneling' exercise, the relevance of the implications for the organisation in each scenario is assessed on a 3-point scale.

By doing this, robust implications, that are relevant in at least two scenarios can be identified. Wind-tunneling is intended to decide which implications must be improved in the next iteration, so that they become as robust as possible. The horizontal scores give an indication for which implications might be more robust than others. The key argument would be that these implications need to be picked up, regardless of what scenario will become more relevant in the future. However, this does not imply that none-robust implications are irrelevant. The implications might need to be more closely monitored or small experiments or investments might be an approach to understand these implications. The vertical score can indicate which scenario might be the most challenging to deal with in terms of having many implications.

Implication	Scenario A	Scenario B	Scenario C	Total
Implication A	5	4	2	11
Implication B	3	2	1	6
Implication C	2	3	3	8
Implication D	1	3	2	6
Implication E	2	3	1	6
Implication F	3	4	3	10
Total	16	19	12	

Figure 7.5 Implication matrix.

7.2.5 After creating scenarios

While not a deliberate part of scenario creation, I want to say a few words about what happens with scenarios once they have been created. While the process of scenario planning builds on the participation of decision-makers, scenarios might also need to be communicated in an organisation. The central question is how scenarios can be communicated and visualised (Müller and Schwarz 2016) in a form that allows others to enter these alternative pictures of the future. The scenarios might guide the scanning and monitoring activities in a firm, pointing out to weak signals or trends that should be watched more closely. It might also be necessary to update the scenarios, for instance when significant changes in the organisational environment have emerged or new trends have been identified.

Concerning the effects of scenario planning on an organisation, Grant (2003: 511) states that Royal Dutch/Shell's scenario planning process was a vehicle for organisational learning by sharing and integrating multiple knowledge bases from both within and outside the Shell group, with emphasis on communication and knowledge sharing: 'Shell's "scenario-to-strategy" framework involved discussion workshops in which scenarios would provide the foundation for an interactive strategy formulation'. Understanding a foresight process as a vehicle for strategic conversation in an organisation helps to define one role of such a process in an organisation. The value of such a strategic conversation is that mental models in an organisation are not only shared but also most likely to be challenged and that an organisation is likely to eliminate blind spots from its sensemaking of the environment and of the future.

References

Augier, M., N. Dew, T. Knudsen, and N. Stieglitz. 2018. "Organizational Persistence in the Use of War Gaming and Scenario Planning." *Long Range Planning* 51 (4): 511–525. https://doi.org/10.1016/j.lrp.2017.12.005.

Bradfield, R., G. Wright, G. Burt, G. Cairns, and K. van der Heijden. 2005. "The Origins and Evolution of Scenario Techniques in Long Range Business Planning." *Futures* 37 (8): 795–812.

Burt, G. 2010. "Revisiting and Extending Our Understanding of Pierre Wack's the Gentle Art of Re-Perceiving." *Technological Forecasting and Social Change* 77 (9): 1476–1484.

Cairns, G., and G. Wright. 2017. *Scenario Thinking: Preparing Your Organization for the Future in an Unpredictable World.* Basel: Springer International Publishing.

Chermack, T. J. 2011. *Scenario Planning in Organizations: How to Create, Use, and Assess Scenarios.* San Francisco: Berrett-Koehler.

———. 2017. *Foundations of Scenario Planning: The Story of Pierre Wack.* Routledge International Studies in Business History. New York: Taylor & Francis.

Eden, C., and F. Ackermann. 1998. *Making Strategy.* London: Sage.

Grant, R. M. 2003. "Strategic Planning in a Turbulent Environment: Evidence from the Oil Majors." *Strategic Management Journal* 24 (6): 491–517. https://doi.org/10.1002/smj.314.

Kahn, H., and A. J. Wiener. 1967. *The Year 2000: A Framework for Speculation on the Next Thirty-Three Years.* New York: Macmillan.

MacKay, R. B., and P. McKiernan. 2018. *Scenario Thinking: A Historical Evolution of Strategic Foresight.* Elements in Business Strategy. Cambridge: Cambridge University Press.

McKiernan, P. 2017. "Prospective Thinking; Scenario Planning Meets Neuroscience." *Technological Forecasting and Social Change.* North-Holland. https://doi.org/10.1016/j.techfore.2016.10.069.

Müller, A. W., and J. O. Schwarz. 2016. "Assessing the Functions and Dimensions of Visualizations in Foresight." *Foresight* 18 (1): 76–90. https://doi.org/10.1108/FS-04-2014-0027.

Schoemaker, P. J. H. 1993. "Multiple Scenario Development: Its Conceptual and Behavioral Foundation." *Strategic Management Journal* 14 (3): 193–213. https://doi.org/10.1002/smj.4250140304.

———. 1995. "Scenario Planning: A Tool for Strategic Thinking." *MIT Sloan Management Review* 36 (2): 25–40.

Schoemaker, P. J. H., and C. A. J. M. van der Heijden. 1992. "Integrating Scenarios into Strategic Planning at Royal Dutch/Shell." *Planning Review* (May/June): 41–46.

Schwartz, P. 2012. *The Art of the Long View: Planning for the Future in an Uncertain World.* New York: Crown.

Schwarz, J. O. 2021. "From Alternative Pictures of the Future to an Organizational Intervention: A Commentary on Rowland and Spaniol." *Futures & Foresight Science* (n/a): e2105. https://doi.org/10.1002/ffo2.105.

Schwenker, B., T. Wulf, and C. Krys. 2013. *Scenario-Based Strategic Planning: Developing Strategies in an Uncertain World.* Roland Berger School of Strategy and Economics. Wiesbaden: Springer Fachmedien.

van der Heijden, K. 1996. *Scenarios: The Art of Strategic Conversation.* Chichester: John Wiley & Sons.

van der Heijden, K., R. Bradfield, G. Burt, G. Crains, and G. Wright. 2002. *The Sixth Sense.* Chichester: John Wiley & Sons.

Wack, P. 1985a. "Scenarios: Uncharted Waters Ahead." *Harvard Business Review* 63 (5): 73–89.

———. 1985b. "Scenarios: Shooting the Rapids." *Harvard Business Review* 63 (6): 139–150.

8

FUTURES LITERACY LABORATORIES

Stefan Bergheim[1]

8.1 History of the method

One of the deepest roots of the Futures Literacy Laboratory can be traced to the early 1980s, when Riel Miller, a young Canadian economist, started to work at the Organization for Economic Cooperation and Development (OECD) in Paris. Miller was tasked with analysing trends and creating forecasts for technology, education, and the economy. But he saw limits to the ways in which humans deal with the future in an uncertain environment. While he noticed a strong tendency of established organisations to plan, predict, and control the future, these efforts often seemed futile, narrow, and sometimes even harmful to themselves. Therefore, he explored additional ways to use the future with a stronger focus on complexity, novelty, and openness.

In 2006 he met sociologist Roberto Poli at a conference in Cardiff. Poli introduced him to the theory of anticipation, devised by the biologist Robert Rosen (Rosen, Rosen, Kineman, and Nadin 2012) in the 1970s. Together they developed the theoretical foundations of the competence that became 'Futures Literacy' (Miller 2007). They also developed, refined, and

DOI: 10.4324/9781003302735-11

experimented with the method 'Futures Literacy Laboratory', initially Know-Lab. When Miller joined UNESCO in 2012, they convened more meetings and conferences. In 2015 they began a book project that brought together all the insights developed over the previous decade and made them accessible to the public. *Transforming the Future – Anticipation in the 21st Century* was published in 2018 in open access. By 2022, hundreds of Futures Literacy Laboratories have been run around the globe by UNESCO and other organisations.

8.2 Description of the method

In line with the method's foundation in complexity theory, learning theories, and the theory of anticipation, three major principles shape the design of Futures Literacy Laboratories (Miller 2018): First, they are action learning or learning-by-doing activities. They engage a large spectrum of participants and invite them to share their individual images of the future and the underlying assumptions. As they do so, participants learn a variety of ways to use the future.

Second, Futures Literacy Laboratories are collective intelligence knowledge creation processes because learning is more likely to occur if it is done with others rather than just by oneself, and because multiple images can be connected in a fruitful way. This is how participants can inspire each other, discuss meaning together and be more creative. The use of a wide range of tools is required, ranging from talking with each other to drawing and sculpting to role-playing.

Perhaps most importantly in comparison with other futures methods, these laboratories use a catalyst to disrupt entrained thinking, move attention beyond the old boxes and stimulate emergence. This is done with an alternative future intended to take participants beyond the frames of likely-unlikely or desirable-undesirable that are ubiquitous in futures work. There are many ways to do this reframing, such as using static or dynamic configurations or games.

8.3 Four phases of a Futures Literacy Laboratory

Every Futures Literacy Laboratory is designed specifically for the host's intentions, topic, participants, length, and other variables. But most laboratories include four sequential phases. Phase 1, the 'reveal', works with the participants' expectations and hopes about the topic under investigation. They reveal their images of probable and desirable futures at a specific time horizon. Phase 1 combines classic elements of trend analysis, forecasting, and visioning, with the purpose of revealing the anticipatory assumptions that led to those images in the present. At the end of this phase, it is possible to discuss the assumptions about the topic that had become visible (or revealed) including their possible sources.

Phase 2, the 'reframe', is a central learning catalyst in which participants imagine the future of the topic through a frame that is unfamiliar and distinct from the ones revealed in Phase 1. By experimenting with this different set of anticipatory assumptions, participants improve their ability to handle uncertainty and ambiguity. Sometimes this alternative future is provided by the design team; sometimes it is developed by participants themselves moving outside of what they had created in the first phase. In either case, participants fill in the details of that unusual future themselves and stretch their imagination.

In Phase 3, the 'rethink', participants come back to the present and look again at the probable, desirable, and alternative futures they explored. They search for new issues that emerged for them, for fissures that opened into new ways of thinking, for powerful new questions that might turn into quests for them later.

Finally, in Phase 4, the 'acting' or next steps, participants select from the wide menu of questions they raised in Phase 3 and focus on next steps. What kinds of activities should follow from the new questions? Which

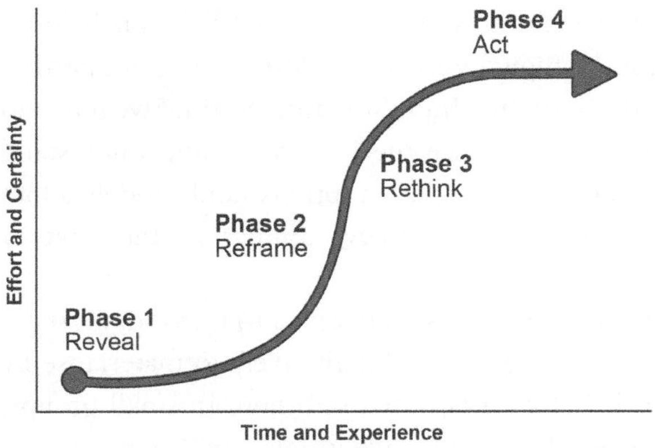

Figure 8.1 Four phases of a Futures Literacy Laboratory.

quests result for individuals or groups? The length of this phase depends on what might happen after the laboratory. Sometimes, ideas for action are briefly alluded to, knowing that change in complex adaptive systems will come when the circumstances are appropriate. In other laboratories, an extended Phase 4 leads to tangible projects, prototypes, action plans, or even briefings to the CEO. Sometimes a shared consensus is sought. In other cases, individual explorations are acceptable.

Research on the evaluation of Futures Literacy Laboratories based on interviews with designers and facilitators shows evidence that this method contributes to four related results (Bergheim 2022). First, laboratories allow participants to acquire the skill of using the future more broadly than before. They understand anticipatory systems and processes better and their competence of Futures Literacy increases. Second, stronger associations develop among participants, which is an ingredient for and a result of successful collective intelligence knowledge creation. Third, participants gain an awareness of novelty, especially in the third phase of the lab. They sense and make sense of new knowledge, new ideas, and questions on the topic of the laboratory that they had not been aware of before. And finally, there is the application of the skills and of the new insights on the topic. This application into practice usually comes towards the end of the laboratory or during the follow-up.

8.4 Example: Futures of media

A colleague and I offered an introductory Futures Literacy Laboratory in May 2022 on the topic of 'The Futures of Media in 2040'. It was held on Zoom over six hours with the digital whiteboard Miro. The eight participants came from Denmark, Germany, Iran, Romania, Spain, Sweden, and Switzerland. They wanted to experience the Laboratory and understand the importance of anticipatory systems, assumptions, and models about the future. They also wanted to discover new content on the topic of media.

In the first part of Phase 1, participants were asked to think about statements about media that they thought were highly likely to materialise by 2040. They wrote down their forecasts, discussed them in small groups, expanded on them, compared them, and made visible a wide spectrum of expectations. For the second part, we took them on a thought journey into their most preferred futures of media in 2040. In their groups, they first shared the images that they took in their minds. Then they jointly created characters, like 'Glassy' and 'The Enabler'. At the end of Phase 1, participants discussed assumptions about media that they had recognised to that point, either their own or those of others. This exercise was not about assessing the validity or veracity of those assumptions but to recognise that most of those assumptions were human-made and therefore changeable. Changed assumptions can lead to changed actions.

In Phase 2, participants were invited to imagine details of a future of media given to them by the facilitators. In this laboratory, the outline of the alternative future was quite short: 'In the year 2040, humans focus solely on their local living environments because this is where they are active and connected'. To add more detail and to exercise their imagination, participants first improvised a day in the life of a person called Alex and how Alex relates to media. Afterward, they imagined the details of that future in a more structured way, exploring actors, technologies, and the content of media. After some initial discomfort and feeling overwhelmed, participants enjoyed the creativity that burst forth.

In Phase 3, participants were led back into the present and asked to compare the three futures they had explored so far. Which new powerful questions emerged for them? Examples included: Is unbiased media possible? What is the role of nature in media? Do we need to know it all?

Finally, and very briefly in this short laboratory with a heterogenous group of participants from different organisations and countries, ideas for action were attached to the questions. What activities might develop from those questions? Some of the results were: Dig deeper into the issue of human cognition! Invite opposing views into all my projects! Focus on more dimensions of any issue than I would naturally do! The application of those ideas into their practice is now up to the participants.

8.5 Challenges of the method

Since Futures Literacy Laboratories are a new method and since the theory of anticipation is not yet widely known, some misunderstandings and challenges remain. For example, the creation of an alternative future – a unique feature of the method – is an art at this stage and not yet well researched. In practice, a just-do-it approach has led to good results possibly exactly because of the creativity and specificity this allows.

In addition, while Futures Literacy Laboratories are often implemented using the four phases, there is a huge variety of possible configurations depending on intentions, resources, and initial conditions. Some phases can be left out, others added, and others split up as needed. Also, the amount of time spent, and the depth explored in the different phases can vary. The creativity techniques in each phase can differ as well. There is no template. This makes the method appealing to experienced designers and facilitators who want to create a unique experience every time. But the absence of a template makes it harder for novices to get started.

Partly because of the versatility of the method, it is not trivial to identify the most appropriate configuration for the context at hand. Here it is helpful to start with a small and modest version. Some sharing of experiences with other designers and facilitators might be similarly beneficial. Chapter 4 in *Transforming the Future* contains 14 case studies that might provide additional inspiration. Bergheim (2021) also includes examples and learnings.

The fourth challenge is the openness to outcomes inherent in this method. Not all hosts and participants are open to the wide variety of possible outcomes. This openness is usually an issue during the co-design of the laboratory and for its framing for participants in the invitation. If hosts want to limit creativity, then the boundaries should be made clear in advance. Ideally, the host should commit to some sort of usage of the results of the laboratories. Otherwise, participant engagement might be affected negatively.

Note

1 Stefan Bergheim works as a futurist since 1995, initially as a forecaster in the financial industry, then as a trend researcher, later focusing on participatory visioning processes for a German think tank he founded, and since 2015 affiliated with UNESCO's Futures Literacy network. He contributed to UNESCO's 2018 book "Transforming the Future" and was one of the curators of the 2019 Futures Literacy Design Forum and the 2020 High-Level Futures Literacy Summit.

References

Bergheim, S. 2021. *Futures - Open to Variety: A Manual for the Wise Use of the Later-than-Now*. Frankfurt: ZgF Publishers.

———. 2022. *On the Evaluation of Futures Literacy Laboratories*. Frankfurt: Zgf Publishers, https://zgf-fortschritt.de/media/pages/methodik/zukuen-ftelabore/1356567106-1666609788/bergheim-2022-evaluationof-flls.pdf.

Miller, R. 2007. "Futures Literacy: A Hybrid Strategic Scenario Method." *Futures* 39 (4): 341–362. https://doi.org/10.1016/j.futures.2006.12.001

———. 2018. "Sensing and Making-Sense of Futures Literacy: Towards a Futures Literacy Framework (FLF)." In *Transforming the Future: Anticipation in the 21st Century*, edited by R. Miller, 15–50. London: Taylor & Francis.

Rosen, R., J. Rosen, J. J. Kineman, and M. Nadin. 2012. *Anticipatory Systems: Philosophical, Mathematical, and Methodological Foundations*. IFSR International Series on Systems Science and Engineering. New York: Springer.

9

BUSINESS WARGAMING

Wargaming is among the oldest tools of strategy formulation and planning. It has been in use for almost 200 years and has enjoyed increased academic interest over the last decade (Augier, Dew, Knudsen, and Stieglitz 2018). Schwarz (2009) and Oriesek and Schwarz (2008, 2020) have tied business wargaming to the field of foresight and have described how foresight in regard to the future competitive dynamics in an industry is developed through a business wargame.

Business wargaming can be traced at least as far back as ancient Greece. It grew out of military wargaming, which was used to prepare generals and officers for unforeseen circumstances on the battlefield. Games about warfare have probably existed as long as warfare itself (Perla 1990). It has been argued that wargaming has been around in the military as long as armed forces have been evaluating plausible defensive and offensive options (Vanderveer and Heasley II 2005). Even though business wargaming was applied to company planning in the 1950s (Bellman, Clark, Malcolm, Craft, and Ricciardi 1957), it was not until the mid-1980s that wargaming

DOI: 10.4324/9781003302735-12

was strategically adapted for business purposes (Ginter and Rucks 1984; Treat, Thibault, and Asin 1996). In the field of competitive business intelligence, wargames have drawn special attention for their focus on competitors (Kurtz 2003).

A business wargame is a role-playing simulation of a dynamic business situation (Kurtz 2003). It involves several teams, each assigned to play a stakeholder (competitor, customer, governing bodies) in a business situation. A business wargame evolves over several moves. Each move represents a defined period and/or scenario. A business wargame should always be prefaced by extensive research and include a review of trends and hypotheses for the industry in which the wargame is taking place. In contrast to 'red teams', where one team is a competitor or other stakeholder to understand interests, intentions and capabilities (Zenko 2015), a business wargame is much more complex and therefore suitable for more complex challenges.

Oriesek and Schwarz (2020) introduce the following typology of wargaming in Figure 9.1. The term 'wargaming' can be broken down by application into military and business wargaming. All non-military applications subsume not only wargaming for public and private corporations, but also non-government and other non-military organisations; these are labelled business wargames. Both military and business wargaming can be conducted at the strategic-, operational-, and tactical leadership level and address questions within a context such as strategy testing, foresight, crisis management, or training.

Oriesek and Schwarz (2020) introduce the following cluster of applications:

- Strategy testing, developing foresight, change management
- Crisis response preparation
- Education, training and recruiting

Strategy testing has been described as the primary application of business wargaming, allowing leaders to test a strategy prior to its implementation, and especially in assessing how it works in the competitive landscape (Schwarz 2011). Alongside the pure strategy testing comes the exploration of future industry developments and therefore the development of foresight. Allowing a group of managers to experience future dynamics of an

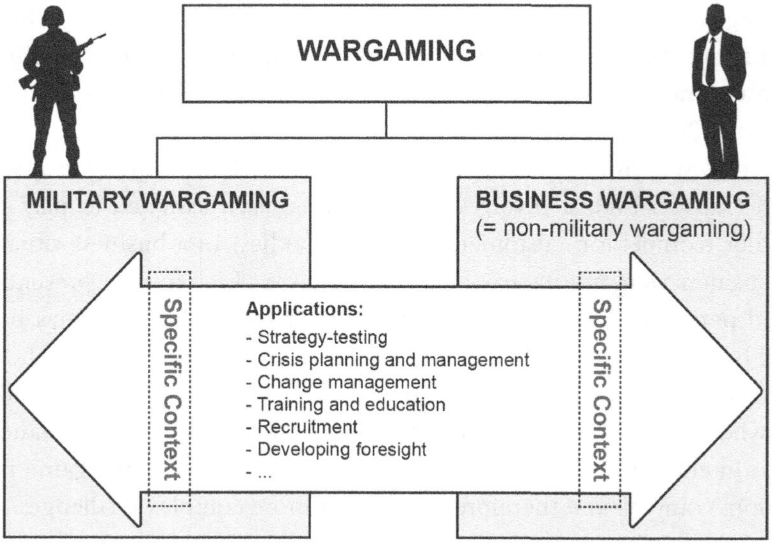

Figure 9.1 Typology of wargaming (Oriesek and Schwarz 2020).

industry will create a sense of urgency for action, which is critical for any change effort (Kotter 2012).

A second application of wargaming is crisis response preparation. In post 9/11 times, wargaming was applied in United States government agencies to anticipate and simulate how to react to crises such as attacks with biological weapons, all in light of how to improve the collaboration among agencies and other key actors. A third application is to use wargames in education and training, which can range from introducing students to strategy, familiarising newly hired managers with a company and an industry, to brainstorming developments in educational settings.

9.1 Application of business wargaming

A business wargame evolves over three moves, representing a certain length of time (e.g., three to ten years) and/or different scenarios (Oriesek and Schwarz 2020). The first move starts in the present and is based on available data. A 'move' is a decision cycle, which begins with the stakeholders or competing companies and the client team taking the initial actions, including competitive offerings, alliances, investments, or lobbying efforts.

In general, as depicted in Figure 9.2, four types of teams exist in a business wargame: a client team, competitive or stakeholder team, market team, and control team.

Competing teams will have to think about strategy, product, pricing, capacity, technology, and the like, while considering the business environment. The members of the client team must execute and adjust their strategy. An essential part of each move is the reaction of the customers, usually played by a market team, consisting of a group of market experts. The market team will provide the reactions of customers, providing figures such as size of the market, market segments, market share, and revenue, and how these figures have changed in the course of the moves. All this data will be passed on to the control team, usually run by the wargaming experts, who use a financial model to calculate profits and losses. In addition, the control team is in charge of supervising the wargame and introducing discontinuities (e.g., technological developments or policy issues) to add real-life dynamics. The control team can also assume the role of other stakeholders, such as regulators who are not represented in the business wargame as competitors.

At the end of each move, the control team calculates the results of the competing teams and the client team and uses these results as the starting point for making the next move. While the market team reacts and the control team calculates the figures, the competitive teams and the client teams plan their next move. They then learn the results of the previous move. As pointed out earlier concerning computer-based simulations, models are supportive, adding reality to the business wargame but not intended to drive the simulation.

Prior to carrying out a business wargame it is essential that the objectives are laid out, and that a database of all information that is potentially useful to the players is created. The model translates the game's data and the players' decisions into game events, quantifying the moves and results of the wargame (Perla 1990). After a business wargame has been played to completion, an analysis is carried out by the wargaming experts and managers of the client company, which is important for discussing lessons learned during the exercise. This analysis allows the managers to describe their experiences, to reflect on what they have learned, and to discuss subsequent steps, which will then result in additional lessons learned.

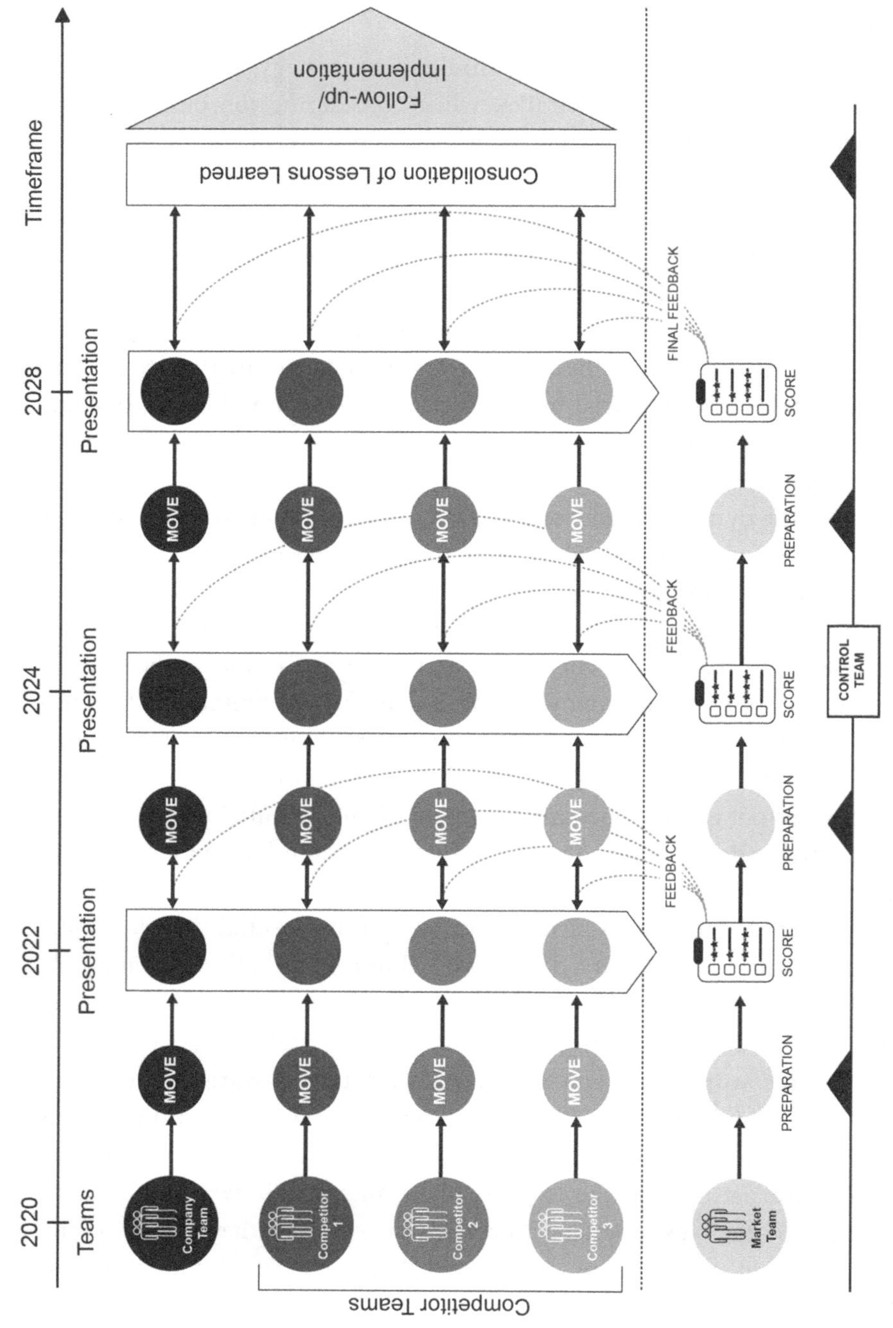

Figure 9.2 Basic structure of a business wargame (Oriesek and Schwarz 2020).

The application of a business wargame is described here to illustrate this approach (Lüchinger 2001; Oriesek and Schwarz 2008, 2020) in the European airline industry. In the late 1990s, European airlines were confronted with dramatic rises in competition. One of the emerging strategies in this environment was to forge alliances, such as Star Alliance or One World, to keep customers and to achieve economies of scale. The strategic questions confronting European airlines included whether they should join one of the alliances, remain independent, or form a new alliance. The objective of this business wargame was to test the strategy. A key finding of this wargame was that remaining outside of one of the major alliances would be difficult for this airline.

However, the CEO of the airline was already convinced, before the business wargame took place, of a different strategic option. While neglecting the outcomes of the business wargame he chose to stick with his original strategic plan that revolved around the formation of a fourth alliance. A few years later the airline went bankrupt and became a part of a larger alliance. While this brief case study suggests that the outcomes of business wargames can be of value, the issue of how the outcomes of a business wargame are handled have also been discussed by Hanley (1992) in a more historic perspective of the military application of wargaming. He points to an aspect that can be deemed responsible for the results of wargames not being translated into action, as described in the European airline case: 'Egocentric constraints were responsible for most cases of active suppression of valid game results' (Hanley 1992: 225).

Military wargaming was used by military planners in response to increasingly dangerous environments and to meet an obvious need to reduce the risks of strategic miscalculations (Ginter and Rucks 1984). In addition to being future-oriented, a business wargame has several advantages. First, a business wargame involves participants. During a business wargame, the dynamics of a market or of competitors will not only be analysed but also anticipated by simulating the future over the course of several days. By simulating an industry, market, or competition over time, managers can 'experience' the dynamics and the competition in their market or industry and see the consequences of their strategy. As the wargaming exercise proceeds, participants will be forced to think about the future. Essentially, early signals of change that might be relevant for the organisation are identified.

The argument can be made that business wargaming is a powerful tool to challenge the mental models of participants. In short, the benefit of business wargaming is that managers of an organisation can think like their competitors, thereby anticipating what those competitors are likely to do.

9.2 Inner mechanics of business wargaming

I reflect here from two angles why business wargaming works and what the inner workings of business wargames are. These two angles are sensemaking and gaming. Sensemaking describes how business wargaming supports understanding the uncertainty imposed by the business environment. Gaming focusses on how participants can be engaged.

9.2.1 Sensemaking

Decision-making in a VUCA world has been researched with the aim of understanding how organisations can best use these attributes to gain insights into environmental changes so that they can take more robust strategic decisions. Previous research has suggested that organisations seek an inter-organisational collaborative sensemaking process (Seidl and Werle 2018) to understand VUCA. The necessity of sensemaking capabilities has been emphasised by the concept of dynamic capabilities (Teece, Pisano, and Shuen 1997; Day and Schoemaker 2016) through which collaborators can 'see different aspects of a problem [...], constructively explore their differences [and] search for solutions that go beyond their own limited vision of what is possible' (Gray 1989: 5).

Many developments and phenomena facing organisations are beyond their 'sensemaking capacity', so they fail to comprehend the scope and consequences of such dynamics. In addition, collaborative sensemaking embraces a variety of perspectives because of the different industries and expertise that participating organisations bring to the process. Hence, the inclusion of this variety of perspectives in exploring new topics, developments, or problems facilitates an extension of sensemaking capacity that no single organisation can generate. It has been argued that the 'variety can improve sensemaking capacity' (Seidl and Werle 2018: 4) and recommended that 'managers actively engage other people in processes of strategic sensemaking in order to gain access to a wider range of frames and thereby develop a better understanding of their world'.

With respect to appreciating the uncertainty of developments that will affect them, organisations and managers draw on inter-organisational (often cross-industry) collaboration, frequently in the form of sensemaking and strategising formats such as scenario planning and business wargaming simulations. Such formats help them anticipate challenges and the need for change. Since initiating and running such processes is resource-intensive, corporations are attentive to ways in which not only decision-makers, but also management can generate and broaden the requisite variety of perspectives that will inform their internal strategising and innovation activities in addition to their decision-making.

The ability to sensitise and alert organisational members and employees to build a change mindset that enables them to consider and appreciate uncertainty and change from different perspectives and often different strategic assumptions is crucial to specify what makes decision-making and strategy more robust. The central argument for business wargaming in this context is that managers of one organisation are enabled, by stepping into the shoes of their competitors for a day or two. Hereby, they develop multiple perspectives on their own industries, core competitors but also their own organisation. Further, the literature on business wargaming (Bracken 2001; Kurtz 2003; Oriesek and Schwarz 2008; Schwarz 2009, 2011, 2013; Krupp and Schoemaker 2014) suggests that the purpose of a business wargame is to support management teams in generating the requisite variety of perspectives and enabling collaborative sensemaking.

9.2.2 *Gaming*

When taking a closer look at business wargaming, it makes sense to consider the gaming part. Video games have become a revenue generator in the same league as films and books (McGonigal 2012; Isbister 2016). Looking at the mechanics of video games in respect to how they emotionally engage their players, not only sheds light on why business wargames work, but also what needs to be considered in designing business wargames to leverage their effects. This emotional activation which attributes to successful computer and video games (McGonigal 2012), is exactly what organisation and its leaders try to achieve: engaging managers and members of an organisation in discussions on strategy, the future, change, or transformation.

The engagement of managers with their organisation is key for the successful transformation of an organisation. I have argued that the emotional engagement of participants in a business wargame is a prerequisite for sustainable learning. Isbister (2016) argues, from the perspective of video game design, that well-designed games engage players in a flow-state. Oriesek and Schwarz (2020) argue that the idea of a flow-state, describing the zone in which a person is fully immersed in an activity, explains what happens in well-designed business wargames. Isbister (2016) refers to the following factors contributing to flow-state in gaming, here related to business wargaming (Oriesek and Schwarz 2020).

1. A challenging activity requiring skill: A business wargame should address a difficult question for the participants to answer. This implies that participants must apply a range of skills to develop solutions for a special situation which is not a typical business problem.
2. A merging of actions and awareness: A business wargame needs to be structured in a way that allows participants to become aware of a new business situation and then act accordingly, for instance, by introducing a new strategy or product or service.
3. Clear goals: The goal of the simulation clearly needs to be defined: what is expected from participants but also what shall be achieved by running the simulation.
4. Direct, immediate feedback: The competing teams must receive feedback from the market and customer team as soon as possible. Providing this feedback is essential for the dynamics of a business wargaming and forces the team members to rethink their strategy if the feedback from the market is disappointing.
5. Concentration on the task at hand: A business wargame needs to focus on what shall be achieved. If for instance too many topics are covered in one simulation, the danger is that the teams lose focus and participants become disengaged.
6. A sense of control: The control team needs to make sure that the business wargame is evolving within the defined boundaries and that for instance the dynamics of the simulation are at a high level.
7. A loss of self-consciousness: Participants should immerse themselves in their role in the business wargame and to be critical of their own perception on how they think their industry might work.

8. An altered sense of time: A business wargame should be structured so that 'time flies' and participants feel so engaged in the simulation that they lose track of the hours.

9.2.3 General learnings and advantages of business wargaming

Besides the above mentioned learnings and benefits business wargaming provides (see sensemaking caption), one of the many advantages of business wargaming is the involvement of participants:

> The problem with many strategy techniques is that they are too cold and bloodless. They fail to capture human emotions, and because of their icy rational character, people don't really pay attention to them. They are soon forgotten, and they make no lasting impact on the organisation. Gaming is profound learning experience, one that is not soon forgotten.
>
> *(Bracken 2001: 18)*

During a wargame the dynamics of a market and of the competitors will be analysed, anticipated, and experienced, by simulating a likely future over the course of several days. As a wargaming exercise evolves, the participants must think creatively about the future, which will also expose them to early signals of imminent change, which may be relevant for the organisation or industry.

Van der Heijden (1998: 351) summarises the relevance of testing a strategy in scenarios, which is also applicable for testing strategies by use of wargaming:

> Strategy is the art of making choices – investing both for current and future success. To understand these choices clearly, organisations should identify a business idea and test it in substantially different scenarios. This process can help an organisation to develop a business idea that will serve it well as the future evolves.

References

Augier, M., N. Dew, T. Knudsen, and N. Stieglitz. 2018. "Organizational Persistence in the Use of War Gaming and Scenario Planning." *Long Range Planning* 51 (4): 511–525. https://doi.org/10.1016/j.lrp.2017.12.005.

Bellman, R., C. E. Clark, D. G. Malcolm, C. J. Craft, and F. M. Ricciardi. 1957. "On the Construction of a Multi-Stage Multi-Person Business Game." *Operations Research* 5 (4): 469–503.

Bracken, P. 2001. "Business War Gaming." *Scenario and Strategy Planning* 3 (2): 15–18.

Day, G. S., and P. J. H. Schoemaker. 2016. "Adapting to Fast-Changing Markets and Technologies." *California Management Review* 58 (4): 59–77.

Ginter, P. M., and A. C. Rucks. 1984. "Can Business Learn from Wargames?" *Long Range Planning* 17 (3): 123–128.

Gray, B. 1989. *Collaborating: Finding Common Ground for Multiparty Problems.* San Francisco: Joey-Bass.

Hanley Jr., J. T. 1992. *On Wargaming: A Critique of Strategic Operational Gaming.* New Haven: Yale University.

Isbister, K. 2016. *How Games Move Us: Emotion by Design.* Cambridge: MIT Press.

Kotter, J. P. 2012. *Leading Change.* Boston: Harvard Business Review Press.

Krupp, S., and P. J. H. Schoemaker. 2014. *Winning the Long Game: How Strategic Leaders Shape the Future.* New York: Public Affairs.

Kurtz, J. 2003. "Business Wargaming: Simulations Guide Crucial Strategy Decisions." *Strategy and Leadership* 31 (6): 12–21.

Lüchinger, R. 2001. *Der Fall Der Swissair.* Zürich: Bilanz Verlag.

McGonigal, J. 2012. *Reality Is Broken: Why Games Make Us Better and How They Can Change the World.* London: Vintage Books.

Oriesek, D. F., and J. O. Schwarz. 2008. *Business Wargaming: Securing Corporate Value.* Aldershot: Gower.

———. 2020. *Winning the Uncertainty Game: Turning Strategic Intent into Results with Wargaming.* London: Taylor & Francis.

Perla, P. P. 1990. *The Art of Wargaming.* Annapolis: Naval Institute Press.

Schwarz, J. O. 2009. "Business Wargaming: Developing Foresight within a Strategic Simulation." *Technological Analysis and Strategic Management* 21 (3): 291–305.

———. 2011. "Ex-Ante Strategy Evaluation: The Case for Business Wargaming." *Business Strategy Series* 12 (3): 122–135.

———. 2013. "Business Wargaming for Teaching Strategy Making." *Futures* 51: 59–66. http://doi.org/10.1016/j.futures.2013.06.002.

Seidl, D., and F. Werle. 2018. "Inter-Organizational Sensemaking in the Face of Strategic Meta-Problems: Requisite Variety and Dynamics of

Participation." *Strategic Management Journal* 39 (3): 830–858. https://doi.org/10.1002/smj.2723.

Teece, D. J., G. Pisano, and A. Shuen. 1997. "Dynamic Capabilities and Strategic Management." *Strategic Management Journal* 18 (7): 509–533. https://doi.org/10.1002/(sici)1097-0266(199708)18:7<509::aid-smj882>3.0.co;2-z.

Treat, J. E., G. E. Thibault, and A. Asin. 1996. "Wargaming - As a Strategic Tool." *Strategy - Management - Competition* 1996 (3): 46–55.

van der Heijden, K. 1998. "Articulating the Business Idea: The Key to Relevant Scenarios." In *Learning from the Future: Competitive Foresight Scenarios*, edited by R. Fahey and R. M. Randall, 335–351. San Francisco John Wiley & Sons.

Vanderveer, R., and J. Heasley II. 2005. "War Gaming: Exercises in Defending Brand Territory." *MM&M* (May): 68–72.

Zenko, M. 2015. *Red Team: How to Succeed By Thinking Like the Enemy*. New York: Basic Books.

10

DESIGN THINKING

Design thinking can play a vital role in the probing phase of the strategic foresight process. In this section I will explore the potential of design thinking and discuss how foresight can be part of design thinking projects.

Several authors (e.g., von der Gracht, Vennemann, and Darkow 2010; Rohrbeck and Gemünden 2011; Jissink, Schweitzer, and Rohrbeck 2018; Sarpong and Meissner 2021) have linked strategic foresight to innovation management and emphasised the relevance of foresight for innovation. The principal argument is that insights from foresight provide input at and beyond the fuzzy front-end of innovation. If we consider the process of strategic foresight, as described in Chapter 4, we will refer to the phase of probing where insights from foresight are funnelled into innovation management.

To be more specific in our discussion on the interrelationship between foresight and innovation, I refer to design thinking as a process for innovation. While this relationship has already been established (Gordon,

DOI: 10.4324/9781003302735-13

Rohrbeck, and Schwarz 2019), I look at this relationship from the perspective of design thinking and foresight.

Design thinking has been described as an integrative framework that brings together creative and analytic modes of reasoning while accompanied by a process, a set of tools, and techniques (Liedtka 2015). The prevailing definition describes design thinking as a human-centred innovation process, emphasising aspects such as observation, collaboration, fast learning, visualisation of ideas, rapid prototyping, and a mix of analytical and intuitive thinking (Micheli, Wilner, Bhatti, Mura, and Beverland 2019). Broadly speaking, design thinking consists of three process steps: inspiration, ideation, and implementation (Seidel and Fixson 2013).

Design thinking and new product development are inevitably concerned with the future (Gordon, Rohrbeck, and Schwarz 2019). Specifically, this is the case if design thinking is applied to future challenges, customer needs, and emerging technologies. Yet, design thinking's inspiration phase is anchored firmly in the present. Based on an empirical investigation of 302 design thinkers (Schwarz 2019) assess the use of foresight methods in design thinking projects.

Schwarz (2019) finds that design thinkers use specific strategic foresight techniques, for instance scenario planning or trend research, in their

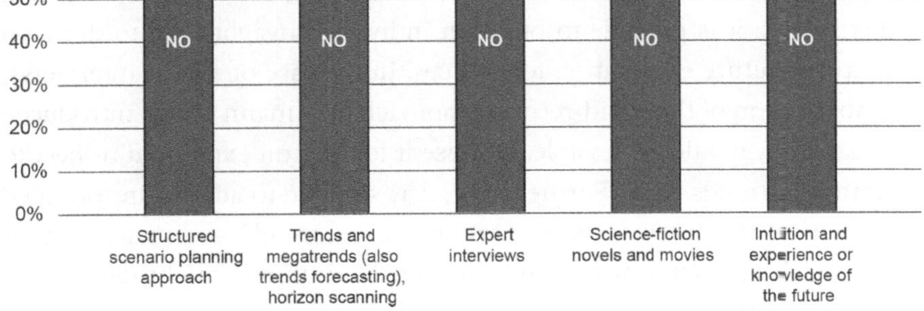

Figure 10.1 Tools used in dealing with the future in design thinking projects (Schwarz 2019).

projects and that these techniques make a major contribution to the success of these projects. Schwarz (2019) suggests that the toolbox of design thinkers should be expanded by including techniques from foresight, such as trend research or scenario planning, and that these should be integrated into design thinking projects. Based on these insights I'm suggesting the Foresight Design Thinking Framework (FDTF), with the aim of combining design thinking with foresight and allowing for a structured approach for incorporating the future into design thinking.

On the basis of the practice at firms such as IDEO, Liedtka (2015: 3) explains that design thinking 'specifies an initial exploratory phase focused on data gathering to identify user needs and define the problem, followed by a second stage of idea generation, followed by a final phase of prototyping and testing…' Seidel and Fixson (2013) have described three elements of the process: needfinding, brainstorming, and prototyping, building models to facilitate the development and selection of concepts. I follow Liedtka's (2015) perspective, focusing on the three stages of exploration, idea generation, and prototyping and testing.

Based on the empirical investigation, Schwarz (2019) identifies two meaningful factors:

1. Individual insights (expert interviews, intuition and experience or knowledge of the future)
2. Structured future exploration (structured scenario planning, trends and megatrends, science-fiction novels, and movies)

As described in Figure 10.2, in the *exploratory phase* in design thinking, in which information is gathered on user needs, both factors are relevant. In other words: it is possible to build on individual insights and to draw on structured future exploration techniques. In the case of expert interviews, the application of the trend-receiver approach (Hofmann 2015), introduced in chapter 5, would be feasible, because it focuses on experts in respect to identifying trends early. Furthermore, it is sensible to add future-oriented perspectives from structured scenario planning, trends, and megatrends as well as science-fiction novels and movies. How trends might develop, as discussed in the processes of foresight (Schwarz 2005; Schwarz, Rohrbeck, and Wach 2020), can be explored in scenarios that allow emerging trends to be linked and probable future schemes to be understood.

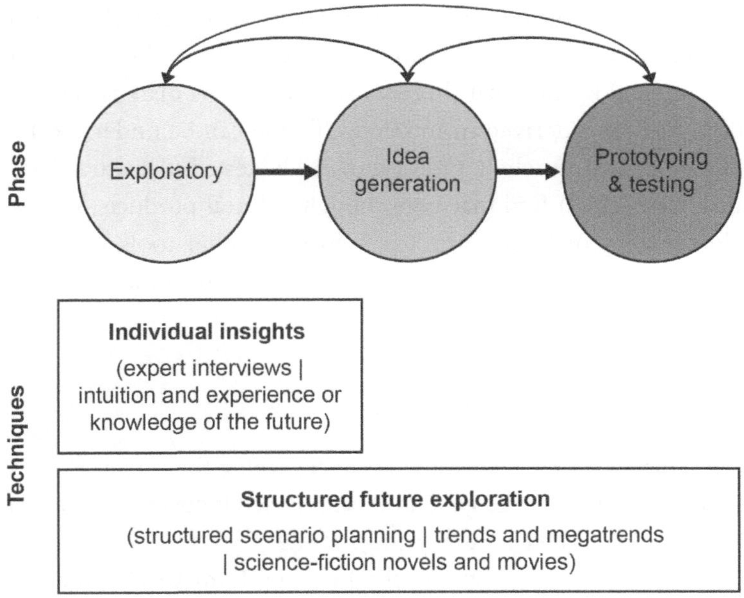

Figure 10.2 Foresight Design Thinking Framework (FDTF).

The factor of structured future exploration continues to be relevant in the *idea generation phase* of design thinking. In this phase, scenario planning can be used as a frame or canvas for innovation. This means that the alternative pictures of the future serve to ideate new products and services. As scenarios and science fiction represent alternative, yet plausible pictures of the future (which are probably distant in time), new products, and services can be ideated in these different futures, meaning also that a large number of ideas can be generated. This notion is in line with the trends and megatrends that are more feasible when compared to the aforementioned techniques but maps a potential direction that is not cast in stone.

In design thinking, new ideas that have been generated eventually need to be *prototyped and tested*. Again, the factor of structured future exploration and, in particular, the scenarios serve as a valuable framework for testing. Testing a prototype within different scenarios might lead to the following result. If the assessment is that a new product or service might fit into multiple scenarios, it is possible to conclude that this is a robust or a future-proof product or service that might develop well in reality. By contrast, if this assessment reveals that a product or service is only suitable for one or a few scenarios, it should be treated with caution. Such outcomes either mean that a new iteration of the design thinking process will be beneficial or, given the detrimental results, that this idea should best be disregarded.

In addition, scenarios derived from science fiction can be understood as a future framework against which prototypes can be tested (Schwarz 2015). Schwarz and Wach (2022) find that even though cultural products, such as science-fiction movies or books, are less used than other tools in design thinking projects, cultural products are positively related to the success of design thinking projects.

References

Gordon, A., R. Rohrbeck, and J. Schwarz. 2019. "Escaping the 'Faster Horses' Trap: Bridging Strategic Foresight and Design-Based Innovation." *Technology Innovation Management Review* 9 (8): 30–42.

Hofmann, R. 2015. "Visionary Competence for Long-Term Development of Brands, Products, and Services: The Trend Receiver Concept and Its First Applications at Audi." *Technological Forecasting and Social Change* 101: 83–98. http://doi.org/10.1016/j.techfore.2014.06.005.

Jissink, T., F. Schweitzer, and R. Rohrbeck. 2018. "Forward-Looking Search during Innovation Projects: Under Which Conditions It Impacts Innovativeness." *Technovation* in press (July). https://doi.org/10.1016/j.technovation.2018.07.001.

Liedtka, J. 2015. "Perspective: Linking Design Thinking with Innovation Outcomes through Cognitive Bias Reduction." *Journal of Product Innovation Management* 32 (6): 925–938. https://doi.org/10.1111/jpim.12163.

Micheli, P., S. J. S. Wilner, S. H. Bhatti, M. Mura, and M. B. Beverland. 2019. "Doing Design Thinking: Conceptual Review, Synthesis, and Research Agenda." *Journal of Product Innovation Management* 36 (2): 124–148. https://doi.org/10.1111/jpim.12466.

Rohrbeck, R., and H. G. Gemünden. 2011. "Corporate Foresight: Its Three Roles in Enhancing the Innovation Capacity of a Firm." *Technological Forecasting and Social Change* 78 (2): 231–243.

Sarpong, D., and D. Meissner. 2021. *Corporate Foresight and Innovation Management.* London: Taylor & Francis Limited.

Schwarz, J. O. 2005. "Pitfalls in Implementing a Strategic Early Warning System." *Foresight* 7 (4): 22–30.

———. 2015. "The 'Narrative Turn' in Developing Foresight Assessing How Cultural Products Can Assist Organisations in Detecting Trends." *Technological Forecasting and Social Change* 90 (Part B): 510–513. http://doi.org/10.1016/j.techfore.2014.02.024.

———. 2019. *The World of the Customer and the Future in Design Thinking: Assessment of the Practice of Design Thinkers.* Munich: Hochschule Fresenius University of Applied Sciences, https://www.researchgate.net/publication/330888489_The_World_of_the_Customer_and_the_Future_in_Design_Thinking_Assessment_of_the_Practice_of_Design_Thinkers.

Schwarz, J. O., and B. Wach. 2022. "The Usage of Cultural Products in Design Thinking: An Assessment of an Underestimated Approach." *The Design Journal* 25 (1): 4–24. https://doi.org/10.1080/14606925.2021.2005879.

Schwarz, J. O., R. Rohrbeck, and B. Wach. 2020. "Corporate Foresight as a Microfoundation of Dynamic Capabilities." *Futures & Foresight Science* 2 (2): e28. https://doi.org/10.1002/ff02.28.

Seidel, V. P., and S. K. Fixson. 2013. "Adopting Design Thinking in Novice Multidisciplinary Teams : The Application and Limits of Design Methods and Reflexive Practices." *The Journal of Product Innovation Management* 30: 19–33.

von der Gracht, H. A., C. R. Vennemann, and I.-L. Darkow. 2010. "Corporate Foresight and Innovation Management: A Portfolio-Approach in Evaluating Organizational Development." *Futures* 42 (4): 380–393.

11

TRANSFORMATION AND OPEN STRATEGY

In this chapter, I discuss how in the phase of transforming, foresight links to change management. However, I want to stress this relationship between transformation and foresight by referring to the concept of open strategy.

If we consider the seminal work of John Kotter (2012) on leading change and the process he formulated (Figure 11.1), two avenues immediately become apparent how foresight plays into this approach. More precisely in step 1 'establish a sense of urgency' and in step 3 'develop a change vision'. Referring to step 1, how can a sense of urgency be established? Scenarios, or involving the management of an organisation into the process of creating scenarios, can create this sense of urgency. This also applies for the other more participative tools, such as the futures literacy lab or business wargaming. Second, creating a change vision (step 3) asks an organisation to create a vision which has the purpose to provide an orientation towards the future. This, however, also implies that an organisation needs to understand how the future could evolve and what possible futures could be.

While in the past strategy processes were an activity for senior managers and their internal or external strategy consultants, and highly confidential,

DOI: 10.4324/9781003302735-14

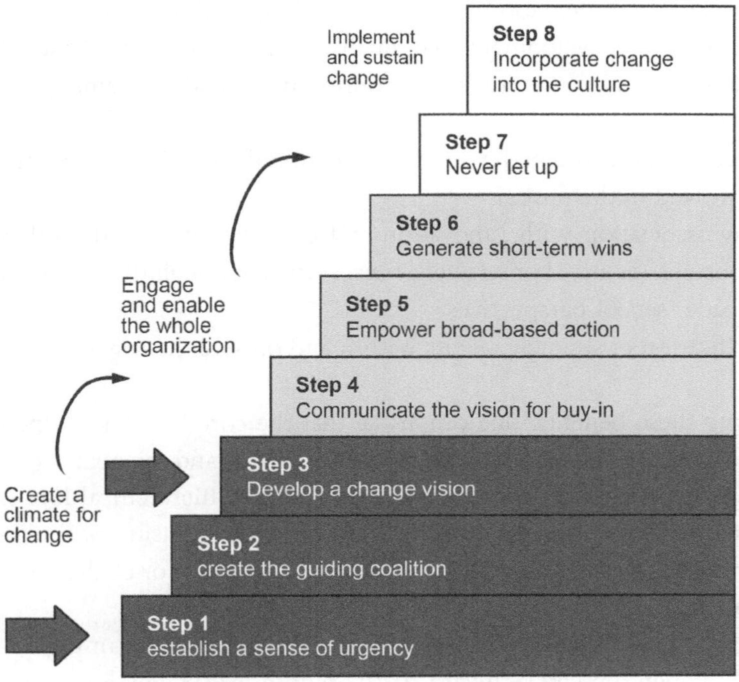

Figure 11.1 Kotter's eight steps to transformation.

open strategy describes approaches that include large groups within an organisation as well as external stakeholders to develop a strategy (Seidl, von Krogh, and Whittington 2019b). Birkinshaw (2017) argues that strategy-making has become much more open, inclusive, and transparent.

The term 'open strategy' was introduced by Whittington, Cailluet, and Yakis-Douglas (2011: 532) (Seidl, von Krogh, and Whittington 2019a) and defined as

> an openness in terms of inclusiveness, in other words, the range of people involved in making strategy; and an openness in terms of transparency, both in the strategy formulation stage and, more commonly, in the communication of strategies once they are formulated.

Seidl, von Krogh, and Whittington (2019a) describe five benefits of open strategy:

1. Greater access to information and knowledge by including a broader set of stakeholders in the process of strategy formulation

2. Improved implementation because including middle managers in the process of development may enable them to better understand the strategy and therefore better communicate and implement the strategy
3. Collective legitimisation by top managers, middle managers, and others involved in the process
4. Strong association with innovation and challenging mental models and business-as-usual or strategic conservatism through the inclusion of broader sets of perspectives
5. Identification of strategic talents within and outside the organisation

Summarising these benefits, one can argue that the core benefits of open strategy are linked to the involvement of its participants and the opening of the process to participants from diverse backgrounds, hierarchical levels, or functions within an organisation. This leads to the inclusion of multiple perspectives with the potential for achieving innovation or challenging mental models.

It has been argued that scenario planning and business wargaming appear useful in open strategy (Schwarz 2020). In both scenario planning and business wargaming, interactive workshop settings enable the future to be explored. While in scenario planning, the focus is on scanning the external environment of a business or industry and exploring alternative plausible pictures of the future, business wargaming focuses on understanding how the competitive landscape of an industry will change in the future. I have mentioned that van der Heijden (1996) referred to scenario planning as the 'art of strategic conversations', and one can argue that at the core of both approaches, participants make sense of their business environment by including the perspectives of those involved.

Several authors (e.g., Schoemaker 1995; van der Heijden 1996; van der Heijden, Bradfield, Burt, Crains, and Wright 2002; Chermack 2011; Ramirez and Wilkinson 2016) highlight the participatory nature of scenario planning or formulate the participation and involvement of top management as a prerequisite. This participation can be achieved by conducting interviews about the perceived challenges in the business environment, discussing trends and selecting key uncertainties, selecting the two key uncertainties for building a scenario matrix, describing the scenarios, and finally deriving the implications. The key argument for the involvement in this process is to strengthen buy-in into the results and, of course, the actions taken based upon the developed scenarios. Oriesek and Schwarz (2020) deliberately position business wargaming as an approach to change management.

While open strategy addresses this issue of opening the strategy development process to all levels in an organisation, the question of how to facilitate such openness also arises. Experience with scenario planning and business wargaming, however, has shown that these types of workshops allow a good integration of participants from different levels of an organisation. In scenario planning workshops, one assessment is that discussing the future is beyond the daily politics of an organisation and therefore fosters an open exchange of ideas. Additionally, in business wargaming, the competitive element between the teams and the dynamics to win, foster teamwork and weaken the perception of hierarchy.

Overall the argument can be made that scenario planning should be understood not only as a tool for identifying pictures of the future, but first and foremost as an approach that will foster strategic conversations in an organisation (van der Heijden 2005). Scenario planning can also be seen as an organisational intervention, actually facilitating open strategising. In this regard, scenario planning is perceived as a means of opening up strategic discussions and including wider groups of stakeholders both inside and outside an organisation with the aim of achieving greater buy-in to corresponding outcomes (Lehr, Lorenz, Willert, and Rohrbeck 2017). What this also implies is that the usefulness of communicating scenarios via a report, whether written or video, might be limited. While there is of course research suggesting what factors might support scenarios in being accepted or rejected by different audiences (Schmidt-Scheele 2020), the open strategy perspective places the focus on the creation of scenarios, understood as an organisational intervention, not on producing reports. One can also think about creating engagement formats which allow those who were not part of a scenario planning process, to immerse themselves into these future worlds in smaller workshop settings.

Conceptualising strategic foresight as an open strategy bears a lot of potential, especially when the purpose of a foresight activity is to contribute to organisational transformation. While I have focused on scenario planning and business wargaming as tools for open strategy and transformation in this chapter, this is not to be understood as a limitation. Also, the phase of prospecting can be opened to the members of an organisation in searching for and evaluating trends. This also applies to the phases of prospecting and probing, and to tools such as scenario planning, futures literacy laboratories or business wargaming.

References

Birkinshaw, J. 2017. "Reflections on Open Strategy." *Long Range Planning* 50 (3): 423–426. https://doi.org/10.1016/j.lrp.2016.11.004.

Chermack, T. J. 2011. *Scenario Planning in Organizations: How to Create, Use, and Assess Scenarios*. San Francisco: Berrett-Koehler.

Kotter, J. P. 2012. *Leading Change*. Boston: Harvard Business Review Press.

Lehr, T., U. Lorenz, M. Willert, and R. Rohrbeck. 2017. "Scenario-Based Strategizing: Advancing the Applicability in Strategists' Teams." *Technological Forecasting & Social Change* 124: 214–224. https://doi.org/10.1016/j.techfore.2017.06.026.

Oriesek, D. F., and J. O. Schwarz. 2020. *Winning the Uncertainty Game: Turning Strategic Intent into Results with Wargaming*. London: Taylor & Francis.

Ramirez, R., and A. Wilkinson. 2016. *Strategic Reframing: The Oxford Scenario Planning Approach*. Oxford: Oxford Publishing.

Schmidt-Scheele, R. 2020. *The Plausibility of Future Scenarios: Conceptualising an Unexplored Criterion in Scenario Planning*. Science Studies. Transcript Verlag.

Schoemaker, P. 1995. "Scenario Planning : A Tool for Strategic Thinking." *Sloan Management Review* 36 (2): 25–40. https://doi.org/10.1016/0024-6301(95)91604-0.

Schwarz, J. O. 2020. "Revisiting Scenario Planning and Business Wargaming from an Open Strategy Perspective." *World Futures Review* 12 (3): 291–303. https://doi.org/10.1177/1946756720953182.

Seidl, D., G. von Krogh, and R. Whittington. 2019a. "Defining Open Strategy: Dimensions, Practices, Impacts, and Perspectives." In *Cambridge Handbook of Open Strategy*, edited by D. Seidl, G. von Krogh, and R. Whittington, 9–26. Cambridge: Cambridge University Press. https://doi.org/10.1017/9781108347921.002.

———. 2019b. "Introduction." In *Cambridge Handbook of Open Strategy*, edited by D. Seidl, G. von Krogh, and R. Whittington, 1–6. Cambridge: Cambridge University Press. https://doi.org/10.1017/9781108347921.001.

van der Heijden, K. 1996. *Scenarios: The Art of Strategic Conversation*. Chichester: John Wiley & Sons.

———. 2005. *Scenarios: The Art of Strategic Conversation*. Chichester: Wiley.

van der Heijden, K., R. Bradfield, G. Burt, G. Crains, and G. Wright. 2002. *The Sixth Sense*. Chichester: John Wiley & Sons.

Whittington, R., L. Cailluet, and B. Yakis-Douglas. 2011. "Opening Strategy: Evolution of a Precarious Profession." *British Journal of Management* 22 (3): 531–544. https://doi.org/10.1111/j.1467-8551.2011.00762.x.

12

STRATEGIC FORESIGHT AND ITS ANTECEDENTS

Strategic foresight has several antecedents. Although this makes it difficult to define the boundaries of strategic foresight, it points out to several fields for which strategic foresight might have relevance or vice versa. This chapter, however, describes only some of those antecedents, while others have been part of previous chapters (see Figure 12.1).

Chapter 12 discussed innovation as an antecedent. Chapter 4 covered issue management, under the assumptions that trends develop from a weak signal to strong signals or trends and that therefore trends can also be perceived as precursors of issues.

12.1 Strategy and dynamic capabilities

A fundamental question in strategic management is how firms achieve and sustain a competitive advantage (Teece, Pisano, and Shuen 1997). Academics and managers have long tried to understand why some firms survive and even prosper in times of environmental changes and uncertainty while

DOI: 10.4324/9781003302735-15

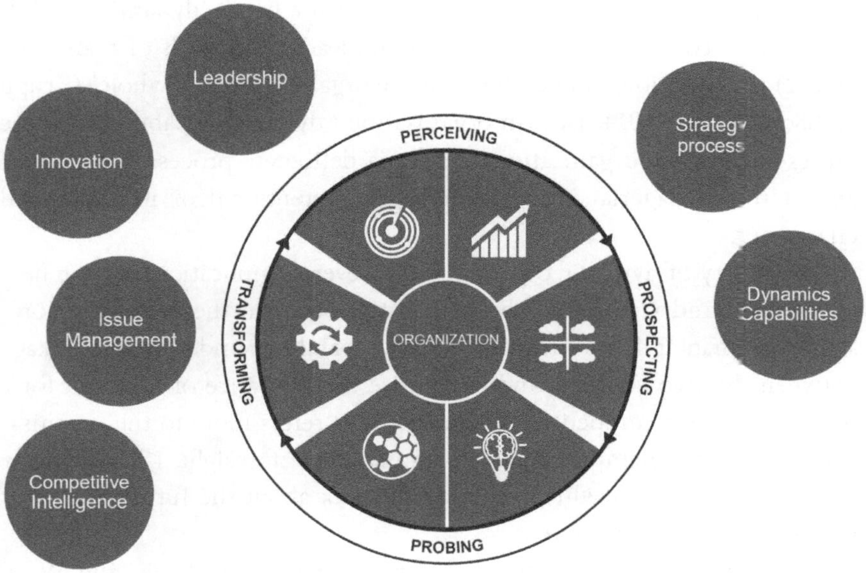

Figure 12.1 Strategic foresight and its antecedents.

others do not (Levinthal 1997; Siggelkow 2002; Birkinshaw, Zimmermann, and Raisch 2016).

In strategic management the theory of dynamic capabilities has gained significant importance since publication of Teece et al.'s (1997) seminal paper. The dynamic capabilities perspective (Teece, Pisano, and Shuen 1997) explains why some firms are adept at anticipating and exploiting opportunities via advances in technology and rapid changes in their market space while others struggle or go out of business. The ways in which firms sense and seize opportunities lie at the core of the dynamic capabilities theory.

Teece, Pisano, and Shuen (1997: 516) suggested that the ability of an organisation to adapt to a changed environment rests on its dynamic capabilities, 'the firm's ability to integrate, build, and reconfigure internal and external competences to address rapidly changing environments'. Their rationale is that the resource-based view has not explained how and why certain firms have a competitive advantage in situations of rapid and unpredictable change. Others argue that dynamic capabilities rest on existing processes such as strategic decision-making, alliances, and research and development (Eisenhardt and Martin 2000).

Teece, Pisano, and Shuen (1997) argued that a firm's 'dynamic capabilities rest on two pillars: (1) the vision and leadership skills of managers, and (2) the cohesion and flexibility of the organisation as a whole' (Krupp and Schoemaker 2014: IX). An organisation's dynamic capabilities can be represented at the organisational level via deliberate processes and tools; and at the managerial level via managers' intentional or unintentional enactment.

The theory of dynamic capabilities has several implications for the field of foresight and vice versa. The similarities between the process of foresight (see Chapter 4) and that of dynamic capabilities underline the linkage between the two concepts and emphasise the relevance of foresight for a firm to achieve a competitive advantage. This refers more to the organisational level of dynamic capabilities and strategic foresight. I have already described the way in which individuals think about the future and what the associated challenges are.

Based on an empirical investigation, Schwarz, Rohrbeck, and Wach (2020) found that corporate foresight practices, which refer to the process of strategic foresight, and the training of individuals in strategic foresight, support dynamic capabilities in a firm. This perception also emphasises that dynamic capabilities build on existing standard processes, such as research and development, alliances, and strategic decision-making, as suggested by Martin and Eisenhardt (Eisenhardt and Martin 2000).

This study (Schwarz, Rohrbeck, and Wach 2020) further suggests that corporate foresight trainings can help leaders to make cognitive leaps by overcoming dominant mental models (Vecchiato, Favato, Maddaloni, and Do 2019). These findings also suggest that while setting up a strategic foresight activity at an organisation it is important to design an appropriate process, and to train managers to address strategic foresight at the organisational and individual levels.

Schwarz, Rohrbeck, and Wach (2020) suggest that individuals be trained in reserving and directing attention towards the periphery and the mid- to long term (Schoemaker 2019); in decision-making under uncertainty (Courtney, Kirkland, and Viguerie 1997); and in reconfiguring resources, a skill also emphasised via the increasing interest in using a business model language to allow firms to survive transitions (Teece, Peteraf, and Leih 2016; Teece 2017; Bidmon and Knab 2018).

12.2 Strategy process

Strategic foresight has the potential to play a vital role in helping a firm to achieve a long-term competitive advantage. One way to do this is to integrate foresight into the strategy process, as used by Royal Dutch Shell's (Figure 12.2) integrating scenarios (Johnson, Scholes, Whittington, and McMillan 2008). Scenarios are developed at a corporate level and then broken down to regional or business unit levels, asking for strategising within the boundaries of the scenarios. The aim of this approach is to begin the strategy process with creating scenarios both to reflect on current trends and to explore alternative pictures as well as to create a basis for decision-makers to choose which future direction an organisation should take.

Here I take a closer look at the ways in which foresight and some of its essential approaches can be incorporated into strategy development. My focus will be on scenario planning and business wargaming. The aim is twofold: (1) to integrate foresight into a strategy process; and (2) to add a competitive perspective to strategising.

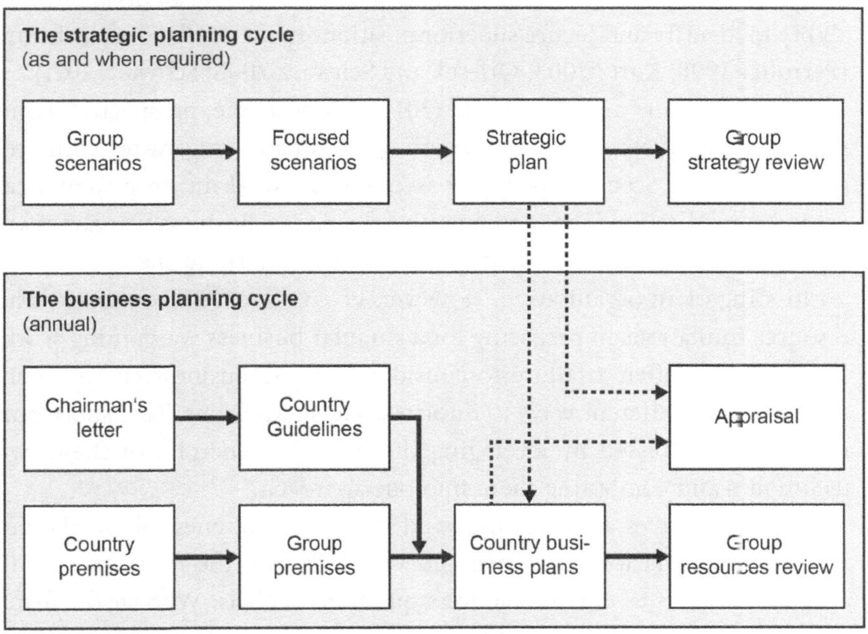

Figure 12.2 Strategy process at Royal Dutch Shell (Johnson, Scholes, Whittington, and McMillan 2008).

Although I have already outlined how scenarios can be incorporated into a strategy progress, the question remains, how to integrate a stronger focus on competitors, or future competitors. In order to identify early warning signals of new rivals (Geroski 1999), it is imperative to understand how an industry might be disrupted and how actors and other stakeholders might behave in different contexts. The scenario technique provides rich insights into a range of futures but leaves management with an ambiguous future space that is difficult to navigate even without the added complexity of potential future moves of competitors (Ramírez and Selsky 2016).

One promising way to address this is to complement scenario analysis with business wargaming. Reibstein and Chussil (1999) have observed that scenario planning does not always capture the impact of competitive dynamics, but that strategy simulations, such as business wargaming, can incorporate insights, for example, from scenario planning. A business wargame takes its appeal from being engaging and from providing a platform for experimenting with different strategic moves and their ability to trigger desired outcomes. Combining the two should allow one to prepare for competition in uncertain environments and thus improve an organisation's ability to identify and secure superior positions in the markets of the future (Perrottet 1998; Kurtz 2003; Oriesek and Schwarz 2008; Schwarz 2011).

Schwarz, Ram, and Rohrbeck (2018) propose the prospective competitive strategy process, which combines business wargaming with scenario planning. Scenario planning is often too weak in proposing clear organisational actions about competitive dynamics. Business wargames are typically directly linked to strategic decisions concerning the competitive positioning of an organisation. However, given the cognitive capacity and resource limitations in preparing for a singular business wargaming workshop, there is often a failure to consider how the business environment may evolve in different ways to inform strategic decisions. This shortcoming can be addressed by leveraging the inherent synergies of these two techniques and combining them into one approach.

Perceiving involves activities designed to create awareness about the environment and factors that have and will influence the focal market. To achieve this, it is essential to address questions such as: Who *are* the firm's competitors? Who *will be* its competitors?

On its own, business wargaming is based on the competitors or stakeholders included in the simulation. Perceiving emphasises that prior to designing a business war game, assumptions about the future course of

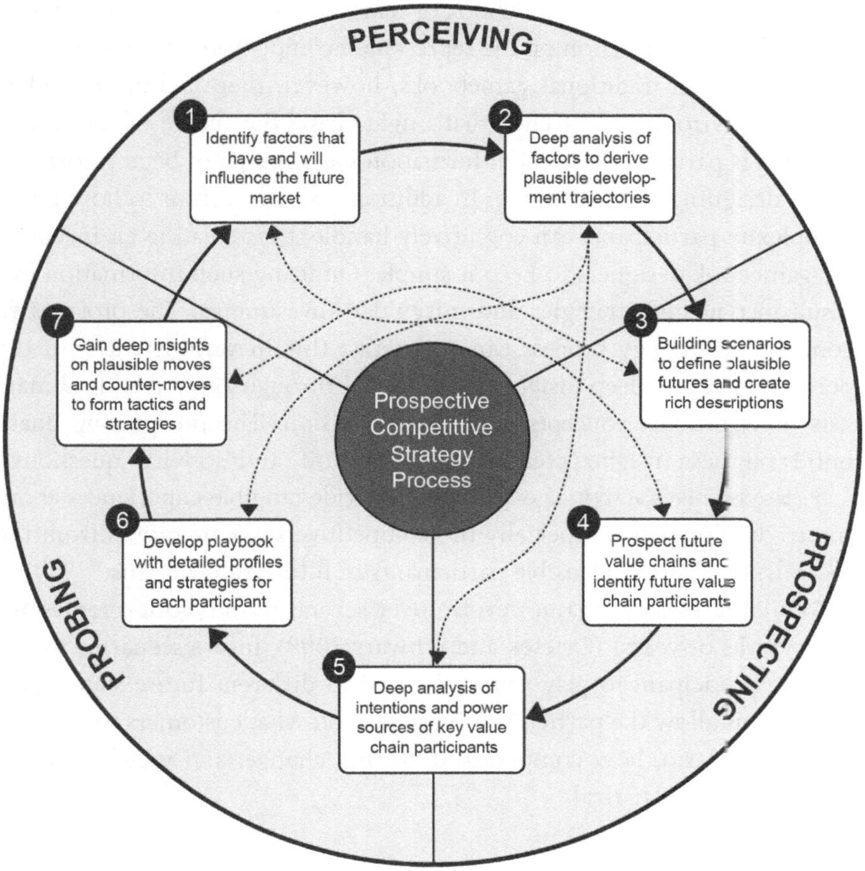

Figure 12.3 Prospective competitive strategy process (Schwarz, Ram, and Rohrbeck 2018).

the industry must be made to include meaningful competitors To do this, the perceiving phase should leverage data collection from several viewpoints (e.g., experience from outside the focal industry, other geographical regions, and lead markets) and different time horizons. Here, firms in fast-moving environments should also consult experts from long-term oriented industries such as the building industry or energy industry, where architects and energy planners must form opinions about future states in a 20- to 40-year time frame.

Prospecting involves activities designed to uncover systemic effects and project developments to define and describe plausible future states and their effects on the focal firm. This requires setting the simulation in an appropriate context. The gamebook or playbook is used to set this context, and contains information on the represented competitors, stakeholder, and the industry. In traditional gamebooks, however, there is little consideration of environmental trends that might shape the future of the industry. This is partly because the information has simply not been prioritised when designing the gamebook. In addition, there is a limit to how much complexity participants can cognitively handle, thus creating an incentive for gamebook designers to keep it simple. Omitting such information can result in outdated strategies and misguided investments. The prospective competitive strategy process can counteract this in various ways. In the perceiving phase, deep insights are collected through analysis of lead markets, new product concepts, and product vision. The prospecting phase builds on these insights, creating sets of 'what if' and 'so what' questions.

Probing involves activities designed to provide tangible experiences about future developments, especially the competitive strategic moves from the focal firm and other plausible participants in future value chains.

While business wargames evolve over several rounds (often representing months or years) (Oriesek and Schwarz 2008), future scenarios would allow a participant to play rounds located in different future states. This would thus allow the participant to understand what customers value most in each scenario, how competitor dynamics change, and which scenario favours a particular rival.

12.3 Competitive intelligence

Competitive intelligence has been perceived as an activity primarily concerned with analysing the competitors of an organisation and as one that considers the environments of that organisation. This means that competitive intelligence professionals must focus on their competitors, on their organisation's stakeholders, and on developments in their environment.

Competitive intelligence is rooted in governmental intelligence agencies, management studies, and market research (Michaeli 2006). Competitive intelligence emerged as a part of market research in the 1970s (Lux and Peske 2002). However, the field experienced a boost with Michael Porter's (1980) *Competitive Strategy: Techniques for Analyzing Industries and Competitors.* The focus of competitive intelligence shifted from an activity concerned primarily with the competitors, to one focusing on the organisational environment. This development is reflected in the definition of 'competitive intelligence' by Society of Competitive Intelligence Professionals, cited by Lux and Peske (2002: 27):

> Competitive intelligence is the process of monitoring the competitive environment. Competitive intelligence enables senior managers... to make informed decisions about everything from marketing, R and D, and investing tactics to long-term business strategies. Effective competitive intelligence is a continuous process involving the legal and ethical collection of information analysis....

A wide variety of analytical tools (Lux and Peske 2002) are used in competitive intelligence activities. However, foresight, in general, is of relevance to the field of competitive intelligence and tools like business wargaming (Chapter 9). Further, the process of strategic foresight (Chapter 3), which also has been labelled as strategic early warning process or system (Schwarz 2005), seems to be of interest for competitive intelligence (Kurtz 2003; Gilad 2004). While business wargaming has been discussed in Chapter 9 and its relevance with the focus on competition for the field of competitive intelligence is obvious, we discuss foresight and its relevance for competitive intelligence

Gilad (2004) describes the implementation of competitive early warning systems as a key component of competitive intelligence. The goal of such a system is to detect weak signals of change before they escalate to a full crisis, or to detect opportunities before competitors do (Fuld 2003). Such systems are meant to provide decision-makers with foresight and insight (Bernhardt 2003). This activity is labelled described here as an early warning system (Tessun 1997; Fuld 2003) or as strategic early warning (Bernhardt 2003; Schwarz 2005).

This entailed that strategic early warning and foresight have the same purpose: to deal with the future or with environments that are turbulent and difficult to predict (Schwarz 2007). In addition, both concepts share an interest in trends. While strategic early warning systems are designed to detect trends, foresight is concerned with creating alternative futures or gaining insight into the future by using trends. Strategic early warning systems are directed more towards a real-time activity, with a diagnosis phase, with the aim of understanding the phenomena.

Linking strategic early warning and competitive intelligence to the process of strategic foresight as discussed in Chapter 4, one can argue that this activity is geared towards the phase of perceiving. However, competitive intelligence provides essential information for instance for designing a business wargame and can therefore enhance foresight in relation to an organisation's competitive landscape.

12.4 Leadership

The link between strategic foresight and leadership is not yet well established. In strategic thinking (Schoemaker, Krupp, and Howland 2013; Krupp and Schoemaker 2014, 2015), the ability to sense changes in the business environment has been pinpointed as a valuable leadership skill. Schoemaker (2019) emphasised that individual leaders need to direct attention to the periphery and ensure sufficient mental slack. Mackay and Burt (2015) described foresight as a means of strategic learning and as a creative process.

In this section I cite research on robust leadership (Schwarz, Wach, von Held, and Werle 2021). The robust leadership framework identifies the challenges leaders need to address and how, and which mechanisms enable them to do so, particularly in VUCA environments. Schwarz, Wach, von Held, and Werle (2021) argue that the need for leaders to provide orientation is key. How can leaders provide orientation in VUCA environments? The answer is in two components of the robust leadership framework:

What vision do leaders have? What purpose should their organisation fulfil? Why does it matter? Leaders need to provide orientation in terms of what the organisation needs to achieve. In this sense, leadership depends on strategic thinking. It is a leader's task to visualise how the organisation's future can be different, to challenge assumptions about how to do business or how the industry operates and to consider new ways of going forward. In making insights and assumptions explicit, leaders define the organisation's vision and tangible goals. This is essential for strategy-making.

Providing orientation on *how* members of an organisation can work together to achieve its *what*. This question relates to the values and more generally to the culture of an organisation and how these are enacted and experienced. Leaders also need to provide orientation in terms of how the organisation will achieve its goals. In other words: 'How do we need to

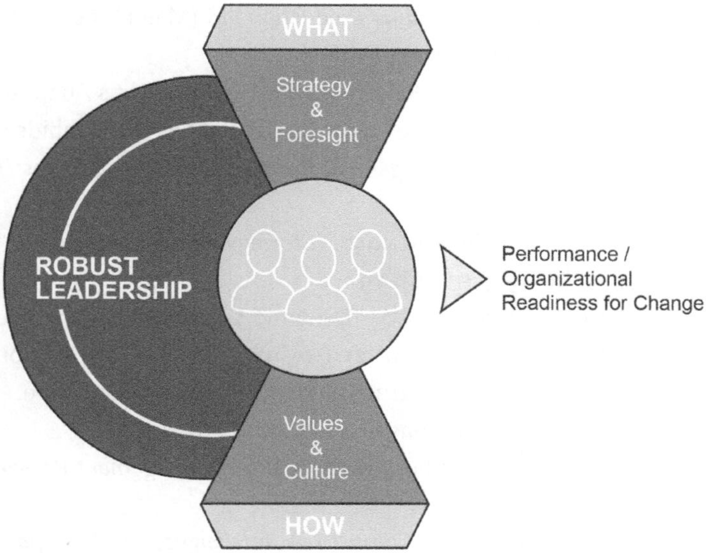

Figure 12.4 Robust leadership framework (Schwarz, Wach, von Held, and Werle 2021).

work together to be successful? Why do we need to change and adapt?' How an organisation pursues its goals in practice relates to the organisation's values and how its members enact and experience them. Organisational values are essential for collaboration across hierarchical levels and functions and to define the organisation's culture.

Schwarz, Wach, von Held, and Werle (2021) found in their empirical investigation that about 20 per cent of the respondents described the leadership of their company as robust and 10 per cent described it as fragile. Around 70 per cent assessed their leadership as neither good nor bad. This category of respondents was labelled as 'stuck in the middle'. Developing foresight capabilities in an organisation on an organisational and individual level has the potential to enable robust leadership.

References

Bernhardt, D. 2003. *Competitive Intelligence: How to Acquire and Use Corporate Intelligence and Counter-Intelligence*. London: Prentice-Hall.

Bidmon, C. M., and S. F. Knab. 2018. "The Three Roles of Business Models in Societal Transitions: New Linkages between Business Model and

Transition Research." *Journal of Cleaner Production* 178 (March): 903–916. https://doi.org/10.1016/j.jclepro.2017.12.198.

Birkinshaw, J., A. Zimmermann, and S. Raisch. 2016. "How Do Firms Adapt to Discontinuous Change? Bridging the Dynamic Capabilities and Ambidexterity Perspectives." *California Management Review* 58 (4): 36–58. https://doi.org/10.1525/cmr.2016.58.4.36.

Courtney, H., J. Kirkland, and P. Viguerie. 1997. "Strategy Under Uncertainty." *Harvard Business Review* 1997 (11): 67–79.

Eisenhardt, K. M., and J. A. Martin. 2000. "Dynamic Capabilities: What Are They?" *Strategic Management Journal* 21 (10–11): 1105–1121. https://doi.org/10.1002/1097-0266(200010/11)21:10/11<1105::aid-smj133>3.0.co;2-e.

Fuld, L. 2003. "Be Prepared." *Harvard Business Review* 81 (11): 20–21.

Geroski, P. A. 1999. "Early Warning of New Rivals." *Sloan Management Review* 40 (3): 107–116.

Gilad, B. 2004. *Early Warning: Using Competitive Intelleignce to Anticipate Market Shifts, Control Risk, and Create Powerful Strategies.* New York: AMACOM.

Johnson, G., K. Scholes, R. Whittington, and K. McMillan. 2008. *Exploring Corporate Strategy: And How to Write Dissertations and Project Reports: Text and Cases.* London: Pearson Education.

Krupp, S., and P. J. H. Schoemaker. 2014. *Winning the Long Game: How Strategic Leaders Shape the Future.* New York: Public Affairs.

———. 2015. "The Power of Asking Pivotal Questions." *MIT Sloan Management Review* 56 (2): 39–47.

Kurtz, J. 2003. "Business Wargaming: Simulations Guide Crucial Strategy Decisions." *Strategy and Leadership* 31 (6): 12–21.

Levinthal, D. A. 1997. "Adaptation on Rugged Landscapes." *Management Science* 43 (7): 934–950.

Lux, C., and T. Peske. 2002. *Competitive Intelligence Und Wirtschaftsspionage.* Wiesbaden: Gabler.

Mackay, D., and G. Burt. 2015. "Strategic Learning, Foresight and Hyperopia." *Management Learning* 46 (5): 546–564. https://doi.org/10.1177/1350507614549120.

Michaeli, R. 2006. *Competitive Intelligence.* Berlin: Springer.

Oriesek, D. F., and J. O. Schwarz. 2008. *Business Wargaming : Securing Corporate Value.* Aldershot: Gower.

Perrottet, C. M. 1998. "Testing Your Strategies in Scenarios." In *Learning from the Future*, edited by L. Fahey and R. M. Randall, 122–139. San Francisco: John Wiley and Sons.

Porter, M. E. 1980. *Competitive Strategy – Techniques for Analyzing Industries and Competitors*. New York: Free Press.

Ramírez, R., and J. W. Selsky. 2016. "Strategic Planning in Turbulent Environments: A Social Ecology Approach to Scenarios." *Long Range Planning* 49 (1): 90–102. https://doi.org/10.1016/j.lrp.2014.09.002.

Reibstein, D. J., and M. J. Chussil. 1999. "Putting the Lesson before the Test: Using Simulation to Analyze and Develop Competitive Strategies." *Competitive Intelligence Review* 10: 34–48.

Schoemaker, P. J. H. 2019. "Attention and Foresight in Organizations." *Futures & Foresight Science* 1:e5. https://doi.org/10.1002/ffo2.5.

Schoemaker, P. J. H., S. Krupp, and S. Howland. 2013. "Strategic Leadership: The Essential Skills." *Harvard Business Review* (January–February), https://hbr.org/2013/01/strategic-leadership-the-esssential-skills.

Schwarz, J. O. 2005. "Pitfalls in Implementing a Strategic Early Warning System." *Foresight* 7 (4): 22–30.

———. 2007. "Competitive Intelligence: A Field for Futurists?" *Futures Research Quarterly* 23 (1): 55–65.

———. 2011. "Ex-Ante Strategy Evaluation: The Case for Business Wargaming." *Business Strategy Series* 12 (3): 122–135.

Schwarz, J. O., B. Wach, F. von Held, and F. Werle. 2021. *A Study on Robust Leadership: Addressing Key Tensions*. Munich: Institute for Innovation and Change Methodologies.

Schwarz, J. O., C. Ram, and R. Rohrbeck. 2018. "Combining Scenario Planning and Business Wargaming to Better Anticipate Future Competitive Dynamics." *Futures* 105: 133–142. https://doi.org/10.1016/j.futures.2018.10.001.

Schwarz, J. O., R. Rohrbeck, and B. Wach. 2020. "Corporate Foresight as a Microfoundation of Dynamic Capabilities." *Futures & Foresight Science* 2 (2): e28. https://doi.org/10.1002/ffo2.28.

Siggelkow, N. 2002. "Evolution toward Fit." *Administrative Science Quarterly* 47 (1): 125. https://doi.org/10.2307/3094893.

Teece, D. J. 2017. "Business Models and Dynamic Capabilities." *Long Range Planning* (July). https://doi.org/10.1016/J.LRP.2017.06.007.

Teece, D. J., G. Pisano, and A. Shuen. 1997. "Dynamic Capabilities and Strategic Management." *Strategic Management Journal* 18 (7): 509–533. https://doi.org/10.1002/(sici)1097-0266(199708)18:7<509::aid-smj882>3.0.co;2-z.

Teece, D., M. Peteraf, and S. Leih. 2016. "Dynamic Capabilities and Organizational Agility: Risk, Uncertainty, and Strategy in the Innovation Economy." *California Management Review* 58 (4): 13–35. https://doi.org/10.1525/cmr.2016.58.4.13.

Tessun, F. 1997. "Scenario Analysis and Early Warning Systems at Daimler-Benz Aerospace." *Competitive Intelligence Review* 8 (4): 30–40.

Vecchiato, R., G. Favato, F. Maddaloni, and H. Do. 2019. "Foresight, Cognition, and Long-term Performance: Insights from the Automotive Industry and Opportunities for Future Research." *Futures & Foresight Science* 1 (3–4): 1–13. https://doi.org/10.1002/ffo2.25.

Part III

STRATEGIC FORESIGHT IN PRACTICE

13

STRATEGIC FORESIGHT

AN EMERGING FIELD

Looking back on five decades of research on corporate foresight, Rohrbeck, Battistella, and Huizingh (2015) describe strategic foresight as an emerging field with a rich tradition. Describing the state of strategic foresight this way captures the status of strategic foresight. While strategic foresight has been around for decades, it appears that the interest in strategic foresight is growing. The factors supporting this rise in interest are the increasing awareness of VUCA or TUNA environments in addition to the perception that more traditional or static planning and strategy tools are not suited for the business environment.

There is a growing awareness that simple approaches are inadequate for a complex business environment. Of course, a barrier to the implementation of strategic foresight is the complexity of its application. Designing and facilitating an intervention which challenges the mental models of leadership teams and allows for strategic conversations on trends in the business environment and probable future scenarios, is a complex undertaking.

However, the intention in dealing with the changes in the environment of a firm is to reduce complexity, this means that complexity needs to be

DOI: 10.4324/9781003302735-17

increased before it can be meaningfully reduced without putting an organisation at risk of increasing its blind spots.

In the following, I look at the practice of strategic foresight from different angles. I consider the value contribution of strategic foresight. Then I examine the evidence on the proliferation of strategic foresight.

13.1 Value contribution of strategic foresight

The issue of the value contribution of strategic foresight is discussed from two angles: how and in what ways strategic foresight can create value in an organisation; and how strategic foresight affects the value of a firm.

13.1.1 How strategic foresight creates value

Based on a literature review and an empirical investigation Rohrbeck and Schwarz (2013) assess the value of strategic foresight for an organisation.

Strategic foresight delivers value to an organisation by developing mechanisms that help to detect weak and interpret signals, then respond to them. Therefore, the first value proposition is to gain insights into environmental changes. Given that the aim of strategic foresight should be to scan for the unknown, the following potential value contribution is proposed: Contributing to a reduction of uncertainty (e.g., through identification of disruptions).

I have discussed the advantages of scenario planning in this book. Scenarios contribute to management's ability to take future-oriented decisions and initiate the configuration and deployment of the firm's strategic resources (Danneels 2011). Scenarios also serve as shared mental images that build emotional capacity. Creating emotional capacity is vital for driving internal change, particularly radical change (Huy 1999). In this respect, the value creation of strategic foresight would be to integrate key stakeholders in the process of strategy formation.

Strategic foresight as a vehicle for strategic conversation involves both value contribution and learning. A strategic foresight process contributes to the learning organisation (Senge 1990) and is a learning process itself (Costanzo 2004). Therefore, the value contribution of strategic foresight is seen in fostering conversations about strategy. It is not enough for a firm to perceive change and develop a strategy to deal with it; the firm also must be able to adapt to change. Further, a value contribution of strategic foresight consists in supporting the adjustment of the company when faced with uncertainty.

Moving from perceiving to responding to external change means that firms need to develop plans and orchestrate actions. Rohrbeck and Schwarz (2013) thus conclude that another potential value contribution of strategic foresight is improving the coordination of business objectives. Another potential value contribution of strategic foresight concerns broadening the scope of perceived alternatives and taking different perspectives to make better strategic choices. Strategic foresight exercises can contribute value by creating the ability to adopt alternative perspectives.

It has been argued that a strategic foresight process can incorporate creativity and innovation and that there is a connection between innovation and strategic foresight. Strategic foresight can support new product development in three ways: (1) exploring new innovation fields; (2) identifying promising innovations; and (3) challenging research and development teams throughout the development of new products. This leads to the following value contribution of strategic foresight: Strategic foresight reduces uncertainty in R&D projects.

Searching for non-articulated needs can be seen as exploring either emerging or future needs, this can be a task of strategic foresight. Strategic foresight could contribute by enhancing the understanding of customer needs. In addition, firms report that they use strategic foresight to identify new customer groups, either for current or future products and services. Strategic foresight will contribute to identifying potential customers, to enhancing the understanding of the market, and to identifying opportunities and threats to the firm's product and technology portfolio.

On a more generic level, the role of an organisation's strategic foresight process organisation is to prepare for the future. Strategic foresight can contribute to an organisation by facilitating organisational learning. Strategic foresight could play an enabling role in systemic innovations, particularly where multiple actors need to work together to create a market or an industry. Thus, strategic foresight contributes value by influencing others to act.

Based on an empirical investigation of 77 European-based corporations, Rohrbeck and Schwarz (2013) assess whether and to what degree these value contributions can be found in practice. All individual value contributions have been confirmed by at least a few firms. This means that, if

Group	Potential value of strategic foresight
Perception	• Gaining insights into changes in the environment • Contributing to a reduction of uncertainty (e.g. through identification of disruptions)
Interpretation and usage (for strategic management)	• Fostering conversations about the overall strategy of the company • Supporting the adjustment of the company in situations of uncertainty • Improving the coordination of business objectives • Creating the ability to adopt alternative perspectives
Interpretation and usage (for innovation management)	• Reducing the level of uncertainty in R&D projects • Enhancing the understanding of customer needs • Identifying potentail customers • Enhancing the understanding of the market • Identification of opportunities and threats regarding our product and technology portfolio
Overall	• Facilitate organizational learning • Shaping the future (e.g. through influencing other parties, such as politics and other companies)

Figure 13.1 Overview value contributions of strategic foresight activities (Rohrbeck and Schwarz 2013).

formalised strategic foresight activities are implemented, then a firm can expect: (1) an enhanced perception, (2) an enhanced ability to interpret change, and (3) an enhanced ability to propose responses, and an enhanced capacity for organisational learning and influencing others. The highest figures were for firms, which report an enhanced perception, either through gaining insights into changes in the environment, or reducing uncertainty. The phase of perceiving in the strategic foresight process is the one that is serviced the best; the others have room for improvement.

13.1.2 *Impact of strategic foresight on firm performance*

Paramount for describing the linkage of strategic foresight to the performance of a firm is the work of Professor René Rohrbeck from the EDHEC Business School in Lille, France (Rohrbeck and Kum 2018; Rohrbeck, Kum, Jissink, and Gordon 2018).

Rohrbeck and Kum (2018) developed a model that judges a firm's future preparedness by assessing the need for corporate foresight in respect to the perceived dynamics in the industry and by comparing it to the maturity of its corporate foresight activities. Corporate foresight activities were measured along five dimensions:

1. Information usage: the information which is collected
2. Method sophistication: methods used to interpret the information
3. People and networks: characteristics of individual employees and networks used by the organisation to acquire and disseminate information on change (Subsequent iterations of the maturity model split these items)
4. Organisation: how information is gathered, interpreted, and used in the organisation
5. Culture: the extent to which the corporate culture supports organisation foresight

A longitudinal research design was applied to measure future preparedness in 2008 and its impact on firm performance in 2015. Surveyed firms were clustered into four groups: vigilant, neurotic, vulnerable, and in danger.

As described in Figure 13.2, Rohrbeck and Kum's (2018) findings show that vigilant firms achieved, on average, 16 per cent profitability, surpassing the industry average profitability of 12 per cent, and made vigilant firms 33 per cent more profitable than the average. The value of future preparedness is even more apparent when looking at the discounts that the firms with deficiencies needed to assume. Neurotic and vulnerable firms have a 37 per cent lower profitability than that of vigilant firms. In-danger firms realised a 44 per cent lower profitability.

Rohrbeck and Kum's (2018) findings suggest that strategic foresight activities, designed and implemented in the right balance in respect to the dynamics in the business environment, lead to performance improvements. Interesting also is that this research reveals that foresight activities might be at the right level in the business environment. Rohrbeck and Kum (2018) found evidence that some firms are too concerned with their strategic foresight activities (neurotic) while others are not concerned enough. Schwarz (2019) discussed a similar effect in the fashion industry.

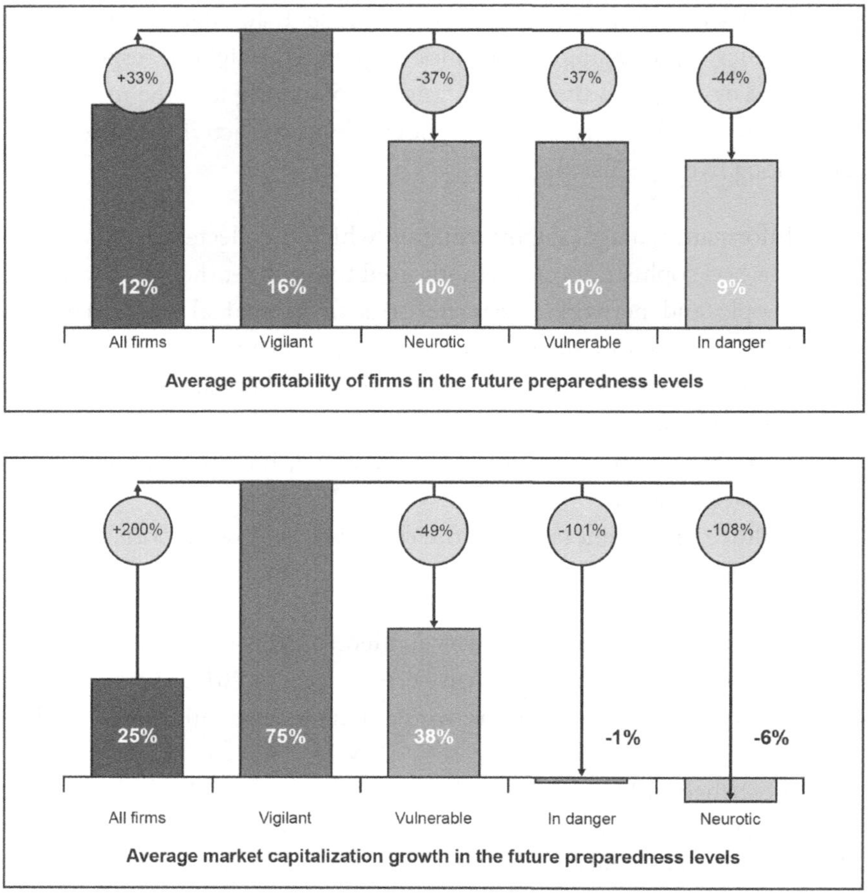

Figure 13.2 Effects of strategic foresight activities on firm performance (Rohrbeck, Kum, Jissink, and Gordon 2018).

13.2 Proliferation of strategic foresight

Here I highlight two studies (Schwarz 2008; Schwarz, Rohrbeck, and Wach 2020) on the proliferation of strategic foresight. The more recent study (Schwarz, Rohrbeck, and Wach 2020), takes a closer look at the stage of strategic foresight and the challenges that arise here. This discussion will also lead to the following chapter on applying strategic foresight.

Schwarz (2008) studied the future of strategic foresight in large German corporates. He cites the examples of German corporates and their strategic foresight activities, at Daimler, Deutsche Bank, BASF, Siemens, Deutsche Bahn, Lufthansa, BMW, Allianz Group, Deutsche Post, EADS, Siemens, Volkswagen, Munich Reinsurance Group, and Deutsche Telekom. While many of these activities have changed over the years, it underlines the fair assessment that the majority of large corporations have implemented strategic foresight activities.

Based on the respondents of large German corporates, defined corporations with an annual turnover of at least €500 million and with more than 10,000 employees, an assessment on their usage of strategic foresight approaches can be made.

Figure 13.3 shows that the approaches used the most are environmental scanning, trend monitoring, trend research, strategic early warning, and the scenario technique, followed by quantitative forecasting and other methods for thinking creatively about the future.

The survey (Schwarz 2008) also investigated how strategic foresight is carried out in terms of its organisational setup. Thirty-six per cent of the respondents reported that strategic foresight is carried out by single employees, followed by 28 per cent in an own department. Only 13 per cent stated that strategic foresight is an integral part of the planning process or that external consultants (17 per cent) are brought in. The results further revealed that 56 per cent of strategic foresight is carried out in departments

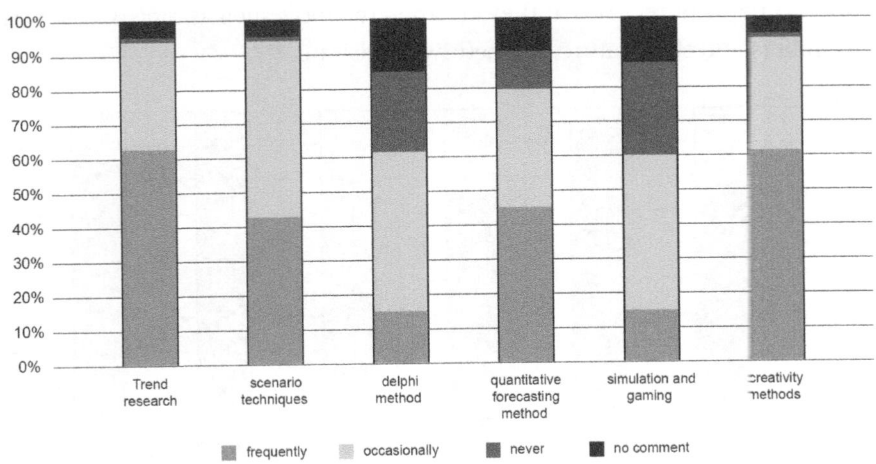

Figure 13.3 Methods used strategic foresight, large German corporates (Schwarz 2008).

such as strategy, planning, corporate development, or marketing and only 23 per cent in technology-related departments.

The study therefore concludes that strategic foresight in the future is likely to increase in relevance.

In a more recent study (Schwarz, Rohrbeck, and Wach 2020) the foresight activities among firms of the Dow Jones Sustainability World Index were assessed. The DJSI World consists of the top 10 per cent of companies from a sustainability perspective per industry, based on the largest 2,500 companies in the Dow Jones Global Stock Market Index. In the 2015 DJSI World, used for this empirical analysis, 317 companies were listed.

While the assessment could be made that all these firms exercise strategic foresight activities, they appear to do so to different degrees. The assessment (Figure 13.4) is along the strategic foresight process, already introduced in this book.

In the *perceiving phase* only 22 per cent of the surveyed firms state that their *scanning* activities ensure that 80 per cent of all trends that will shape their industry in the next five to ten years are detected. Only 4 per cent in the survey state that they are using cultural products, such as *science fiction* novels or movies as a source in their scanning activities.

In respect to the *prospecting* phase, 25 per cent state that they used approaches such as *scenario planning*. In respect to the probing phase, however, only 15 per cent report that they are applying tools such as *business wargaming* to understand their future competitive landscape.

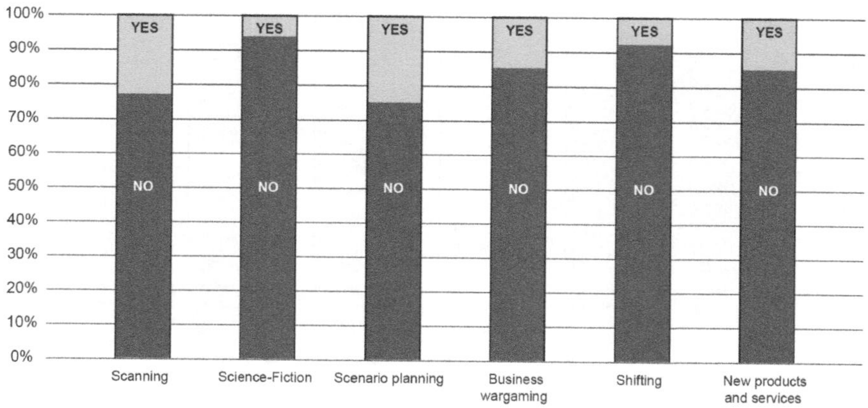

Figure 13.4 Assessment of strategic foresight practices (Schwarz 2019).

Concerning the *transforming phase*, only 15 per cent of the firms in our sample report that identified trends and/or developed scenarios regularly support them in *shifting* resources of the firm into new strategies. A mere 14 per cent state that longer-term future trends or scenarios are often the starting point for the development of *new products and services*.

While interesting to see that the majority of firms in this sample exercises strategic foresight to some degree, the level of sophistication, summarised in Figure 13.4, calls for improvement. Throughout this book, numerous challenges and the approaches to strategic foresight have been highlighted. In the following chapter I will suggest some ways of working with strategic foresight.

References

Costanzo, L. A. 2004. "Strategic Foresight in a High-Speed Environment." *Futures* 36 (2): 219–235.

Danneels, E. 2011. "Trying to Become a Different Type of Company: Dynamic Capability at Smith Corona." *Strategic Management Journal* 32 (1): 1–31. https://doi.org/10.1002/smj.863.

Huy, Q. N. 1999. "Emotional Capability, Emotional Intelligence, and Radical Change." *Academy of Management Review* 24 (2): 325–345. https://doi.org/10.5465/amr.1999.1893939.

Rohrbeck, R., C. Battistella, and E. Huizingh. 2015. "Corporate Foresight: An Emerging Field with a Rich Tradition." *Technological Forecasting & Social Change* 101: 1–9. https://doi.org/10.1016/j.techfore.2015.11.002.

Rohrbeck, R., and J. O. Schwarz. 2013. "The Value Contribution of Strategic Foresight: Insights from an Empirical Study of Large European Companies." *Technological Forecasting and Social Change* 80 (8) 1593–1606. http://doi.org/10.1016/j.techfore.2013.01.004.

Rohrbeck, R., and M. E. Kum. 2018. "Corporate Foresight and Its Impact on Firm Performance: A Longitudinal Analysis." *Technological Forecasting and Social Change* in press (January): 1–30. https://doi.org/10.1016/j.techfore.2017.12.013.

Rohrbeck, R., M. E. Kum, T. Jissink, and A. V. Gordon. 2018. "Corporate Foresight Benchmarking Report 2018: How Leading Firms Build a Superior Position in Markets of the Future." *SSRN Electronic Journal*. Aarhus, Denmark. https://doi.org/10.2139/ssrn.3178562.

Schwarz, J. O. 2008. "Assessing the Future of Futures Studies in Management." *Futures* 40 (3): 237–246.

———. 2019. "Strategy Orientation in The Fashion Industry: Short- Or Long-Term?" *Journal of Futures Studies* 24 (1): 77–90.

Schwarz, J. O., R. Rohrbeck, and B. Wach. 2020. "Corporate Foresight as a Microfoundation of Dynamic Capabilities." *Futures & Foresight Science* 2 (2): e28. https://doi.org/10.1002/ff02.28.

Senge, P. M. 1990. *The Fith Discipline*. London: Century Business.

14

WORKING WITH STRATEGIC FORESIGHT

SEVEN PRINCIPLES

I end this book with principles of working with foresight. These principles are based on my practical experience in working with companies on strategic foresight and on my academic research.

In the context of analysing the strategic foresight process, Schwarz (2005) has used a case study to identify some pitfalls in implementing this process, which is also relevant for our discussion:

1. **Starting with scenarios**: If scenarios are used as a starting point for a strategic foresight process, these scenarios must be alternative pictures of the future, in contrast to variations of a favourable business case. This pitfall limits further scanning and monitoring activities.
2. **The lonely foresight-manager**: Being well networked is essential for the foresight manager, both to detect weak signals in the business environment and to remain abreast of informed about ongoing changes in the organisation. The inclusion of different perspectives is an imperative. It has been argued that a variety of perspectives is non-negotiable in making sense of changes in VUCA-dominated business environments (Seidl and Werle 2018).

DOI: 10.4324/9781003302735-18

3. **A trend is a trend**: Attention must be devoted to conceptualising a trend and how that trend evolves from a weak to a strong signal (Liebl and Schwarz 2010; von Groddeck and Schwarz 2013). If this is not done, there is a risk of missing blind spots, and the organisation will be flooded with irrelevant information.
4. **The quantitative temptation**: The inclusion of quantitative analysis in a strategic foresight process must be done carefully because the future cannot be predicted.
5. **Organisational early warning disability**: Here a critical issue is raised in terms of the purpose of a strategic foresight process in an organisation. Is it about only providing trend input? Is it part of the strategic planning process? Or is it deliberately designed to contribute to changing an organisation? Here the emphasis is on how open and willing an organisation is to challenge its assumptions and mental models about the future of its business or industry. Schwarz (2005) observed several situations in his case where trends critical to the strategy of the organisation were avoided, to prevent conflicts between the foresight manager and senior management. This attempt to avoid or ignore information that does not conform to perceptions is of little to know use to individuals and to organisations. This phenomenon is known as 'cognitive dissonance' (Festinger 1957).

The seven principles of working with strategic foresight, listed here, will be explored in the next chapter.

1. Theory is key
2. Engagement of leadership
3. Starting small
4. Overall engagement
5. Cross-industry collaboration
6. Training of leadership
7. Facilitation skills

14.1 Theory is key

Although it might seem counterintuitive in a corporate setting, it is essential for an organisation to have a sound theoretical basis for doing foresight. In corporate settings, I often hear complaints that 'this is too academic'.

However, I understand this to mean that the people at the organisation either do not understand it or think that it is too complicated.

I enjoyed watching a conversation at MIT between strategy guru Henry Mintzberg and the Ricardo Semler, CEO, and majority owner of Semco Partners, on the usefulness of MBAs. Semler described conversation with a French CEO who had asked a team to solve a problem. The team came back with the successful implementation of a new process that solved the problem. The CEO then said to his managers, 'Well, it seems to work in practice, but does it work in theory?'

This is an elegant twist to the often-asked question 'Will this work in practice?' The question 'does it work in theory?' indicates the capability of reflecting what has been done and why something is working. I need a context to reflect practice; this is where theory comes into play. Reflecting practice in light of theory means creating a deeper understanding of practice and leas to a better understanding of the levers of a practice for its continued improvement.

Developing foresight is a complex task. To do this, organisations are tasked with identifying trends, making sense out of them, creating pictures of and understanding the future, and making decisions. This entails dealing with uncertainty, complexity and challenging the beliefs or mental models of the members of an organisation. Of course, foresight is the exploration of something that does not exist yet: the future. A sound theoretical understanding, as described in this book, is therefore necessary. Further, the growing body of research on foresight allows for multiple insights to understand how to apply foresight or what the challenges might be.

Before engaging with strategic foresight, one needs to understand what the cognitive and organisational challenges are most likely to arise. Further, it is imperative to clarify how trends should be conceptualised and perceived. Applying the tools described in this book, requires an actual understanding of the challenges involved. In this sense, the theoretical backgrounds to foresight discussed here have very practical implications for an organisation's development of foresight.

14.2 Engagement of leadership

From the beginning, the engagement of leadership is critical. In particular, as the outcomes of a strategic foresight process are most likely to be at one point part of the decision-making of a company's leadership team. Further, challenging the mental models of a leadership team and discussing alternative picture of the future is a challenge by itself for leadership teams.

I will now describe one of the greatest lessons that I learned from doing foresight work. As an internal strategy consultant at a large corporation, I had the exciting task of developing scenarios, applying Royal Dutch Shell's scenario planning methodology. Although I knew the literature and had access to training by Kees van der Heijden (Schwarz 2021), I had no exposure to the decision makers who were the ultimate customers of this project. After a lengthy process of developing the scenarios with input from many other stakeholders inside and outside of the organisation, my boss gave a 20-minute presentation of the scenarios to the leadership team. No feedback.

In hindsight, I would argue that a 20-minute presentation would go nowhere without involvement of the leadership team in the project. My master's students and I have had similar experiences when doing scenario planning exercises for corporations. The more engaged the corporate representatives were in the process, the more they could engage with its results.

In contrast, all of my most successful scenario planning exercises have had strong engagement by the leadership of the organisation. However, engagement is not involvement in every step of process or attendance at every workshop. But wisely choosing how to engage leadership either through interviews or through small and well-designed workshops can make a huge difference.

14.3 Starting small

In this book strategic foresight has been predominantly discussed in the context of large corporations. These often have the resources to dedicate entire teams to developing foresight or even have their own think tanks. However, strategic foresight can also be developed on a much smaller scale.

A good starting point might be to discuss the frames of the leadership team in respect to the future of an industry. Establishing a small-scale future literacy laboratory might be a first step. Another approach might be to hold a simple trend workshop in which participants review several trends, while asking questions such as 'do we have a common understanding on what this trend is?' and 'what could this trend imply for our business?'

As a starting point one can refer to freely available trend collections such as Future Agenda (https://www.futureagenda.org) or the Megatrend Hub of the EU Commission (https://knowledge4policy.ec.europa.eu/foresight/tool/megatrends-hub_en). Of course, these high-level trends might not be industry specific, but they might be a good start into having strategic conversations in an organisation.

At this point, I would warn against depending on software tools for strategic foresight, but they can be useful in obtaining buy-in from leadership.

14.4 Overall engagement

As we discussed in Chapter 13, the overall engagement of the members of an organisation is fruitful in many ways. Including many members of an organisation in detecting trends, for instance, both uses the intelligence in an organisation and ensures that multiple perspectives are reflected. At the same time, it creates buy-in into strategic foresight activities.

The same applies in the strategic foresight process for the phase of prospecting or probing. Tools such as the futures literacy laboratories, scenario planning, business wargaming, or design thinking can be easily made available to the members of an organisation, ideally fostering cross-hierarchical and cross-functional collaboration.

And while not widely used, doing parts of the strategic foresight process online or scenario planning interventions online (Schwarz 2020), also has the potential to include large and diverse groups, for instance, by inviting international colleagues without incurring travel costs. Here the COVID-19 pandemic surely has necessitated more strategic foresight work to be conducted online.

It is also possible to identify an implementation gap in strategic foresight (Schwarz, Rohrbeck, and Wach 2020). Organisations are much better at perceiving and prospecting than at probing and transforming. This underlines the argument in this book that foresight needs to be perceived as part of a change journey, rather than an analytical exercise that produces reports on the future.

A last aspect can be mentioned here. The existence of the strategic foresight process might signal to others that the 'future is taking care of' (Schwarz 2009). By creating overall engagement the danger of this

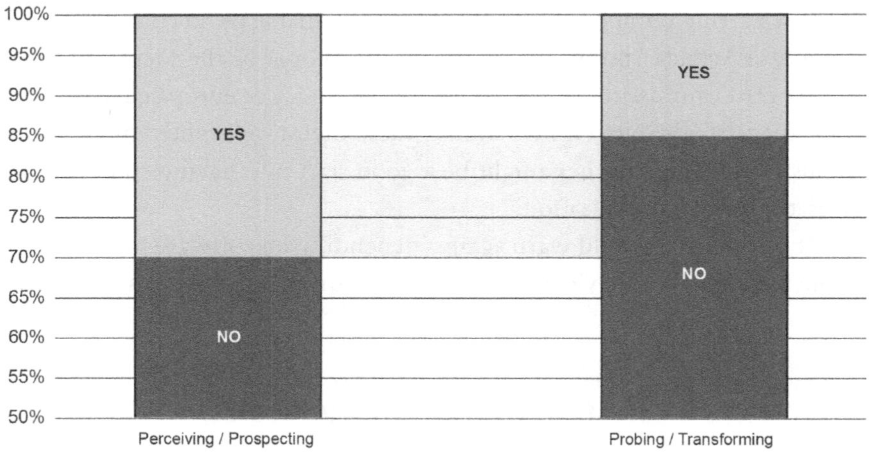

Figure 14.1 Implementation gap in strategic foresight.

perception is averted, and it reiterates that these types of activities are relevant to the entire organisation.

But engagement can also be perceived in a different way. From my perspective, diversity, the engagement of many different perspectives, is crucial in foresight work. It has been argued that diverse perspectives are crucial when it comes to decision-making processes on innovations and that diverse teams tend to be better at identifying opportunities and risks (Hill, Tedards, and Swan 2021).

14.5 Cross-industry collaboration

Statements like 'We all live in the same future' or 'We all have the same customer' reveal another important principle of strategic foresight work. If the aims of strategic foresight are to understand possible current trends and weak signals and to think about alternatives futures, then how can we include many perspectives? While some organisations might already have this kind of diversity, it often makes sense to engage a mode of cross-industry collaboration to include multiple perspectives but also expertise in understanding trends and possible futures.

A prime example of cross-industry collaboration is the Foresight Academy (https://www.foresightacademy.com). The Foresight Academy consists of several companies that partner to explore how the future could look. To do so, however, these companies have to trust each other. One principle of the Foresight Academy is thus that only one player per industry can join, to avoid competitiveness.

14.6 Training of leadership

Schwarz, Rohrbeck, and Wach (2020) have found, based on an empirical investigation, that corporate foresight training for managers has a positive impact on the dynamic capabilities of a firm. The implication is that leaders need to be trained in sensing change or developing peripheral vision, making decisions under uncertainty, being confronted with an uncertain future, and how to reconfigure strategic resources accordingly.

It has also been argued that foresight is not something that people are born with but that it is a way of thinking, of gathering information, and updating beliefs. These habits can be cultivated and learned (Tetlock and Gardner 2015). I have described how both organisations and individuals can be trained to develop foresight capabilities.

From my perspective, having a leadership team with this capability is one side of the strategic foresight coin. Having a good process in place, perhaps with the assistance of software tools or consultants, does not guarantee that the results of strategic foresight processes will be useful for a leadership team. However, and this is the other side of the coin, if that element of training is in place, then one can ensure that the phases of perceiving, prospecting, and probing actually translate into the phase of transforming.

14.7 Facilitation skills

A theme of this book is that strategic foresight is closely linked to the capability of those running such a process to facilitate discussions and workshops. This is especially true for tools such as futures literacy laboratories, scenario planning or business wargaming. Bowman and MacKay (2020) insist on the facilitator's role in managing people and process in a scenario planning workshop. They add that it takes a rare skill to take a group of people from a position of perceived uncertainty concerning the future of their industry to one where they have the confidence to take on strategic challenges.

Scharmer (2009) has argued, in developing Theory U, that there are two ways of learning: learning from the past, which is well developed; and learning from the future, which is not. However, as he observed the manner in which 'impressive leaders' use future possibilities, he wondered how one could be better at sensing and connecting with future possibilities. I have described in Section 2.4 how assumptions concerning the future of individuals can be made explicit through a structured process. Further, as I discussed in Chapter 9, challenging mental models and creating alternative pictures for the future follow similar ideas.

Strategic foresight tools can be considered as facilitators as well. Hence, an approach such as scenario planning must be less understood as a many-step process that needs to be followed but more as an organisational intervention. Moreover, this puts an emphasis not only on the individuals involved in such an exercise but also on how to facilitate these interventions.

Adam Kahane (2021) has worked in scenario planning at Royal Dutch Shell and in the facilitation of the Mont Fleur scenarios. The purpose of the Mont Fleur scenarios in South Africa during 1991–1992 was to think creatively about the future of the country, highlighting a number of challenges for facilitation. He points out that the task of a facilitator is to remove obstacles to contribution, to create opportunities for participants to voice their ideas, skills, and resources. The facilitator should also remove obstacles to connection. Kahane (2021) argues that all collaboration requires connection. This implies engaging participants emotionally, requiring them to get to know each other better. His argument is that as soon as participants in a workshop know each other well, they can more easily engage in the kind of discussion that than can enable an organisation to transform. Removing the obstacles to equity means holding workshops that are open to all participants.

References

Bowman, G., and R. B. MacKay. 2020. "Scenario Planning as Strategic Activity: A Practice-Orientated Approach." *Futures & Foresight Science* n/a (n/a). https://doi.org/10.1002/ffo2.32.

Festinger, L. 1957. *A Theory of Cognitive Dissonance*. Palo Alto: Stanford University Press.

Hill, L. A., E. Tedards, and T. Swan. 2021. "Drive Innovation with Better Decision-Making: Don't Let Old Habits Undermine Your Organization's Creativity." *Harvard Business Review* (November-December): 70–79.

Kahane, A. 2021. *Facilitating Breakthrough: How to Remove Obstacles, Bridge Differences, and Move Forward Together.* Oakland: Berrett-Koehler Publishers.

Liebl, F., and J. O. Schwarz. 2010. "Normality of the Future: Trend Diagnosis for Strategic Foresight." *Futures* 42 (4): 313–327.

Scharmer, C. O. 2009. *Theory U: Learning from the Future as It Emerges*. San Francisco: Berrett-Koehler Publishers.

Schwarz, J. O. 2005. "Pitfalls in Implementing a Strategic Early Warning System." *Foresight* 7 (4): 22–30.

———. 2009. "The Symbolism of Foresight Processes in Organizations." In *The Handbook of Research on Strategy and Foresight*, edited by L. A. Costanzo and R. B. MacKay, 82–89. Cheltenham: Edward Elgar.

———. 2020. "Revisiting Scenario Planning and Business Wargaming from an Open Strategy Perspective." *World Futures Review* 12 (3): 291–303. https://doi.org/10.1177/1946756720953182.

———. 2021. "From Alternative Pictures of the Future to an Organizational Intervention: A Commentary on Rowland and Spaniol." *Futures & Foresight Science* n/a (n/a): e2105. https://doi.org/10.1002/ffo2.105.

Schwarz, J. O., R. Rohrbeck, and B. Wach. 2020. "Corporate Foresight as a Microfoundation of Dynamic Capabilities." *Futures & Foresight Science* 2 (2): e28. https://doi.org/10.1002/ffo2.28.

Seidl, D., and F. Werle. 2018. "Inter-Organizational Sensemaking in the Face of Strategic Meta-Problems: Requisite Variety and Dynamics of Participation." *Strategic Management Journal* 39 (3): 830–858. https://doi.org/10.1002/smj.2723.

Tetlock, P., and D. Gardner. 2015. *Superforecasting: The Art and Science of Prediction*. London: Random House.

von Groddeck, V., and J. O. Schwarz. 2013. "Perceiving Megatrends as Empty Signifiers: A Discourse-Theoretical Interpretation of Trend Management." *Futures* 47: 28–37.

15

CONCLUSION

The relevance of strategic foresight is most likely to increase in the future. In the near future, new acronyms will be developed to show that VUCA is not only not going away but increasing. Strategic foresight supports individuals and organisations alike in dealing with VUCA.

For me however one aspect of strategic foresight is central, and I hope this book has demonstrated this. While it is essential to follow a process in strategic foresight, this is only one side of the coin. Strategic foresight needs to change the minds of leaders and decision-makers. It needs to challenge mental models, assumptions and redirect our attention. Many drivers of VUCA, for instance digitalisation and sustainability, are trends or megatrends that have been around for decades. Nonetheless, organisations are still struggling to meet these challenges.

A strategic foresight intervention needs to ensure that not only reports are produced, but that the stakeholders in an organisation are engaged in a way that is conducive to generating, understanding, and making decisions based on new insights. A foresight intervention must achieve a common

DOI: 10.4324/9781003302735-19

understanding in an organisation on what the changes, the trends, in the respective environment are, what they mean, how possible futures could look like, and of course what this implies for an organisation.

This implies the positioning of strategic foresight not only to inject future into a strategy, which already is difficult, but to understand strategic foresight as a starting point for the transformation of an organisation. In other words, foresight and change/transformation are the two sides of the same coin. If an organisation develops foresight, it most likely will also need to change or even to transform itself to steer to a desired future. And while understanding and implementing strategic foresight in this way is a challenge, it is also the most fascinating aspect of foresight.

And in the end dealing with the future is what organisations and individuals do and how they act on what they perceive.

> I read the future like a picture book
> I have always believed in what I see when you're near
> Each day I look in your eyes
> I see a lifetime with you so crystal clear
>
> 'So Much More Together', The Solsonics, 1994

INDEX

Note: *Italic* page numbers refer to figures.

For Product Safety Concerns and Information please contact our EU representative GPSR@taylorandfrancis.com Taylor & Francis Verlag GmbH, Kaufingerstraße 24, 80331 München, Germany